PERTURBATION METHODS FOR ENGINEERS AND SCIENTISTS

CRC Press Library of Engineering Mathematics

SERIES EDITOR

PROFESSOR ALAN JEFFREY

University of Newcastle upon Tyne
and University of Delaware

Linear Algebra and Ordinary Differential Equations
ALAN JEFFREY, *University of Newcastle upon Tyne, UK, and University of Delaware*

Nonlinear Ordinary Differential Equations
R. GRIMSHAW, *University of New South Wales, Australia*

Complex Analysis and Applications
ALAN JEFFREY, *University of Newcastle upon Tyne, UK, and University of Delaware*

Perturbation Methods for Engineers and Scientists
ALAN W. BUSH, *Teesside Polytechnic, UK*

PERTURBATION METHODS FOR ENGINEERS AND SCIENTISTS

Author
Alan W. Bush
School of Computing and Mathematics
Teesside Polytechnic
Middlesbrough, U.K.

CRC Press
Boca Raton Ann Arbor London Tokyo

Library of Congress Cataloging-in-Publication Data

Bush, Alan W.
Perturbation methods for engineers and scientists / Alan W. Bush.
p. cm.
Includes bibliographical references and index.
ISBN 0-8493-8608-X (hardcover). — 0-8493-8614-4 (softcover).
1. Perturbation (Mathematics) I. Title.
QA871.B94 1992
531′.11′0151535—dc20 91-39626
CIP

Direct all inquiries to CRC Press, Inc., 2000 Corporate Blvd., N.W., Boca Raton, Florida, 33431.

International Standard Book Number 0-8493-8608-X (Hardcover)
International Standard Book Number 0-8493-8614-4 (Softcover)

Library of Congress Card Number 91-39626

Printed in the United States 1 2 3 4 5 6 7 8 9 0

To my Wife
SUSAN

and my Mother
ANNE

Contents

Preface

The purpose of this book is to describe the application of perturbation expansion techniques to the solution of differential equations and the approximation of integrals. Approximate expressions are generated in the form of asymptotic series. These may not and often do not converge but, in a truncated form of only two or three terms, provide a useful approximation to the original problem.

The techniques, being analytical rather than numerical, provide an alternative to a direct computer solution. An awareness of the perturbation approach is sometimes essential even when a direct numerical approach is adopted. An example of this occurs in boundary layer problems where there are regions of rapid change of quantities such as fluid velocity, temperature or concentration. Appropriate scaling of the boundary layer dimension is required before a numerical solution can be generated which will capture the behavior in the rapidly changing region.

The material is based on lectures given at Teesside Polytechnic as part of the Master's Degree in Applicable Mathematics. The main techniques with some of the applications are taught in a course of sixty hours duration. The book can be used at advanced undergraduate or postgraduate level.

Students, teachers and researchers in applied and engineering mathematics should find this book of help in gaining an understanding of most of the powerful perturbation techniques along with their applications. The essential prerequisite knowledge is usually gained by the end of the first year mathematical methods course in an engineering or mathematics degree. Specifically the reader is assumed to be familiar with the elementary techniques for solving first order differential equations and second order constant coefficient equations.

The perturbation methods for differential equations are explained using simple 'model' ordinary differential equations. The motivation for developing the methods is the need to deal with more complicated, often nonlinear, ordinary and partial differential equations. Applications to these problems are presented. To fully appreciate some of the partial differential equation examples a knowledge of the governing equations of fluid mechanics and heat transport are required. Readers unfamiliar with these subjects may either take the formulation of the problems for granted and follow the mathematical techniques, or the material may be omitted without impeding their understanding of later material.

The idea of a perturbation expansion is introduced informally in Chapter 1 with a study of algebraic equations, differential equations and integrals. Formal definitions and the terminology of asymptotics are presented in Chapter 2. The first two chapters contain essential material on which the rest of the book is based. The remaining chapters are self-contained although links between subjects are made when appropriate. It is natural for Chapter 3 on strained coordinates to be read before the following chapter on the method of multiple scales.

Chapter 5 on boundary layers can, if required, be studied without reference to the previous two chapters. The subject is particularly rich in applications, and it is intended that the lengthy section on practical applications should reflect this.

Chapter 6 is devoted to a study of the asymptotic behavior of solutions of differential equations for large values of the independent variable and large values of a parameter.

The final chapter is concerned with the asymptotic approximation of integrals. This material is independent of most of the earlier work requiring only a review of the two introductory chapters.

Some computer program listings have been included when they were considered helpful for the reader in exploring an application.

I first acquired my knowledge of the subject of perturbation expansions from the work of other authors. I would like to gratefully acknowledge the contribution to my understanding from the books listed in the bibliography.

I am grateful to my colleagues, Alan Lowdon, Michael Priddis, Sally and Alan Cook, for providing helpful comments on various sections of the work. I am grateful to Professor W. F. Ames for his encouragement and helpful comments during the writing of the book.

Finally I wish to thank Lynne Rhucroft for her patience and care in the preparation of the typed version of the book.

ALAN BUSH
December 1990

Chapter 1
Introduction to Perturbation Expansions

The governing equations of physical, biological and economic models often involve features which make it impossible to obtain their exact solution. Examples of such features are:

the occurrence of a complicated algebraic equation
the occurrence of a complicated integral
varying coefficients in a differential equation
an awkwardly shaped boundary
a nonlinear term in a differential equation

When a large or small parameter occurs in a mathematical model of a process there are various methods of constructing perturbation expansions for the solution of the governing equations. Often the terms in the perturbation expansions are governed by simpler equations for which exact solution techniques are available. Even if exact solutions cannot be obtained, the numerical methods used to solve the perturbation equations approximately are often easier to construct than the numerical approximation for the original governing equations.

This chapter is an introduction to the study of perturbation expansions. We will first consider a model problem for which an exact solution is available against which the perturbation expansion can be compared. Next we obtain expansions for the solution of polynomial equations involving small parameters. Then various initial value problems are considered. This is followed by a study of expansions involving powers and inverse powers of the independent variable. These provide approximations for small and large values respectively of the independent variable. A feature of perturbation expansions is that they often form divergent series. The concept of an asymptotic expansion will be introduced and the value of a truncated divergent series will be demonstrated.

1.1 Model problem. Resisted motion of a particle

Our first example is a study of the effect of small damping on the motion of a particle. Consider a particle of mass M which is projected vertically upward with an initial speed U_0. Let U denote the speed at some general time T. If air resistance is neglected then the only force acting on the particle is gravity, $-Mg$, where g is the acceleration due to gravity and the minus sign occurs because the upward direction is chosen to be the positive direction. Newton's second law governs the motion of the projectile, i.e.

$$M\frac{dU}{dT} = -Mg. \qquad 1.1.1$$

The solution, $U = C - gT$, is obtained by direct integration. The constant of integration, C, is determined from the initial condition, $U(0) = U_0$, so that

$$U = U_0 - gT. \qquad 1.1.2$$

On defining the nondimensional velocity, v, and time, t, by the expressions $v = U/U_0$ and $t = gT/U_0$, the governing equation becomes

$$\frac{dv}{dt} = -1, \quad v(0) = 1, \qquad 1.1.3$$

with solution $v = 1 - t$.

Whenever nondimensional variables are introduced, upper case letters will be used to denote dimensional variables and lower case letters to denote nondimensional variables. If possible this convention will be followed for parameters as well. However, in the case of gravitational acceleration it is standard to use the lower case letter, g, to represent the dimensionalized quantity.

Air resistance is included in Newton's second law as a force dependent on the velocity. The simplest model is a linear law. This is reasonable for low speeds but must be modified to a quadratic or a more complex polynomial in powers of the velocity at higher speeds. The governing equation for a linear resistance model is

$$M\frac{dU}{dT} = -Mg - KU, \qquad 1.1.4$$

where the drag constant K has the dimensions of mass/time. This equation expressed in terms of the previously introduced nondimensional variables becomes,

$$\frac{dv}{dt} = -1 - (KU_0/Mg)v. \qquad 1.1.5$$

The combination KU_0/Mg is a dimensionless drag constant which we will denote by the symbol ε. The nondimensional form of the governing equation is

$$\frac{dv}{dt} = -1 - \varepsilon v, \qquad 1.1.6$$

with the initial condition $v(0) = 1$.

The care which has been taken over the transformation to nondimensional form emphasizes the importance of the procedure. We will be considering the effect of *small disturbances*. The word 'small' must be used with caution. It would not, for example, be sufficient to simply state that the damping constant, K, which appears in equation 1.1.4 is small, since K has the dimensions of mass/time and a small quantity in units of kilograms per second could turn out to be a large quantity in other units such as micrograms per year. The important consideration is the size of K in comparison with another quantity having the same dimensions as K. To properly compare terms, appropriate velocity and time scales must be used for the problem. In this case they are respectively the initial speed, U_0, and the ratio U_0/g. It then becomes clear that the nondimensional combination, ε $(=KU_0/Mg)$, measures the importance of the resistance force. If we revert to dimensionalized quantities we see that the drag constant K is to be compared with the quantity Mg/U_0 in order to decide whether K is or is not a small quantity.

In the case of small resistance the nondimensional term, ε, is small. By this we mean that ε is significantly less than unity. Although we will usually not be specific about the size of small nondimensional terms we will expect them to be no greater than 0.1. Intuitively the small term, εv, in equation 1.1.6 may be expected to have a small effect on the solution. In fact this is not always the case as we shall discover when the solutions of the unperturbed problem (equation 1.1.3) and the perturbed problem (equation 1.1.6) are compared. The solutions are found to have similar values (the differences are of the magnitude of ε) for values of the independent variable, t, which are not large. On the other hand, for large t, the solutions of the perturbed and unperturbed problem differ greatly. This behavior is described as a nonuniformity associated with the unbounded domain of the independent variable. The problem of nonuniformities will be dealt with in detail in later chapters.

A perturbation expansion

It is possible to solve equation 1.1.6 exactly since it is of variables separable form. However, unlike equation 1.1.3 it cannot be solved by direct integration. We temporarily suppose that the process of direct integration is all that we have at our disposal and develop a solution of equation 1.1.6 by an iterative process which only requires direct integration at each stage. The iterative procedure will create what is called a *perturbation expansion* for the solution.

The Ith iterate is denoted by $v^{(I)}$. The brackets around I indicate that I numbers the iteration and is not to be confused with an exponent. The Ith iterate is obtained by solving the equation

$$\frac{dv^{(I)}}{dt} = -1 - \varepsilon v^{(I-1)}, \qquad 1.1.7$$

with $v^{(I)}(0) = 1$.

The justification for this iterative scheme is that the term εv involves the small multiplying coefficient, ε, and so the term itself may be expected to be small. Thus the term $\varepsilon v^{(I)}$ which should appear on the right-hand side of 1.1.7 to make it exact, may be replaced by $\varepsilon v^{(I-1)}$ with an error which is expected to be small.

The first iterate $v^{(0)}$ is obtained by neglecting the perturbation, thus

$$\frac{dv^{(0)}}{dt} = -1, \quad v^{(0)}(0) = 1.$$

This is the unperturbed problem (equation 1.1.3). The solution, which is obtained by direct integration, is $v^{(0)} = 1 - t$.

The next iterate, $v^{(1)}$, satisfies

$$\frac{dv^{(1)}}{dt} = -1 - \varepsilon(1 - t), \quad v^{(1)}(0) = 1,$$

and direct integration yields

$$v^{(1)} = 1 - t(1 + \varepsilon) + \varepsilon t^2/2.$$

Similarly, $v^{(2)}$, satisfies

$$\frac{dv^{(2)}}{dt} = -1 - \varepsilon[1 - t(1 + \varepsilon) + \varepsilon t^2/2], \quad v^{(2)}(0) = 1.$$

Direct integration yields the solution

$$v^{(2)} = 1 - t(1 + \varepsilon) + \varepsilon(1 + \varepsilon)t^2/2 - \varepsilon^2 t^3/6.$$

It is helpful to rearrange the terms in these iterates in ascending powers of ε,

$$\begin{aligned} v^{(0)} &= 1 - t \\ v^{(1)} &= 1 - t + \varepsilon(t^2/2 - t) \\ v^{(2)} &= 1 - t + \varepsilon(t^2/2 - t) + \varepsilon^2(t^2/2 - t^3/6). \end{aligned} \qquad 1.1.8$$

Clearly as the iteration proceeds the expressions are refined by terms which involve increasing powers of ε. These terms become progressively smaller since ε is a small parameter. This is an example of a perturbation expansion. It will often be the case that perturbation expansions involve ascending integer powers of the small parameter i.e. $\{\varepsilon^0, \varepsilon^1, \varepsilon^2, \varepsilon^3, \ldots\}$. Such a sequence is called an *asymptotic sequence*. Although this is the most common sequence which we shall meet, it is by no means unique. Examples of other asymptotic sequences are $\{\varepsilon^{1/2}, \varepsilon, \varepsilon^{3/2}, \varepsilon^2, \ldots\}$ and $\{\varepsilon^0, \varepsilon^2, \varepsilon^4, \varepsilon^6, \ldots\}$. In each case the essential feature is that subsequent terms tend to zero faster than previous terms as ε tends to zero. A precise definition of an asymptotic sequence and further examples will be presented in Chapter 2.

An alternative procedure to that of developing the expansion by iteration is to assume the form of the expansion at the outset. Thus if we assume that the perturbation expansion involves the standard asymptotic sequence $\{\varepsilon^0, \varepsilon^1, \varepsilon^2, \varepsilon^3, \ldots\}$ then the solution v, which depends on the variable t and the parameter ε, is expressed in the form

$$v(t;\varepsilon) = \varepsilon^0 v_0(t) + \varepsilon^1 v_1(t) + \varepsilon^2 v_2(t) + \cdots. \qquad 1.1.9$$

The coefficients $v_0(t)$, $v_1(t)$, ... of powers of ε are functions of t only. This means that the dependence of v on the parameter ε is assumed to be of the above power series form. If this assumption is wrong then an inconsistency is to be expected to occur in the process of determining the coefficient functions v_0, v_1, etc. It is important to notice the distinction between the subscripted quantities v_I which are coefficients of the powers of ε in equation 1.1.9, whereas the bracketed superscripts used in the quantities $v^{(I)}$ denote the full solution at the Ith stage of iteration in the earlier approach.

The expression ε^0 has been used to emphasize that it is a member of the family ε^n but it will subsequently be replaced by its numerical value of unity.

Substituting the expansion 1.1.9 into the governing equation 1.1.6 yields the following,

$$\frac{dv_0}{dt} + \varepsilon\frac{dv_1}{dt} + \varepsilon^2\frac{dv_2}{dt} + \cdots = -1 - \varepsilon v_0 - \varepsilon^2 v_1 - \cdots, \qquad 1.1.10$$

where for convenience the dependence on t of the coefficients v_0, v_1 ... is not shown explicitly. The *single* equation 1.1.10 along with the *single* initial condition

$$v_0(0) + \varepsilon v_1(0) + \varepsilon^2 v_2(0) + \cdots = 1, \qquad 1.1.11$$

are to be used to determine *all* the functions $v_I(t)$. The extra condition which is imposed is that these equations are true for all values of the parameter ε in some continuous range, (say $0 \leqslant \varepsilon \leqslant 0.1$, for example, although it is not necessary to specify the range). Thus coefficients of powers of ε can be equated on the left- and right-hand sides of equation 1.1.10.

$$\begin{aligned} \varepsilon^0&: \frac{dv_0}{dt} = -1 \\ \varepsilon^1&: \frac{dv_1}{dt} = -v_0 \\ \varepsilon^2&: \frac{dv_2}{dt} = -v_1 \text{ etc.} \end{aligned} \qquad 1.1.12$$

The proof of the validity of this fundamental procedure can be developed by first setting $\varepsilon = 0$ in equation 1.1.10 which yields the first of the equations 1.1.12. Using this result allows the first members of the left- and right-hand sides of equation 1.1.10 to be removed. Then after dividing the remaining terms by ε we obtain the equation

$$\frac{dv_1}{dt} + \varepsilon\frac{dv_2}{dt} + \cdots = -v_0 - \varepsilon v_1 + \cdots.$$

This is valid for all nonzero values of ε so that on taking the limit as ε tends to zero we obtain the second of equations 1.1.12 namely $dv_1/dt = -v_0$. Repeating the procedure leads to $dv_2/dt = -v_1$ and so on.

The equations 1.1.12 each require an initial condition. A set of initial conditions is generated from equation 1.1.11 by rewriting the right-hand side in the power series form $1 = 1 + 0\varepsilon + 0\varepsilon^2 + \cdots$ Then the initial condition becomes

$$v_0(0) + \varepsilon v_1(0) + \varepsilon^2 v_2(0) + \cdots = 1 + 0\varepsilon + 0\varepsilon^2 + \cdots$$

which is valid for all ε in some continuous range so that again we may equate coefficients of powers of ε on the left- and right-hand sides of the equation to obtain

$$v_0(0) = 1, \quad v_1(0) = 0, \quad v_2(0) = 0, \text{ etc.}$$

The functions $v_I(t)$ are determined by combining the initial conditions with the equations 1.1.12 to yield the following set of initial value problems,

$$\varepsilon^0: \quad \frac{dv_0}{dt} = -1, \quad v_0(0) = 1 \qquad 1.1.13a$$

$$\varepsilon^1: \quad \frac{dv_1}{dt} = -v_0, \quad v_1(0) = 0 \qquad 1.1.13b$$

$$\varepsilon^2: \quad \frac{dv_2}{dt} = -v_1, \quad v_2(0) = 0 \text{ etc.} \qquad 1.1.13c$$

The solution of 1.1.13a, obtained by direct integration, is $v_0 = 1 - t$. Then on substituting for v_0 in 1.1.13b the solution, again obtained by direct integration, is $v_1 = t^2/2 - t$. Similarly 1.1.13c yields

$$v_2 = t^2/2 - t^3/6.$$

In this way the coefficients in the perturbation expansion 1.1.9 are found and we have

$$v(t; \varepsilon) = 1 - t + \varepsilon(t^2/2 - t) + \varepsilon^2(t^2/2 - t^3/6) + \cdots. \qquad 1.1.14$$

This is the same as the expansion 1.1.8 which was generated by iteration.

The exact solution

We are able to obtain the exact solution of equation 1.1.6 by the variables separable technique. This can be used for comparison with the perturbation expansion. Starting from

$$\frac{dv}{dt} = -1 - \varepsilon v,$$

the following steps are standard:

$$\int \frac{1}{1+\varepsilon v}\frac{\mathrm{d}v}{\mathrm{d}t}\,\mathrm{d}t = -\int \mathrm{d}t$$

$$\frac{1}{\varepsilon}\int \frac{\varepsilon}{1+\varepsilon v}\,\mathrm{d}v = -\int \mathrm{d}t$$

$$\frac{1}{\varepsilon}\ln|1+\varepsilon v| = -t + c,$$

where c is a constant of integration. Exponentiating yields

$$|1+\varepsilon v| = \mathrm{e}^{\varepsilon c}.\mathrm{e}^{-\varepsilon t}.$$

We may omit the modulus function if the constant $\mathrm{e}^{\varepsilon c}$ is allowed to take positive or negative values. On denoting this constant by a we have

$$1+\varepsilon v = a\mathrm{e}^{-\varepsilon t}.$$

The initial condition $v(0) = 1$ determines the constant A to be $1 + \varepsilon$ and on rearranging we obtain the exact solution

$$v = [(1+\varepsilon)\mathrm{e}^{-\varepsilon t} - 1]/\varepsilon. \qquad 1.1.15$$

The perturbation expansion can be obtained from 1.1.15 by replacing the exponential function by its Maclaurin expansion, i.e.

$$\mathrm{e}^{-\varepsilon t} = 1 - \varepsilon t + \varepsilon^2 t^2/2 - \varepsilon^3 t^3/6 + \cdots.$$

The expression 1.1.15 becomes

$$v = \frac{1}{\varepsilon}\left(1 - \varepsilon t + \frac{\varepsilon^2 t^2}{2} - \frac{\varepsilon^3 t^3}{6} + \cdots + \varepsilon - \varepsilon^2 t + \frac{\varepsilon^3 t^2}{2} - \cdots - 1\right)$$

$$= 1 - t + \varepsilon\left(\frac{t^2}{2} - t\right) + \varepsilon^2\left(\frac{t^2}{2} - \frac{t^3}{6}\right) + \cdots.$$

This is the same as the expansion 1.1.14. Thus the perturbation expansion approach is justified in this case.

Although the exact solution for this example was easily found, it required the use of the variables separable technique whereas direct integration was all that the perturbation expansion required. By analogy, more complicated problems may be governed by equations for which no analytical solution technique exists but, if a perturbation expansion is used, each coefficient function of the expansion may be governed by a relatively simple equation for which analytical solution techniques do exist.

The expansion 1.1.14 of the solution of equation 1.1.6 provides an example of one of the interesting complications which often arise in perturbation expansions, namely that a small perturbation in the governing equation may have a large effect on the solution. In this example the unperturbed problem (with $\varepsilon = 0$) has the solution $v = 1 - t$. The small perturbation has a small effect on the solution provided that the range of the independent variable, t, is restricted. Comparing the first two terms of 1.1.14, namely $1 - t$ and $\varepsilon(t^2/2 - t)$ we see that the second term is a small correction to the first provided t is not large. However, the two terms are of the same order if t is of order $1/\varepsilon$. Furthermore, all terms are of the same order as the first term if t is of order $1/\varepsilon$. The expansion is only useful if subsequent terms are small corrections to previous terms. We must therefore restrict the range of values of the independent variable to be significantly less than $1/\varepsilon$.

In Fig. 1.1 the behavior of the exact solution,

$$v_{\mathrm{ex}} = [(1 + \varepsilon)e^{-\varepsilon t} - 1]/\varepsilon,$$

is compared with the solution of the unperturbed problem

$$v_0 = 1 - t,$$

along with the two and three term expansions,

$$v_{2T} = 1 - t + \varepsilon(t^2/2 - t)$$

$$v_{3T} = 1 - t + \varepsilon(t^2/2 - t) + \varepsilon^2(t^2/2 - t^3/6).$$

The value of 0.1 has been chosen for ε.

Clearly the perturbation expansion fails to predict the occurrence of the terminal velocity $v_\infty = -1/\varepsilon$ as $t \to \infty$ and the expansion becomes invalid when t is of order $1/\varepsilon$ (i.e. t of order 10 in this example with $\varepsilon = 0.1$). In fact the scaling of the velocity by the initial velocity U_0 (i.e. defining the nondimensional velocity $v = U/U_0$) is not appropriate for large values of t since the velocity approaches the terminal velocity $U_\infty = -Mg/K$ which is independent of U_0. A scaling is 'appropriate' if it causes the resulting function and its derivatives to be of order unity. Then small terms are identified by the occurrence of a small coefficient. Thus with the scaling used in this

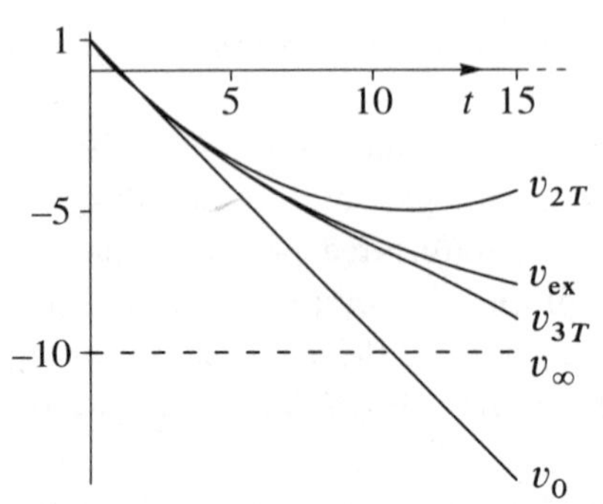

Fig. 1.1 Comparison of the exact solution and perturbation expansions for equation 1.1.6 ($\varepsilon = 0.1$)

problem the term εv is small for times of order unity since then v is order unity but for times of order $1/\varepsilon$, v is of this order so that εv is no longer a small term.

The failure of this expansion for large values of t is an example of what is called a nonuniformity. The region of the failure is called the *region of nonuniformity*. The problem of nonuniformities in perturbation expansions will be considered in detail in subsequent chapters.

In the previous discussion the word 'order' was introduced. That it has the meaning of 'size' can be gained from the context in which it was used. A precise mathematical meaning will be developed in Chapter 2 along with an associated notation. We anticipate this development and use the notation for the rest of the chapter. The order symbol 'big O' will be used so that $O(\varepsilon)$ loosely speaking means of size ε while $O(1)$ means of size unity and $O(1/\varepsilon)$ means of size $1/\varepsilon$. The subtlety which has not been introduced at this stage is that these expressions are to be understood in an *asymptotic* sense as ε tends to zero.

1.2 Roots of polynomials

In this section the problem of obtaining roots of polynomials containing a small parameter, ε, will be considered. The perturbation approach is useful if the roots of the unperturbed polynomial (obtained by setting $\varepsilon = 0$) are known. The study is both of practical value in providing a means of dealing with some algebraic equations and of instructional value in providing easily constructed examples of perturbation expansions which are nonstandard in one way or another. Thus we will see examples of singular perturbations where inverse powers of the small parameter occur in the expansion and we will see fractional powers of the small parameter appear.

Consider the equation

$$\varepsilon x^2 - x + 1 = 0, \qquad 1.2.1$$

where ε is small and positive. This is a quadratic equation which can be solved exactly. However, the construction of the exact solution will be postponed until a perturbation expansion has been obtained. The point of this is to develop a technique which can be tested for a simple problem against the exact solution before applying the technique to more complex cases where exact solutions may not be available.

The following iterative scheme is suggested by the presence of the small parameter,

$$x^{(I)} - 1 = \varepsilon(x^{(I-1)})^2. \qquad 1.2.2$$

It is supposed that the term involving ε as a multiplying coefficient is small so that the earlier iterate may be used for x in this term. The leading approximation $x^{(0)}$ is provided by the unperturbed problem (obtained by setting $\varepsilon = 0$) i.e. $x^{(0)} = 1$. The subsequent iterates are:

$$x^{(1)} = 1 + \varepsilon(x^{(0)})^2 = 1 + \varepsilon$$

$$x^{(2)} = 1 + \varepsilon(1 + \varepsilon)^2$$

$$= 1 + \varepsilon + 2\varepsilon^2 + \varepsilon^3$$

$$x^{(3)} = 1 + \varepsilon(1 + \varepsilon + 2\varepsilon^2 + \varepsilon^3)^2$$

$$= 1 + \varepsilon + 2\varepsilon^2 + 5\varepsilon^3 + 6\varepsilon^4 + 6\varepsilon^5 + 4\varepsilon^6 + \varepsilon^7.$$

The iterative approach shows that a sequence of terms involving integer powers of ε is developed, i.e. the asymptotic sequence is $\{\varepsilon^0, \varepsilon^1, \varepsilon^2, \varepsilon^3, \ldots\}$. A difficulty with the iteration is that of determining when the coefficients of the powers of ε are fixed. Thus for example $x^{(2)}$ involves the term ε^3 with a coefficient of unity while $x^{(3)}$ involves this term with a coefficient 5. A preferable method is to assume the above asymptotic sequence, i.e.

$$x = x_0 + \varepsilon x_1 + \varepsilon^2 x_2 + \varepsilon^3 x_3 + \cdots. \qquad 1.2.3$$

Substituting the assumed expansion into equation 1.2.1 yields the equation

$$\varepsilon(x_0^2 + 2\varepsilon x_0 x_1 + \varepsilon^2(x_1^2 + 2x_0 x_2) + \cdots) - x_0 - \varepsilon x_1 - \varepsilon^2 x_2 - \varepsilon^3 x_3 - \cdots + 1 = 0. \qquad 1.2.4$$

This equation is valid for all values of ε in a continuous range so that equations may be generated by equating coefficients of powers of ε on the left- and right-hand sides of 1.2.4. The zero on the right-hand side may be expressed as $0 + 0\varepsilon + 0\varepsilon^2 + 0\varepsilon^3 + \cdots$ to obtain the following:

$$\varepsilon^0:\ -x_0 + 1 = 0 \qquad \therefore\ x_0 = 1$$

$$\varepsilon^1:\ -x_1 + x_0^2 = 0 \qquad \therefore\ x_1 = 1$$

$$\varepsilon^2:\ -x_2 + 2x_0 x_1 = 0 \qquad \therefore\ x_2 = 2$$

$$\varepsilon^3:\ -x_3 + x_1^2 + 2x_0 x_2 = 0 \qquad \therefore\ x_3 = 5 \text{ etc.} \qquad 1.2.5$$

Thus the following perturbation expansion for the solution of equation 1.2.1 is obtained,

$$x = 1 + \varepsilon + 2\varepsilon^2 + 5\varepsilon^3 + \cdots.$$

This agrees with the third iteration $x^{(3)}$ up to terms of order ε^3.

Regular and singular perturbation expansions

Equation 1.2.1 is a quadratic equation and therefore has two solutions. The perturbation expansion has yielded only one solution. No matter how small ε becomes in 1.2.1 it remains a quadratic equation with two solutions. Whereas when ε equals zero 1.2.1

becomes a linear equation with one solution. Clearly the problem behaves differently in the limit as $\varepsilon \to 0$ than it does when ε actually equals zero. This is an example of what is called a *singular perturbation.*

To explore the cause of the difficulty consider the exact solution,

$$x = \frac{1 \pm \sqrt{1 - 4\varepsilon}}{2\varepsilon}. \qquad 1.2.6$$

An expression involving powers of ε can be obtained by replacing the square root by its binomial expansion,

$$x = \frac{1 \pm (1 - 2\varepsilon - 2\varepsilon^2 + \cdots)}{2\varepsilon}$$

$$= \frac{1}{\varepsilon} - 1 - \varepsilon + \cdots \quad ; \quad 1 + \varepsilon + \cdots. \qquad 1.2.7$$

The assumed form of the perturbation expansion, 1.2.3, has produced the second of the solutions 1.2.7, (i.e. the solution associated with the minus sign in equation 1.2.6). The leading term in the expansion of the first solution is $1/\varepsilon$. No such term was anticipated in the trial form 1.2.3 and consequently only one of the solutions was created by the perturbation approach. If we consider the behavior of the two expansions 1.2.7 as $\varepsilon \to 0$ we see that the second expansion tends to the solution $x = 1$ of the unperturbed problem whereas the first expansion tends to infinity. There is no counterpart to the first expansion in the unperturbed problem.

The expansion associated with the solution $x = 1$ of the unperturbed problem is called a *regular perturbation expansion.* The regular expansion continuously approaches the unperturbed solution as $\varepsilon \to 0$. The new solution for which there is no counterpart in the unperturbed problem has what is called a *singular perturbation expansion.*

It is reasonable to anticipate that singular perturbation expansions will always be generated for some of the solutions of polynomial equations when the highest power is multiplied by the small parameter. A similar effect occurs in differential equations when the highest derivative is multiplied by a small parameter. In this case the unperturbed problem involves lower order derivatives than the full problem so that the unperturbed problem cannot satisfy all of the boundary conditions of the full problem. This leads to regions of rapid change in the solution where a boundary condition is to be obeyed. These regions are called *boundary layers.* The theory of boundary layers will be considered in detail in Chapter 5.

We next consider an example of a cubic equation,

$$\varepsilon x^3 + x^2 - 1 = 0. \qquad 1.2.8$$

Substituting the standard expansion

$$x = x_0 + \varepsilon x_1 + \varepsilon^2 x_2 + \cdots,$$

into the equation leads to

$$\varepsilon(x_0^3 + 3\varepsilon x_0^2 x_1 + \cdots) + x_0^2 + 2\varepsilon x_0 x_1 + 2\varepsilon^2 x_0 x_2 + \varepsilon^2 x_1^2 + \cdots - 1 = 0.$$

Equating coefficients of powers of ε on the left- and right-hand sides yields the following:

$$\varepsilon^0: \quad x_0^2 - 1 = 0 \qquad \therefore\ x_0 = \pm 1$$

$$\varepsilon^1: \quad x_0^3 + 2x_0 x_1 = 0 \qquad \therefore\ x_1 = -\frac{x_0^2}{2} = -\frac{1}{2}$$

$$\varepsilon^2: \quad 3x_0^2 x_1 + 2x_0 x_2 + x_1^2 = 0 \qquad \therefore\ x_2 = \frac{-1}{2x_0}(x_1^2 + 3x_0^2 x_1) = \frac{5}{8x_0}.$$

Two perturbation expansions are generated associated with the two solutions for x_0, and they are

$$x = 1 - \varepsilon/2 + 5\varepsilon^2/8 + \cdots \quad \text{and} \quad x = -1 - \varepsilon/2 - 5\varepsilon^2/8 + \cdots.$$

The third solution is expected to be associated with large x values such that the term εx^3 in equation 1.2.8 is not small. If x is large, the term x^2 dominates the third member (-1), so that the leading term in the expansion of the third solution is expected to satisfy the equation $\varepsilon x^3 + x^2 = 0$. The nontrivial solution is $x = -1/\varepsilon$. Thus we are led to try an expansion of the form

$$x = \frac{x_0}{\varepsilon} + x_1 + \varepsilon x_2 + \cdots,$$

where x_0 is expected to be equal to -1. Substituting into 1.2.8 leads to

$$\varepsilon\left(\frac{x_0^3}{\varepsilon^3} + \frac{3x_0^2 x_1}{\varepsilon^2} + \frac{3x_0^2 x_2}{\varepsilon} + \frac{3x_0 x_1^2}{\varepsilon} + \cdots\right)$$

$$+ \frac{x_0^2}{\varepsilon^2} + \frac{2x_0 x_1}{\varepsilon} + 2x_0 x_2 + x_1^2 + \cdots - 1 = 0.$$

Equating coefficients yields

$$O(\varepsilon^{-2}): \quad x_0^3 + x_0^2 = 0 \qquad \therefore\ x_0 = -1$$

$$O(\varepsilon^{-1}): \quad 3x_0^2 x_1 + 2x_0 x_1 = 0 \qquad \therefore\ x_1 = 0$$

$$O(\varepsilon^0): \quad 3x_0^2 x_2 + 3x_0 x_1^2 + 2x_0 x_2 + x_1^2 - 1 = 0 \qquad \therefore\ x_2 = 1.$$

Thus the third solution has the expansion

$$x = -\frac{1}{\varepsilon} + \varepsilon + \cdots.$$

The absence of the O(1) term in this expansion suggests that subsequent terms involve powers of ε which are increased by two. This becomes apparent if we make the substitution $x = z/\varepsilon$ and seek a perturbation expansion for z. Equation 1.2.8 becomes

$$z^3 + z^2 - \varepsilon^2 = 0. \qquad 1.2.9$$

Clearly the small parameter is ε^2 so that the perturbation expansion for z is likely to involve powers of ε^2. If we assume the expansion

$$z = z_0 + \varepsilon^2 z_1 + \varepsilon^4 z_2 + \cdots \qquad 1.2.10$$

and substitute into 1.2.9 we obtain the expression

$$z_0^3 + 3\varepsilon^2 z_0^2 z_1 + 3\varepsilon^4 z_0 z_1^2 + 3\varepsilon^4 z_0^2 z_2 + \cdots$$
$$+ z_0^2 + 2\varepsilon^2 z_0 z_1 + \varepsilon^4 z_1^2 + 2\varepsilon^4 z_0 z_2 + \cdots - \varepsilon^2 = 0.$$

Equating coefficients yields

$O(1)$: $z_0^3 + z_0^2 = 0 \qquad \therefore z_0 = -1$

(Note the solution $z_0 = 0$ is rejected since we are searching for the solution of order $1/\varepsilon$.)

$O(\varepsilon^2)$: $3z_0^2 z_1 + 2z_0 z_1 - 1 = 0 \qquad \therefore z_1 = 1$

$O(\varepsilon^4)$: $3z_0 z_1^2 + 3z_0^2 z_2 + z_1^2 + 2z_0 z_2 = 0 \qquad \therefore z_2 = -2.$

The fact that no inconsistency has arisen in determining the coefficients in the expansion 1.2.10 justifies the assumption of using ascending powers of ε^2 rather than of ε.

We have obtained the expansion of the third solution of 1.2.8, $x = z/\varepsilon$, namely

$$x = -\frac{1}{\varepsilon} + \varepsilon - 2\varepsilon^3 + \cdots.$$

This result was based on the reasoning that the leading order term had the form x_0/ε so that to leading order εx^3 and x^2 dominate the equation. Had we assumed that the leading order term in the expansion of the third solution was of a more singular form than x_0/ε, say x_0/ε^2 then to leading order εx^3 dominates the equation having order ε^{-5} while the term x^2 has order ε^{-4}. This forces the choice $x_0 = 0$ showing that our choice of leading order term was wrong. On the other hand had we chosen a less singular form than x_0/ε, say $x_0/\varepsilon^{1/2}$ then to leading order x^2 dominates the equation having order ε^{-1} while the term εx^3 has order $\varepsilon^{-1/2}$. This again forces the choice $x_0 = 0$ showing that our choice of leading order term was wrong. The only choice of leading order term which allows εx^3 to be matched to dominant order by the term x^2 is x_0/ε. This method of determining the form of the singular behavior has an analogy in boundary layer theory where the process of preserving the maximum number of terms in the leading order equation is called *the principle of least degeneracy.*

Our next example has two solutions which involve singular perturbation expansions. Consider the equation

$$\varepsilon x^3 - x + 1 = 0, \quad 0 < \varepsilon \ll 1. \qquad 1.2.11$$

The unperturbed problem is the linear equation $x - 1 = 0$ with, of course, the single solution $x = 1$. We expect there to be only one regular expansion having leading order term $x_0 = 1$. On substituting the standard expansion $x_0 + \varepsilon x_1 + \varepsilon^2 x_2 + \cdots$ into equation 1.2.11 we obtain

$$\varepsilon(x_0^3 + 3\varepsilon x_0^2 x_1 + \cdots) - x_0 - \varepsilon x_1 - \varepsilon^2 x_2 + \cdots + 1 = 0.$$

Equating coefficients leads to

$$\begin{aligned} &O(1): && -x_0 + 1 = 0 && \therefore\ x_0 = 1 \\ &O(\varepsilon): && x_0^3 - x_1 = 0 && \therefore\ x_1 = 1 \\ &O(\varepsilon^2): && 3x_0^2 x_1 - x_2 = 0 && \therefore\ x_2 = 3 \text{ etc.} \end{aligned}$$

Thus one solution of 1.2.11 is generated, namely

$$x = 1 + \varepsilon + 3\varepsilon^2 + \cdots. \qquad 1.2.12$$

There are two solutions of 1.2.11 which we expect to be associated with large values of x so that the term εx^3 is present to leading order. If we follow the previous example and assume an expansion with dominant term x_0/ε, i.e.

$$x = \frac{x_0}{\varepsilon} + x_1 + \varepsilon x_2 + \cdots,$$

then substituting into 1.2.11 yields,

$$\varepsilon\left(\frac{x_0^3}{\varepsilon^3} + \frac{3x_0^2 x_1}{\varepsilon^2} + \cdots\right) - \frac{x_0}{\varepsilon} + \cdots + 1 = 0.$$

The dominant term is of order ε^{-2} but merely leads to $x_0 = 0$. This again shows the need to preserve at least two terms in the dominant order equation. In this case if x is large it dominates the third member of the left-hand side of 1.2.11 so the terms εx^3 and x must have the same order.

If we let the dominant term in the expansion be x_0/ε^p then we require

$$\varepsilon\frac{x_0^3}{\varepsilon^{3p}} - \frac{x_0}{\varepsilon^p} = 0.$$

Now x_0 is a number independent of ε so the coefficients of ε must be the same, i.e. $1 - 3p = -p$ so that $p = \frac{1}{2}$. Then x_0 satisfies $x_0^3 - x_0 = 0$ with solutions $x_0 = 0$ and ± 1. We are only interested in nonzero values of the leading coefficient; thus two possible leading terms have been obtained, $1/\varepsilon^{1/2}$ and $-1/\varepsilon^{1/2}$. The appearance of half powers suggests that the asymptotic sequence should consist of ascending powers of $\varepsilon^{1/2}$, i.e.

$$\{\varepsilon^{-1/2}, 1, \varepsilon^{1/2}, \varepsilon, \varepsilon^{3/2}, \ldots\}.$$

On substituting the expansion

$$x = \frac{x_0}{\varepsilon^{1/2}} + x_1 + \varepsilon^{1/2} x_2 + \varepsilon x_3 + \cdots$$

into 1.2.11 we obtain

$$\varepsilon\left(\frac{x_0^3}{\varepsilon^{3/2}} + \frac{3x_0^2 x_1}{\varepsilon} + \frac{3x_0^2 x_2}{\varepsilon^{1/2}} + \frac{3x_0 x_1^2}{\varepsilon^{1/2}} + \cdots\right) - \frac{x_0}{\varepsilon^{1/2}} - x_1 - \varepsilon^{1/2} x_2 - \cdots + 1 = 0.$$

Equating coefficients of powers of ε yields;

$$\begin{aligned}
&O(\varepsilon^{-1/2}): && x_0^3 - x_0 = 0 && \therefore\ x_0 = \pm 1\ (x_0 = 0 \text{ is rejected})\\
&O(1): && 3x_0^2 x_1 - x_1 + 1 = 0 && \therefore\ x_1 = -1/2\\
&O(\varepsilon^{1/2}): && 3x_0^2 x_2 + 3x_0 x_1^2 - x_2 = 0 && \therefore\ x_2 = \mp 3/8 \text{ etc.}
\end{aligned}$$

Thus the two singular expansions for the solutions of 1.2.11 are

$$x = \frac{1}{\varepsilon^{1/2}} - \frac{1}{2} - \frac{3\varepsilon^{1/2}}{8} + \cdots$$

and

$$x = -\frac{1}{\varepsilon^{1/2}} - \frac{1}{2} + \frac{3\varepsilon^{1/2}}{8} + \cdots.$$

The third solution is given by the regular expansion 1.2.12.

So far we have considered examples of singular perturbations where at least one of the solutions of the algebraic equation is absent in the unperturbed problem. The following are examples of regular perturbations where the unperturbed solutions are continuously approached as $\varepsilon \to 0$.

Consider the equation

$$x^3 - x + \varepsilon = 0, \quad 0 < \varepsilon \ll 1. \tag{1.2.13}$$

Assume an expansion of the form

$$x = x_0 + \varepsilon x_1 + \varepsilon^2 x_2 + \cdots$$

and substitute into 1.2.13 we obtain

$$x_0^3 + 3\varepsilon x_0^2 x_1 + 3\varepsilon^2 x_0^2 x_2 + 3\varepsilon^2 x_0 x_1^2 + \cdots - x_0 - \varepsilon x_1 - \varepsilon^2 x_2 + \cdots + \varepsilon = 0.$$

Equating coefficients of powers of ε leads to

$$\begin{aligned}
&O(1): && x_0^3 - x_0 = 0 && \therefore\ x_0 = 0, 1, -1\\
&O(\varepsilon): && 3x_0^2 x_1 - x_1 + 1 = 0 && \therefore\ x_1 = 1/(1 - 3x_0^2)\\
&O(\varepsilon^2): && 3x_0^2 x_2 + 3x_0 x_1^2 - x_2 = 0 && \therefore\ x_2 = 3x_0 x_1^2/(1 - 3x_0^2).
\end{aligned}$$

Thus when $x_0 = 0$, $\quad x_1 = 1$ and $x_2 = 0$
when $x_0 = 1$, $\quad x_1 = -1/2$ and $x_2 = -3/8$
when $x_0 = -1$, $\quad x_1 = -1/2$ and $x_2 = 3/8$.
The solutions of 1.2.13 are therefore given by

$$x = 1 - \varepsilon/2 - 3\varepsilon^2/8 + O(\varepsilon^3)$$

$$x = -1 - \varepsilon/2 + 3\varepsilon^2/8 + O(\varepsilon^3)$$

$$x = \varepsilon + O(\varepsilon^3).$$

The next example involves a quartic equation

$$x^4 + \varepsilon x^3 - 5x^2 + 4 = 0, \quad 0 < \varepsilon \ll 1.$$

Substitution of a standard expansion $x = x_0 + \varepsilon x_1 + \cdots$ leads to

$$x_0^4 + 4\varepsilon x_0^3 x_1 + \cdots + \varepsilon x_0^3 + \cdots - 5x_0^2 - 10\varepsilon x_0 x_1 + 4 = 0,$$

$$O(1): \quad x_0^4 - 5x_0^2 + 4 = 0 \qquad 1.2.14a$$

$$O(\varepsilon): \quad 4x_0^3 x_1 + x_0^3 - 10x_0 x_1 = 0. \qquad 1.2.14b$$

Although the leading order equation, 1.2.14a, is of the same degree as the original problem, the solutions are easily found since it factorizes, i.e.

$$(x_0^2 - 4)(x_0^2 - 1) = 0.$$

Thus $x_0 = 1, -1, 2, -2$ and equation 1.2.14b yields $x_1 = x_0^2/(10 - 4x_0^2)$ so that $x_1 = 1/6, 1/6, -2/3, -2/3$ respectively. The four solutions of the quartic to $O(\varepsilon)$ are

$$x = 1 + \varepsilon/6 + O(\varepsilon^2)$$

$$x = -1 + \varepsilon/6 + O(\varepsilon^2)$$

$$x = 2 - 2\varepsilon/3 + O(\varepsilon^2)$$

$$x = -2 - 2\varepsilon/3 + O(\varepsilon^2).$$

In the following example of a regular perturbation the unperturbed problem has repeated solutions. This leads to a nonstandard form of perturbation expansion. Consider the equation

$$x^3 - 2x^2 + x(1 + \varepsilon) - 2\varepsilon = 0, \quad 0 < \varepsilon \ll 1. \qquad 1.2.15$$

Assuming a standard expansion $x = x_0 + \varepsilon x_1 + \cdots$ and substituting leads to

$$O(1): \quad x_0^3 - 2x_0^2 + x_0 = 0 \qquad 1.2.16a$$

$$O(\varepsilon): \quad 3x_0^2 x_1 - 4x_0 x_1 + x_1 + x_0 - 2 = 0. \qquad 1.2.16b$$

The solutions of 1.2.16a are $x_0 = 0$ and $x_0 = 1$ (repeated). When $x_0 = 0$ equation 1.2.16b gives $x_1 = 2$. However, when $x_0 = 1$ equation 1.2.16b cannot be satisfied. Thus only one of the solutions has an expansion of standard form namely the solution with $x_0 = 0$, the expansion is

$$x = 2\varepsilon + O(\varepsilon^2).$$

The fact that assuming the standard form of expansion leads to an inconsistency when $x_0 = 1$ indicates that a different asymptotic sequence is involved. Let the small term λ be the difference between the solution x and the leading term $x_0 = 1$. Its order is determined as follows.

Substitute $x = 1 + \lambda$ into equation 1.2.15, to obtain

$$1 + 3\lambda + 3\lambda^2 + \lambda^3 - 2 - 4\lambda - 2\lambda^2 + 1 + \lambda + \varepsilon + \varepsilon\lambda - 2\varepsilon = 0$$

which simplifies to yield

$$\lambda^2 + \lambda^3 - \varepsilon + \varepsilon\lambda = 0.$$

The dominant terms in this equation are λ^2 (which dominates λ^3 for small λ) and ε (which dominates $\varepsilon\lambda$ for small λ). Thus λ^2 is of order ε and so half integer powers of ε will appear in the perturbation expansion.

We therefore assume the expansion

$$x = 1 + x_1\varepsilon^{1/2} + x_2\varepsilon + x_3\varepsilon^{3/2} + \cdots$$

and substitute into equation 1.2.15. Working to $O(\varepsilon^{3/2})$ leads to

$$1 + 3x_1\varepsilon^{1/2} + 3x_2\varepsilon + 3x_3\varepsilon^{3/2} + 3x_1^2\varepsilon + 6x_1x_2\varepsilon^{3/2} + x_1^3\varepsilon^{3/2} - 2 - 4x_1\varepsilon^{1/2}$$
$$- 4x_2\varepsilon - 4x_3\varepsilon^{3/2} - 2x_1^2\varepsilon - 4x_1x_2\varepsilon^{3/2} + 1 + x_1\varepsilon^{1/2} + x_2\varepsilon + x_3\varepsilon^{3/2} + \varepsilon$$
$$+ x_1\varepsilon^{3/2} - 2\varepsilon = 0.$$

After the canceling of terms we obtain

$$O(\varepsilon) \quad : \quad x_1^2 - 1 = 0 \qquad \therefore\ x_1 = \pm 1$$
$$O(\varepsilon^{3/2}) : \quad x_1^3 + 2x_1x_2 + x_1 = 0 \qquad \therefore\ x_2 = -1.$$

Thus the two roots associated with $x_0 = 1$ have the expansions

$$x = 1 + \varepsilon^{1/2} - \varepsilon + O(\varepsilon^{3/2})$$

and

$$x = 1 - \varepsilon^{1/2} - \varepsilon + O(\varepsilon^{3/2}).$$

Exercises

Obtain two-term expansions for the solutions of the following equations:

(i) $(x - 1)(x - 2)(x - 3) + \varepsilon = 0$
(ii) $x^3 + x^2 - \varepsilon = 0$
(iii) $x^4 - 6x^2 + \varepsilon x + 8 = 0$
(iv) $x^3 - 4x + 2\varepsilon = 0$
(v) $(x^2 - 1)(x^2 - 4)(x - 3) + \varepsilon = 0$
(vi) $\varepsilon x^3 + x^2 + 3x + 2 = 0$
(vii) $\varepsilon x^3 + x^2 + 2x + 1 = 0$
(viii) $\varepsilon x^4 - x^2 + 3x - 2 + \varepsilon = 0$

Selected answers

(i) $1 - \varepsilon/2 + \cdots$
$2 + \varepsilon + \cdots$
$3 - \varepsilon/2 + \cdots$

(ii) $\sqrt{\varepsilon} - \varepsilon/2 + \cdots$
$-\sqrt{\varepsilon} - \varepsilon/2 + \cdots$
$-1 + \varepsilon + \cdots$

(vii) $-1 + \sqrt{\varepsilon} + \cdots$
$-1 - \sqrt{\varepsilon} + \cdots$
$-\dfrac{1}{\varepsilon} + 2 + \cdots$

1.3 Initial value problems

We return to the study of differential equations which began in Section 1. Consider the nonlinear differential equation

$$\frac{df}{dt} + f = \varepsilon f^2, \qquad 0 < \varepsilon \ll 1, \tag{1.3.1}$$

with initial condition $f(0) = 1$.

The solution $f(t; \varepsilon)$ depends on the independent variable t and the parameter ε. We will assume that the ε dependence is of a standard power series form

$$f(t; \varepsilon) = f_0(t) + \varepsilon f_1(t) + \varepsilon^2 f_2(t) + \cdots, \tag{1.3.2}$$

where the coefficient functions $f_0(t)$, $f_1(t)$ ... are independent of ε. Substituting the expansion 1.3.2 into the differential equation 1.3.1 leads to

$$\frac{df_0}{dt} + \varepsilon\frac{df_1}{dt} + \varepsilon^2\frac{df_2}{dt} + \cdots + f_0 + \varepsilon f_1 + \varepsilon^2 f_2 + \cdots$$
$$= \varepsilon(f_0^2 + 2\varepsilon f_0 f_1 + \cdots). \tag{1.3.3}$$

The initial condition yields

$$f_0(0) + \varepsilon f_1(0) + \varepsilon^2 f_2(0) + \cdots = 1 + 0\varepsilon + 0\varepsilon^2 + \cdots. \tag{1.3.4}$$

Equating coefficients of powers of ε on the left- and right-hand sides of equations 1.3.3 and 1.3.4 leads to the following set of equations,

$$O(1): \quad \frac{df_0}{dt} + f_0 = 0, \qquad f_0(0) = 1 \tag{1.3.5a}$$

$$O(\varepsilon): \quad \frac{df_1}{dt} + f_1 = f_0^2, \qquad f_1(0) = 0 \tag{1.3.5b}$$

$$O(\varepsilon^2): \quad \frac{df_2}{dt} + f_2 = 2f_0 f_1, \quad f_2(0) = 0. \tag{1.3.5c}$$

Equation 1.3.5a is a linear differential equation with constant coefficients. It has the general solution $f_0 = Ae^{-t}$ and the initial condition requires that

$$A = 1, \quad \text{so} \quad f_0 = e^{-t}.$$

Equation 1.3.5b involves a nonlinear term on the right-hand side but this is a known function. The equation is linear in the unknown function f_1. This constant coefficient form can be solved by the complementary function plus particular integral technique, so that we have

$$\frac{df_1}{dt} + f_1 = e^{-2t}, \quad f_1(0) = 0.$$

The complementary function is Ae^{-t} and a particular integral is $-e^{-2t}$, thus

$$f_1 = Ae^{-t} - e^{-2t},$$

and the initial condition yields $A = 1$, thus

$$f_1 = e^{-t} - e^{-2t}.$$

Equation 1.3.5c is again linear in the unknown function f_2,

$$\frac{df_2}{dt} + f_2 = 2e^{-t}(e^{-t} - e^{-2t}).$$

The complementary function is again Ae^{-t} and a particular integral has the form $\alpha e^{-2t} + \beta e^{-3t}$ where

$$(-2\alpha + \alpha)e^{-2t} + (-3\beta + \beta)e^{-3t} = 2e^{-2t} - 2e^{-3t} \quad \text{leading to} \quad \alpha = -2, \beta = 1,$$

so that

$$f_2 = Ae^{-t} - 2e^{-2t} + e^{-3t}.$$

The initial condition $f_2(0) = 0$ yields $A = 1$.

Thus the three-term expansion for the solution of 1.3.1 is

$$f = e^{-t} + \varepsilon(e^{-t} - e^{-2t}) + \varepsilon^2(e^{-t} - 2e^{-2t} + e^{-3t}) + \cdots. \tag{1.3.6}$$

It is interesting to note that, unlike the expansion 1.1.14 for the introductory problem, the expansion 1.3.6 is uniform for all positive values of the independent variable. This is so because each coefficient function is of the same order for positive t, namely $O(e^{-t})$. Consequently the ascending powers of ε ensure that subsequent terms are small corrections to the previous terms. The perturbation expansion is said to be *uniformly valid* for all positive t.

It has been emphasized that the governing equations for the terms in the expansion 1.3.2 are of linear constant coefficient form and are consequently straightforward to solve. As it happens, the original equation, while nonlinear, is of variables separable form and can be solved exactly. This solution is obtained below to provide a validation of the perturbation approach. In some of the subsequent examples, exact solutions of the original equations cannot be obtained whereas the equations determining the coefficient functions in the perturbation expansion can be solved exactly.

Equation 1.3.1 can be written in the form

$$\frac{df}{dt} = f(\varepsilon f - 1),$$

so that

$$\int \frac{1}{f(\varepsilon f - 1)} \frac{df}{dt}\, dt = \int dt.$$

Using partial fractions we obtain

$$\int \left(-\frac{1}{f} + \frac{\varepsilon}{\varepsilon f - 1} \right) df = \int dt.$$

Integrating gives the expression

$$-\ln|f| + \ln|\varepsilon f - 1| = t + c,$$

which becomes

$$\ln\left|\frac{\varepsilon f - 1}{f}\right| = t + c.$$

After exponentiating and removing the modulus sign by allowing the constant $A(=e^c)$ to be positive or negative we obtain

$$\frac{\varepsilon f - 1}{f} = Ae^t.$$

The initial condition requires $A = \varepsilon - 1$ and on solving for f we obtain

$$f = \frac{e^{-t}}{1 - \varepsilon(1 - e^{-t})}. \qquad 1.3.7$$

The perturbation expansion can be obtained from the exact solution by use of the binomial expansion

$$\frac{1}{1 - x} = 1 + x + x^2 + \cdots,$$

with $x = \varepsilon(1 - e^{-t})$. This yields

$$f = e^{-t}(1 + \varepsilon(1 - e^{-t}) + \varepsilon^2(1 - e^{-t})^2 + \cdots),$$

which is the same as the expansion 1.3.6.

The following example is nonlinear and not of variables separable type so that a perturbation expansion is of genuine value. Consider then the initial value problem,

$$\frac{df}{dt} + f = \varepsilon t f^2, \quad 0 < \varepsilon \ll 1, \quad f(0) = 1. \qquad 1.3.8$$

We will obtain the two-term expansion $f_0 + \varepsilon f_1$ by working to order ε. Thus

$$\frac{df_0}{dt} + \varepsilon \frac{df_1}{dt} + f_0 + \varepsilon f_1 + O(\varepsilon^2) = \varepsilon t f_0^2 + O(\varepsilon^2),$$

and

$$f_0(0) + \varepsilon f_1(0) + O(\varepsilon^2) = 1 + O\varepsilon + O(\varepsilon^2).$$

Equating coefficients of powers of ε leads to

$$O(1): \quad \frac{df_0}{dt} + f_0 = 0, \qquad f_0(0) = 1$$

$$O(\varepsilon): \quad \frac{df_1}{dt} + f_1 = t f_0^2, \quad f_1(0) = 0.$$

The $O(1)$ equation has solution $f_0 = e^{-t}$.

The $O(\varepsilon)$ equation is

$$\frac{df_1}{dt} + f_1 = t e^{-2t}.$$

This may be solved by the complementary function plus particular integral technique but it is more convenient to multiply throughout by the integrating factor, e^t. Then we have

$$\frac{d}{dt}(e^t f_1) = t e^{-t},$$

so that

$$e^t f_1 = -t e^{-t} - e^{-t} + c.$$

The constant c is found to equal unity from the initial condition. Thus

$$f_1 = e^{-t} - e^{-2t} - t e^{-2t},$$

and the two-term expansion for the solution of equation 1.3.8 is

$$f = e^{-t} + \varepsilon(e^{-t} - e^{-2t} - t e^{-2t}).$$

In the concluding example of this section we consider a second order differential equation with a nonlinear perturbation,

$$\frac{d^2 f}{dt^2} + f = \varepsilon f^2, \quad 0 < \varepsilon \ll 1$$

$$f(0) = 1, \quad \frac{df}{dt}(0) = -1. \qquad 1.3.9$$

We work to $O(\varepsilon)$ to obtain the two-term expansion $f_0 + \varepsilon f_1$.

$$O(1): \quad \frac{d^2 f_0}{dt^2} + f_0 = 0, \quad f_0(0) = 1, \quad \frac{df_0}{dt}(0) = -1$$

$$O(\varepsilon): \quad \frac{d^2 f_1}{dt^2} + f_1 = f_0^2, \quad f_1(0) = 0, \quad \frac{df_1}{dt}(0) = 0.$$

The general solution of the $O(1)$ equation is

$$f_0 = A \cos t + B \sin t,$$

and the initial conditions yield $A = 1$, $B = -1$.

The $O(\varepsilon)$ equation becomes

$$\frac{d^2 f_1}{dt^2} + f_1 = \cos^2 t - 2 \cos t \sin t + \sin^2 t = 1 - \sin 2t.$$

The complementary function is $A \cos t + B \sin t$. A particular integral has the form $\alpha + \beta \sin 2t$ where

$$\alpha - 4\beta \sin 2t + \beta \sin 2t = 1 - \sin 2t.$$

Therefore $\alpha = 1$, $\beta = 1/3$ and the general solution for f_1 is given by

$$f_1 = A \cos t + B \sin t + 1 + \sin 2t/3.$$

The initial conditions lead to $A = -1$ and $B = -2/3$, thus the two-term expansion is

$$f = \cos t - \sin t + \varepsilon\left(1 - \cos t - \frac{2}{3}\sin t + \frac{1}{3}\sin 2t\right).$$

Exercises

1 A projectile is moving vertically under the action of gravity and a resistance force which is proportional to the cubed power of the speed. The governing equation in nondimensional form is

$$\frac{dv}{dt} = -1 - \varepsilon v^3, \quad v(0) = 1.$$

Obtain a two-term perturbation expansion $v_0 + \varepsilon v_1$.

2 Obtain a two-term expansion for the solution of

$$\frac{df}{dt} - f = \varepsilon f^2 e^{-t}, \quad f(0) = 1.$$

3 Obtain a two-term expansion for the solution of

$$\frac{d^2 f}{dt^2} + f = \varepsilon \frac{df}{dt}, \quad f(0) = 1, \quad \frac{df}{dt}(0) = 0.$$

Compare this expansion with that obtained by expanding the exact solution.

4 A mass spring system is subject to a small damping proportional to the cubed power of the speed

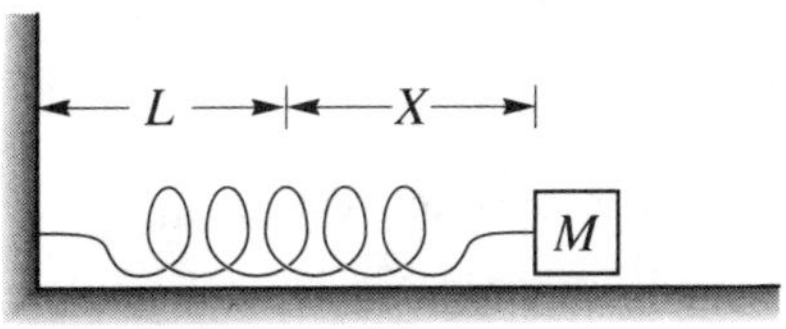

The governing equation is

$$M \frac{d^2 X}{dT^2} = -\frac{\Lambda}{L} X - K \left(\frac{dX}{dT} \right)^3,$$

where X is the extension, L the natural length, T is time, Λ is a spring constant and K a resistance constant. The mass starts from rest with an initial extension X_0. Introduce the nondimensional extension $x = X/X_0$ and time $t = T\sqrt{\Lambda/ML}$ and obtain the nondimensional governing equation

$$\frac{d^2 x}{dt^2} + x = -\varepsilon \left(\frac{dx}{dt} \right)^3, \quad x(0) = 1, \quad \frac{dx}{dt}(0) = 0.$$

Express ε in terms of the constants of the system. Obtain a two-term expansion $x_0 + \varepsilon x_1$.

1.4 Expansions involving the independent variable

In the previous sections we have constructed perturbation expansions using powers of a small parameter. It is often useful to use powers of the independent variable to construct a perturbation expansion. A commonly used technique for investigating the behavior of functions is the construction of its Maclaurin expansion or more generally its Taylor expansion about a chosen point. For example, the behavior of the function e^x near $x = 0$ is described by the first few terms of the Maclaurin expansion

$$e^x = 1 + x + x^2/2 + x^3/6 + \cdots.$$

This can be used to estimate $e^{0.1}$ as

$$e^{0.1} = 1 + 0.1 + 0.005 + 0.00016 + \cdots = 1.105167 \text{ to 6 decimal places.}$$

(The exact value of $e^{0.1}$ is 1.105171 to 6 decimal places.)

Similarly $\sin(31^\circ)$ can be obtained from the Taylor series of $\sin x$ about the value of $\sin(30^\circ)$. The values of $\sin 30^\circ$ and $\cos 30^\circ$ will be needed; they are 1/2 and $\sqrt{3}/2$ respectively. Then, using radians we have

$$\sin(a + x) = \sin a + x \cos a - \frac{x^2}{2} \sin a + \cdots,$$

so that

$$\begin{aligned}\sin\left(\frac{\pi}{6} + \frac{\pi}{180}\right) &= \sin\frac{\pi}{6} + \frac{\pi}{180}\cos\frac{\pi}{6} - \left(\frac{\pi}{180}\right)^2 \frac{1}{2}\sin\frac{\pi}{6} + \cdots \\ &= \frac{1}{2} + \frac{\pi}{180}\frac{\sqrt{3}}{2} - \left(\frac{\pi}{180}\right)^2 \frac{1}{2}\cdot\frac{1}{2} + \cdots = 0.515039 \text{ to 6 decimal places.}\end{aligned}$$

(The exact value of $\sin(31^\circ)$ is 0.515038 to 6 decimal places.)

Of course, tables of the exponential and trigonometric functions exist and these functions are available on most pocket calculators. However, the function

$$f(x) = \int_0^x \exp(-s^3)\,\mathrm{d}s,$$

for example, is not commonly tabulated so that an expansion in powers of x is of practical use. The integrand is first replaced by its Maclaurin expansion

$$f(x) = \int_0^x \left(1 - s^3 + \frac{s^6}{2} - \frac{s^9}{6} + \cdots\right)\mathrm{d}s.$$

Integrating each term leads to

$$f(x) = x - \frac{x^4}{4} + \frac{x^7}{14} - \frac{x^{10}}{60} + \cdots$$

The error associated with this integrated truncated series can be shown to be bounded by the next term in the series. Thus the expansion provides an accurate means of evaluating the function for small values of x.

If the value of a function is required for large values of the independent variable it is often possible to obtain an expansion in inverse powers. Consider for example the function

$$f(x) = \frac{1 + x}{1 - 2x}.$$

Its value as $x \to \infty$ is $-1/2$. To investigate the way in which this limit is reached we expand using powers of $1/x$. On dividing the numerator and denominator by $-2x$ we have

$$f(x) = -\frac{\frac{1}{2}+\frac{1}{2x}}{1-\frac{1}{2x}} = -\left(\frac{1}{2}+\frac{1}{2x}\right)\left(1+\frac{1}{2x}+\frac{1}{4x^2}+\cdots\right)$$

$$= -\frac{1}{2}-\frac{1}{4x}-\frac{1}{8x^2}+\cdots-\frac{1}{2x}-\frac{1}{4x^2}+\cdots = -\frac{1}{2}-\frac{3}{4x}-\frac{3}{8x^2}+\cdots.$$

This expansion shows the way in which the limiting value of $-1/2$ is approached as $x \to \infty$. If we test the accuracy of the three-term expansion for $x = 10$ we have

$$f(x) \simeq -0.5 - 0.075 - 0.00375 = -0.57875,$$

while the exact value is $-11/19 = 0.57894$ (to 5 decimal places.)

Expansions in powers or inverse powers of the independent variable are helpful when the dependent variable is given in the form of a differential equation or an integral. Consider the solution of the following differential equation for both small and large values of the independent variable,

$$(1-x^2)\frac{dy}{dx} = 2y. \qquad 1.4.1$$

First, for small x, we assume an expansion

$$y = a_0 + a_1 x + a_2 x^2 + a_3 x^3 + \cdots$$

where the coefficients $a_0, a_1, a_2, \ldots$ are constants. Substituting the expansion into equation 1.4.1 leads to

$$a_1 + 2a_2 x + 3a_3 x^2 + 4a_4 x^3 + \cdots - a_1 x^2 - 2a_2 x^3 + \cdots$$
$$= 2a_0 + 2a_1 x + 2a_2 x^2 + 2a_3 x^3 + \cdots.$$

Equating coefficients of powers of x yields

$$\begin{aligned}
O(1) &: a_1 &&= 2a_0 \\
O(x) &: 2a_2 &&= 2a_1 && \therefore\ a_2 = 2a_0 \\
O(x^2) &: 3a_3 - a_1 &&= 2a_2 && \therefore\ a_3 = 2a_0 \\
O(x^3) &: 4a_4 - 2a_2 &&= 2a_3 && \therefore\ a_4 = 2a_0.
\end{aligned}$$

Thus

$$y = a_0[1 + 2(x + x^2 + x^3 + x^4 + \cdots)] \qquad 1.4.2$$

where a_0 is determined from the initial value of the function, $y(0) = a_0$.

Next consider large x and assume an expansion

$$y = b_0 + \frac{b_1}{x} + \frac{b_2}{x^2} + \frac{b_3}{x^3} + \cdots$$

where the b_i are constants. Substituting into 1.4.1 leads to

$$-\frac{b_1}{x^2} - \frac{2b_2}{x^3} + \cdots + b_1 + \frac{2b_2}{x} + \frac{3b_3}{x^2} + \frac{4b_4}{x^3} + \cdots$$

$$= 2b_0 + \frac{2b_1}{x} + \frac{2b_2}{x^2} + \frac{2b_3}{x^3} + \cdots.$$

Equating coefficients of inverse powers of x yields,

$$\begin{aligned}
&O(1) &:\quad & b_1 &= 2b_0 & \\
&O(1/x) &:\quad & 2b_2 &= 2b_1 &\qquad \therefore\ b_2 = 2b_0 \\
&O(1/x^2) &:\quad & -b_1 + 3b_3 &= 2b_2 &\qquad \therefore\ b_3 = 2b_0 \\
&O(1/x^3) &:\quad & -2b_2 + 4b_4 &= 2b_3 &\qquad \therefore\ b_4 = 2b_0.
\end{aligned}$$

Thus

$$y = b_0\left[1 + 2\left(\frac{1}{x} + \frac{1}{x^2} + \frac{1}{x^3} + \frac{1}{x^4} + \cdots\right)\right] \tag{1.4.3}$$

where b_0 equals the value of the function as $x \to \infty$, $y(\infty) = b_0$.

Convergent and divergent series

So far no mention has been made of convergence. In fact we will discover that the convergence or divergence of infinite series is not important in the study of perturbation expansions because the expressions we deal with are always finite series with a remainder. The important property is the behavior of the remainder in the region of some point of interest, typically the behavior as x tends to zero or infinity. However, it is of interest to briefly consider the convergence of infinite series at this stage.

The ratio test for convergence states that a series of terms $u_0 + u_1 + u_2 + u_3 + \cdots + u_N + u_{N+1} + \cdots$ is convergent if

$$\lim_{N\to\infty}\left|\frac{u_{N+1}}{u_N}\right| < 1,$$

when applied to the series 1.4.2 this gives

$$\lim_{N\to\infty}\left|\frac{2a_0x^{N+1}}{2a_0x^N}\right| < 1,$$

thus the series converges for $|x| < 1$.

Similarly applying the ratio test to the series 1.4.3 gives

$$\lim_{N\to\infty}\left|\frac{2b_0/x^{N+1}}{2b_0/x^N}\right| < 1,$$

i.e. $|1/x| < 1$. Thus the series of inverse powers converges for $|x| > 1$.

In the next example a series of inverse powers of x is constructed which *diverges* for all x and yet in truncated form the series provides a useful approximation to the function it represents. Consider the behavior of the solution of

$$\frac{dy}{dx} - y = -\frac{1}{x}, \qquad 1.4.4$$

for large values of x with the boundary condition $y(\infty) = 0$. Assume the expansion

$$y = b_0 + \frac{b_1}{x} + \frac{b_2}{x^2} + \frac{b_3}{x^3} + \frac{b_4}{x^4} + \cdots,$$

where, from the boundary condition, $b_0 = 0$.

Substitute into 1.4.4 to obtain

$$-\frac{b_1}{x^2} - \frac{2b_2}{x^3} - \frac{3b_3}{x^4} - \cdots - \frac{b_1}{x} - \frac{b_2}{x^2} - \frac{b_3}{x^3} - \frac{b_4}{x^4} + \cdots = -\frac{1}{x}.$$

Equating coefficients of inverse powers of x gives

$$O\left(\frac{1}{x}\right) \ : -b_1 = -1 \qquad \therefore\ b_1 = 1$$

$$O\left(\frac{1}{x^2}\right) : -b_1 - b_2 = 0 \qquad \therefore\ b_2 = -1$$

$$O\left(\frac{1}{x^3}\right) : -2b_2 - b_3 = 0 \qquad \therefore\ b_3 = 2!$$

$$O\left(\frac{1}{x^4}\right) : -3b_3 - b_4 = 0 \qquad \therefore\ b_4 = -3!$$

Clearly $b_N = (-1)^{N-1}(N-1)!$ so that

$$y = \frac{1}{x} - \frac{1}{x^2} + \frac{2!}{x^3} - \frac{3!}{x^4} + \cdots + \frac{(-1)^{N-1}(N-1)!}{x^N} + \cdots. \qquad 1.4.5$$

The ratio test applied to the series 1.4.5 requires for convergence,

$$\lim_{N\to\infty}\left|\frac{(-1)^N N!}{x^{N+1}} \cdot \frac{x^N}{(-1)^{N-1}(N-1)!}\right| < 1$$

i.e. $$\frac{1}{|x|} < \frac{1}{\lim_{N\to\infty}(N)} = 0.$$

Thus the series 1.4.5 diverges for all x. Nevertheless, the first few terms of the series provide a good approximation to the function for large values of x. To see why this is so we must investigate the behavior of the remainder R_N when the function is approximated by the partial sum of the series S_N, that is

$$y = S_N + R_N$$

with

$$S_N = \frac{1}{x} - \frac{1}{x^2} + \frac{2!}{x^3} - \frac{3!}{x^4} + \cdots + \frac{(-1)^{N-1}(N-1)!}{x^N}.$$

An expression for R_N can be obtained by repeated integration by parts. First multiply equation 1.4.4 throughout by the integrating factor e^{-x} to obtain

$$\frac{d}{dx}(e^{-x}y) = -\frac{e^{-x}}{x}.$$

Integration yields

$$[e^{-s}y(s)]_x^\infty = -\int_x^\infty \frac{e^{-s}}{s}ds,$$

$$e^{-\infty}y(\infty) - e^{-x}y(x) = -\int_x^\infty \frac{e^{-s}}{s}ds,$$

so that

$$y = e^x \int_x^\infty \frac{e^{-s}}{s}ds. \qquad 1.4.6$$

Then on integrating by parts we obtain

$$y = e^x\left\{\left[-\frac{e^{-s}}{s}\right]_x^\infty - \int_x^\infty \frac{e^{-s}}{s^2}ds\right\} = \frac{1}{x} - e^x \int_x^\infty \frac{e^{-s}}{s^2}ds.$$

Integrating by parts again yields

$$y = \frac{1}{x} - e^x\left\{\left[-\frac{e^{-s}}{s^2}\right]_x^\infty - \int_x^\infty 2\frac{e^{-s}}{s^2}ds\right\} = \frac{1}{x} - \frac{1}{x^2} + 2e^x \int_x^\infty \frac{e^{-s}}{s^3}ds.$$

Repeated integration by parts leads to

$$y = \frac{1}{x} - \frac{1}{x^2} + \frac{2}{x^3} - \frac{3!}{x^4} + \cdots + \frac{(-1)^{N-1}(N-1)!}{x^N} + (-1)^N N! e^x \int_x^\infty \frac{e^{-s}}{s^{N+1}}ds. \qquad 1.4.7$$

The truncated series of N terms is the partial sum S_N and the remainder R_N is given by the integral

$$R_N = (-1)^N N! e^x \int_x^\infty \frac{e^{-s}}{s^{N+1}}ds. \qquad 1.4.8$$

The function y is bounded for $x > 0$. This can be shown by investigating the integral expression 1.4.6

$$y = e^{x} \int_{x}^{\infty} \frac{e^{-s}}{s} ds < e^{x} \int_{x}^{\infty} \frac{e^{-s}}{x} ds = \frac{e^{x}}{x} [-e^{-s}]_{x}^{\infty} = \frac{1}{x}.$$

Thus $0 < y < 1/x$ for all positive x. (In practice we are interested in x large and positive.)

Asymptotic series

The ratio test has shown that the partial sum, S_N, in equation 1.4.7 diverges as $N \to \infty$. The remainder, R_N, also diverges as $N \to \infty$ but the combination is the bounded quantity y where $0 < y < 1/x$. Although the limit of the partial sums S_∞ diverges for any value of x, the sum S_N provides a valid approximation of the function for sufficiently large x. To see this we must consider the alternative limit as $x \to \infty$ for N fixed. First we bound R_N as follows:

$$|R_N| = N! e^{x} \int_{x}^{\infty} \frac{e^{-s}}{s^{N+1}} ds \leqslant N! e^{x} \int_{x}^{\infty} \frac{e^{-s}}{x^{N+1}} ds = \frac{N!}{x^{N+1}}.$$

The truncated series involves inverse powers of x up to and including $1/x^N$, whereas the remainder is bounded by the inverse power $1/x^{N+1}$. The numerator, $N!$, of the remainder bound becomes arbitrarily large as N increases but if N is fixed the remainder can be made as small as we like relative to the truncated series by taking sufficiently large values of x.

The question of convergence or divergence is not important since we will always use a truncated series of terms (usually only two or three). The important consideration is the behavior of the remainder for large x. In terms of limits we do not ask
'does the limit of R_N as $N \to \infty$ equal zero for x fixed?'
Instead we ask
'does the limit of R_N as $x \to \infty$ equal zero for N fixed and, if so, does R_N approach zero faster than the terms in the truncated series?'
Series which satisfy the latter condition are called *asymptotic series.*

When asymptotic series are used to obtain approximate values of a function the independent variable will take finite but large values. In practice 'large' often only means $x > 4$. The remainder, R_N, will be nonzero for finite x and judgment must be used to determine the optimum number of terms in the partial sum S_N such that the bound for R_N is minimized.

Optimum truncation rule

Consider the expansion 1.4.7

$$y = \frac{1}{x} - \frac{1}{x^2} + \frac{2!}{x^3} - \frac{3!}{x^4} + \cdots + (-1)^{N-1}\frac{(N-1)!}{x^N} + R_N$$

with $|R_N| \leqslant N!/x^{N+1}$.

Using the series to approximate $y(4)$ we have

Truncated series	Remainder bound
$y_{1\,\text{Term}} = \frac{1}{4}$	$\frac{1}{16}$
$y_{2\,\text{Term}} = \frac{1}{4} - \frac{1}{16}$	$\frac{2}{64}$
$y_{3\,\text{Term}} = \frac{1}{4} - \frac{1}{16} + \frac{2}{64}$	$\frac{6}{256}$
$y_{4\,\text{Term}} = \frac{1}{4} - \frac{1}{16} + \frac{2}{64} - \frac{6}{256}$	$\frac{6 \times 4}{256 \times 4}$
$y_{5\,\text{Term}} = \frac{1}{4} - \frac{1}{16} + \frac{2}{64} - \frac{6}{256} + \frac{6}{256}$	$\frac{6 \times 4 \times 5}{256 \times 4 \times 4}$.

The sign of the remainder is opposite that of the last term in the truncated series; thus we have

1-term approximation

$$\frac{1}{4} - \frac{1}{16} < y(4) < \frac{1}{4}$$

2-term approximation

$$\frac{1}{4} - \frac{1}{16} < y(4) < \frac{1}{4} - \frac{1}{16} + \frac{2}{64}$$

3-term approximation

$$\frac{1}{4} - \frac{1}{16} + \frac{2}{64} - \frac{6}{256} < y(4) < \frac{1}{4} - \frac{1}{16} + \frac{2}{64}$$

4-term approximation

$$\frac{1}{4} - \frac{1}{16} + \frac{2}{64} - \frac{6}{256} < y(4) < \frac{1}{4} - \frac{1}{16} + \frac{2}{64} - \frac{6}{256} + \frac{6}{256}$$

5-term approximation

$$\frac{1}{4}-\frac{1}{16}+\frac{2}{64}-\frac{6}{256}+\frac{6}{256}-\frac{6}{256}\times\frac{5}{4}<y(4)<\frac{1}{4}-\frac{1}{16}+\frac{2}{64}-\frac{6}{256}+\frac{6}{256}.$$

The remainder bounds and the estimates for $y(4)$ are shown in Fig. 1.2 for various values of N.

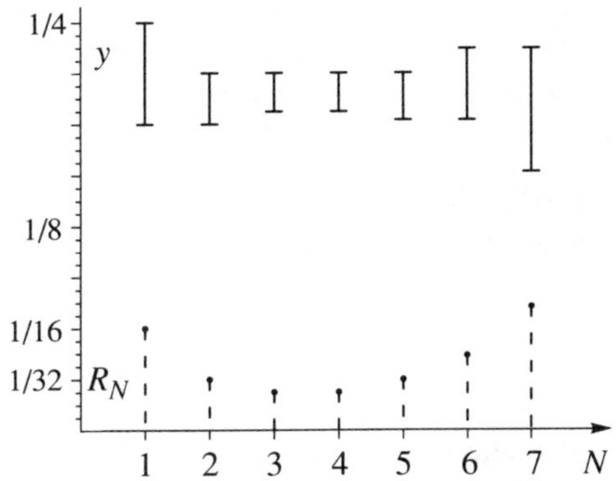

Fig. 1.2 The variation of the bound for $|R_N|$ ---- and the estimate for y ├─┤

The minimum remainder bound occurs for the three-term expansion. The same bound occurs for $|R_4|$ after which the remainder bounds increase tending ultimately to infinity as $N \to \infty$. Thus the best estimate for $y(4)$ is obtained from the three-term expansion.

$$\frac{1}{4}-\frac{1}{16}+\frac{2}{64}-\frac{6}{256}<y(4)<\frac{1}{4}-\frac{1}{16}+\frac{2}{64}$$

$$0.1953125 < y(4) < 0.2187500. \qquad 1.4.9$$

More terms may be included in the truncated series for larger values of the independent variable before the remainder bound starts to increase. For example when $x = 6$ we have

$$y(6)=\frac{1}{6}-\frac{1}{6^2}+\frac{2!}{6^3}-\frac{3!}{6^4}+\frac{4!}{6^5}-\frac{5!}{6^6}+\frac{6!}{6^7}-\frac{7!}{6^8}+\cdots$$

↑ (under $\frac{5!}{6^6}$)

|⟨ Best estimate ⟩| Minimum error bound

i.e.

$$\frac{1}{6}-\frac{1}{6^2}+\frac{2!}{6^3}-\frac{3!}{6^4}+\frac{4!}{6^5}-\frac{5!}{6^6}<y(6)<\frac{1}{6}-\frac{1}{6^2}+\frac{2!}{6^3}-\frac{3!}{6^4}+\frac{4!}{6^5}$$

$$0.1440329 < y(6) < 0.1466049. \qquad 1.4.10$$

For the case $x = 10$ the best estimate is given by the nine-term series with the remainder bounded by the tenth term. This yields

$$0.0915456 < y(10) < 0.0915819. \qquad 1.4.11$$

In the case of noninteger values of the independent variable let $x = M + t$ where M is the nearest integer from below and t lies in the interval $0 < t < 1$. Then the best estimate is determined by the M term series.

$$y(M+t) \simeq \frac{1}{M+t} - \frac{1}{(M+t)^2} + \frac{2!}{(M+t)^3} + \cdots + \frac{(-1)^{M-1}(M-1)!}{(M+t)^M}$$

with the minimum remainder bound $M!/(M+t)^{M+1}$.

The integral 1.4.6 is related to the exponential integral $E_1(x)$,

$$y = e^x \int_x^\infty \frac{e^{-s}}{s}\, ds = e^x E_1(x),$$

which is a tabulated function (Abramowitz and Stegun[1a]).

The accurate values of y along with the average of the expressions 1.4.9–11 and error bounds are presented in Table 1.1 which shows that both the true relative error and the estimated relative error R_e/y_e decrease as x increases.

Table 1.1

x	y_{accurate}	$y_{\text{estimate}} \pm$ Error bound $y_e \pm R_e$	R_e/y_e
4	0.2063456	0.2070363 ± 0.0117238	0.056627
6	0.1452676	0.1453189 ± 0.0012860	0.008853
10	0.0915633	0.0915638 ± 0.0000182	0.000199

In this and subsequent examples the remainder turns out to be bounded by the first neglected term in the expansion. We will consider asymptotic expansions in more detail in Chapter 2 where we make precise the definition of an asymptotic expansion. The essential property is that the remainder, after any chosen number of terms, should tend to zero more rapidly than the previous terms as x tends to the limit value x_0 (currently we are considering $x_0 = \infty$). It is a bonus if the remainder is bounded by the first neglected term of the series as in this and subsequent examples.

Consider next the integral expression

$$f(x) = \int_0^\infty \frac{e^{-xt}}{\sqrt{1+t}}\, dt$$

for large values of x. An asymptotic expansion in inverse powers of x can be obtained using integration by parts as follows:

$$f(x) = \left[\frac{-e^{-xt}}{x(1+t)^{1/2}}\right]_0^\infty - \frac{1}{2x}\int_0^\infty \frac{e^{-xt}}{(1+t)^{3/2}}\,dt$$

$$= \frac{1}{x} - \frac{1}{2x}\left[\frac{-e^{-xt}}{x(1+t)^{3/2}}\right]_0^\infty + \frac{3}{2}\cdot\frac{1}{2}\cdot\frac{1}{x^2}\int_0^\infty \frac{e^{-xt}}{(1+t)^{5/2}}\,dt$$

$$= \frac{1}{x} - \frac{1}{2x^2} + \frac{3}{2}\cdot\frac{1}{2}\cdot\frac{1}{x^3} - \frac{5}{2}\cdot\frac{3}{2}\cdot\frac{1}{2}\cdot\frac{1}{x^4}$$

$$+ \cdots (-1)^{N-1}\frac{[2(N-1)-1]}{2}\cdots\frac{5}{2}\cdot\frac{3}{2}\cdot\frac{1}{2}\cdot\frac{1}{x^N} + R_N$$

where $R_N = (-1)^N\dfrac{[2(N+1)-1]}{2}\cdots\dfrac{5}{2}\cdot\dfrac{3}{2}\cdot\dfrac{1}{2}\,\dfrac{1}{x^N}\displaystyle\int_0^\infty \frac{e^{-xt}}{(1+t)^{[2(N+1)+1]/2}}\,dt.$

The integral can be bounded by replacing $1 + t$ in the denominator by unity; then, since $\int_0^\infty e^{-xt}\,dt = 1/x$, we have

$$|R_N| \leqslant \frac{2(N+1)-1}{2}\cdots\frac{5}{2}\cdot\frac{3}{2}\cdot\frac{1}{2}\cdot\frac{1}{x^{N+1}}.$$

This proves that the series is asymptotic since the remainder tends to zero faster than the last term in the truncated series as $x \to \infty$. In fact just as in the earlier example, the remainder is bounded by the first neglected term in the expansion.

The expansion can be used to provide the best estimate of the function along with an error bound. For example if $x = 4$ then

$$f(4) = \underbrace{\frac{1}{4} - \frac{1}{2}\cdot\frac{1}{4^2} + \frac{3}{2}\cdot\frac{1}{2}\cdot\frac{1}{4^3} - \frac{5}{2}\cdot\frac{3}{2}\cdot\frac{1}{2}\cdot\frac{1}{4^4}}_{\text{Best estimate}} + \underbrace{\frac{7}{2}\cdot\frac{5}{2}\cdot\frac{3}{2}\cdot\frac{1}{2}\cdot\frac{1}{4^5}}_{\text{Minimum remainder bound}} + \cdots$$

Thus $0.2231445 < f(4) < 0.2295532$.

In using expansions to obtain the best estimate of a function the series should be truncated so that the minimum remainder bound occurs. If a bound for the remainder is not available but the expansion is known to be (or more often hoped to be) an asymptotic expansion then the best estimate is obtained by terminating the series when the terms begin to increase in magnitude. This is called the 'optimum truncation rule'.

The error function

The error function, erf(x), is defined by the expression

$$\operatorname{erf}(x) = \frac{2}{\sqrt{\pi}} \int_0^x \exp(-t^2)\,\mathrm{d}t. \qquad 1.4.12$$

The integral can be evaluated exactly for $x = \infty$ and has the value $\sqrt{\pi}/2$ so that erf$(\infty) = 1$.

The complementary error function, erfc(x), is defined by the expression

$$\operatorname{erfc}(x) = \frac{2}{\sqrt{\pi}} \int_x^\infty \exp(-t^2)\,\mathrm{d}t. \qquad 1.4.13$$

The sum of the two functions is unity,

$$\operatorname{erf}(x) + \operatorname{erfc}(x) = \frac{2}{\sqrt{\pi}} \int_0^\infty \exp(-t^2)\,\mathrm{d}t = 1.$$

An asymptotic expansion for erfc(x) can be developed by integrating by parts. The integrand on the right-hand side of equation 1.4.13 must first be multiplied and divided by $-2t$ so that the numerator $-2te^{-t^2}$ can be integrated.

Then we have

$$\begin{aligned}
\frac{2}{\sqrt{\pi}} \int_x^\infty \exp(-t^2)\,\mathrm{d}t &= \frac{2}{\sqrt{\pi}} \int_0^\infty \frac{-2t\exp(-t^2)}{-2t}\,\mathrm{d}t \\
&= \frac{2}{\sqrt{\pi}} \left\{ \left[\frac{\exp(-t^2)}{-2t} \right]_x^\infty - \int_x^\infty \exp(-t^2) \frac{1}{2t^2}\,\mathrm{d}t \right\} \\
&= \frac{2}{\sqrt{\pi}} \left(\frac{\exp(-t^2)}{2x} - \frac{1}{2} \int_x^\infty \frac{-2t\exp(-t^2)}{2t^3}\,\mathrm{d}t \right).
\end{aligned}$$

Repeated integration by parts leads to

$$\begin{aligned}
\operatorname{erfc}(x) &= \frac{\exp(-x^2)}{\sqrt{\pi}} \left(\frac{1}{x} - \frac{1}{2x^3} + \frac{1.3}{2^2x^5} - \frac{1.3.5}{2^3x^7} + \cdots \right. \\
&\left. + (-1)^{N-1} \frac{1.3.5.\ldots.(2N-3)}{2^{N-1}\,x^{2N-1}} \right) + R_N
\end{aligned} \qquad 1.4.14$$

where the remainder after N terms is given by

$$R_N = \frac{(-1)^N}{\sqrt{\pi}} \frac{1.3.5.\ldots.(2N-3)(2N-1)}{2^{N-1}} \int_x^\infty \frac{\exp(-t^2)}{t^{2N}}\,\mathrm{d}t.$$

The remainder can be bounded using the following inequality,

$$\int_x^\infty \frac{\exp(-t^2)}{t^{2N}}\,\mathrm{d}t = \int_x^\infty \frac{2t\exp(-t^2)}{2t^{2N+1}}\,\mathrm{d}t < \int_x^\infty \frac{2t\exp(-t^2)}{2x^{2N+1}}\,\mathrm{d}t = \frac{\exp(-x^2)}{2x^{2N+1}}.$$

Therefore

$$|R_N| < \frac{1.3.5.\ldots.(2N-3)(2N-1)}{2^N \quad x^{2\ \ +1} \quad \sqrt{\pi}} \exp(-x^2),$$

which proves that 1.4.14 is an asymptotic expansion since R_N tends to zero faster than the truncated terms in the series as $x \to \infty$. In fact as in previous examples the first neglected term in the series bounds the remainder.

The factorial behavior of the numerator of the terms in the expansion 1.4.14 shows that the series diverges. It is essential to truncate the series before R_N starts to increase. Quite accurate estimates can be obtained for erfc(x) even for relatively small values of x. For example, if $x = 2$ then

$$\text{erfc}(2) = \frac{1}{\sqrt{\pi}} e^{-4}\left(\frac{1}{2} - \frac{1}{2.2^3} + \frac{1.3}{2^2.2^5} - \frac{1.3.5}{2^3.2^7} + \frac{1.3.5.7}{2^4.2^9} - \frac{1.3.5.7.9}{2^5.2^{11}} + \cdots\right).$$

↑ Smallest term (at $\frac{1.3.5.7}{2^4.2^9}$) ⟼ Terms begin to increase in magnitude

The denominator of subsequent terms increases by a factor of 8; so the series must be truncated before the numerator increases by the factor 9. The remainder is bounded by and has the sign of the first neglected term, so the best estimate for erfc(2) is that its value lies between the four- and five-term truncated series, i.e.

$$\frac{e^{-4}}{\sqrt{\pi}}(0.5 - 0.0625 + 0.0234375 - 0.0146484),$$

and

$$\frac{e^{-4}}{\sqrt{\pi}}(0.5 - 0.0625 + 0.0234375 - 0.0146484 + 0.0128173).$$

Thus

$$0.004612 < \text{erfc}(2) < 0.004744,$$

or equivalently

$$\text{erfc}(2) = 0.004678 \pm 0.000066. \qquad 1.4.15$$

In fact the accurate, tabulated value of erfc(2) happens to be 0.004678 to 6 decimal places.

Comparison with a convergent series

It is possible to estimate the value of erfc(x) using a convergent Maclaurin series. This is generated for erf(x) by replacing the integrand $\exp(-t^2)$ by its Maclaurin series and integrating,

$$\operatorname{erfc}(x) = 1 - \operatorname{erf}(x)$$

$$= 1 - \frac{2}{\sqrt{\pi}} \int_0^x \left(1 - t^2 + \frac{t^4}{2!} - \frac{t^6}{3!} + \cdots\right) dt$$

$$= 1 - \frac{2}{\sqrt{\pi}} \left(x - \frac{x^3}{3} + \frac{x^5}{5.2!} - \frac{x^7}{7.3!} + \cdots\right).$$

The factorial term in the denominator ensures that the series converges for all values of x. Thus we may use the series for the case $x = 2$,

$$\operatorname{erfc}(2) = 1 - \frac{2}{\sqrt{\pi}} \left(2 - \frac{2^3}{3} + \frac{2^5}{5.2} - \frac{2^7}{7.3.2} + \cdots\right).$$

The value of the truncated series is presented in Table 1.2 for various numbers of terms. Clearly far more terms must be used in the convergent Maclaurin series to generate an estimate which is as good as that obtained from only a few terms in the divergent asymptotic expansion. In this sense asymptotic expansions can be more efficient methods of representing functions than their convergent power series when they exist. Of course, convergent series do have the advantage that in principle arbitrary accuracy can be achieved, whereas divergent asymptotic series have a minimum error bound which cannot be reduced.

Table 1.2

Number of terms in truncated series	Resultant sum of terms
4	1.580309
8	0.157841
12	0.007130
16	0.004689
20	0.004678

Exercises

1 Use integration by parts to obtain expansions for large x of the following functions. Show that these are asymptotic expansions by constructing an appropriate bound for the remainder. Use the expansions to obtain the best estimate of the function value when $x = 5$.

(i) $f(x) = \displaystyle\int_0^\infty \frac{e^{-t}}{(x+t)^2} dt$

(ii) $f(x) = e^x \displaystyle\int_x^\infty \frac{e^{-t}}{t^{1/2}} dt$

(iii) $f(x) = \displaystyle\int_0^\infty e^{-xt} \ln(1+t) dt.$

2 Show that the cosine integral

$$\mathrm{Ci}(x) = \int_x^\infty \frac{\cos t}{t}\,\mathrm{d}t$$

has the following asymptotic expansion for large x,

$$\left(\frac{1}{x^2} - \frac{3!}{x^4} + \cdots\right)\cos x - \left(\frac{1}{x} - \frac{2!}{x^3} + \cdots\right)\sin x.$$

Chapter 2
Asymptotics

In this chapter some of the more formal aspects of perturbation expansions will be considered. The rate at which functions approach limit values will be described by comparison with reference functions. Order symbols will be introduced to allow the limiting behavior to be expressed concisely. The behavior of asymptotic expansions will be explored. The occurrence of nonuniformities in expansions will be discussed and the common sources of nonuniformities identified.

2.1 Order symbols

The letters 'O' (upper case) and 'o' (lower case) are order symbols. They are used to describe the rate at which functions approach limit values. We will consider three types of limit value namely, zero, a finite but nonzero value and infinity:

I $\lim_{x\to x_0} f(x) = 0$

II $\lim_{x\to x_0} f(x) = L \quad (L \neq 0 \text{ and finite})$

III $\lim_{x\to x_0} f(x) = \pm\infty.$

Limits which are indefinite such as $\lim_{x\to 0} [\sin(1/x)]$ will not be considered.

Examples of these three types of limit are:

I $\lim_{x\to 0} \sin x = 0, \qquad \lim_{x\to 0} \frac{x^3}{x^2 - x} = 0,$

$\lim_{x\to \pi/2} \cos x = 0, \qquad \lim_{x\to \infty} e^{-x} = 0.$

II $\lim_{x\to 0} \cos x = 1, \qquad \lim_{x\to 1} \frac{x^2 - 1}{x - 1} = 2,$

$\lim_{x\to 0} \frac{1 + x}{1 - x} = 1, \qquad \lim_{x\to \infty} \frac{1 + x}{1 - x} = -1.$

$$\text{III}\quad \lim_{x\to 1^-} \frac{1}{1-x} = \infty, \qquad \lim_{x\to 0^+} \ln(x) = -\infty,$$

$$\lim_{x\to 0^+} e^{1/x} = \infty, \qquad \lim_{x\to 0} \frac{1}{1-\cos x} = \infty.$$

When it is necessary to distinguish the direction from which x approaches x_0 the superscripts + or − are used. Thus, for example, $x \to 1^-$ means x approaches the value +1 through values which are less than +1.

We will be interested for the most part in the behavior of functions either as $x \to 0$ or as $x \to \infty$. In fact we could choose to transform the independent variable so that $x \to x_0$ corresponds to $y \to 0$ by setting $y = x - x_0$. In the case when x_0 is infinite the transformation $y = 1/x$ makes $x \to \infty$ correspond to $y \to 0$. It is therefore possible to restrict our study to the limit as the independent variable tends to zero without loss of generality. For this reason most of our attention will be paid to this case.

Consider the following two limits,

$$\lim_{x\to 0^+} \sqrt{x} = 0 \quad \text{and} \quad \lim_{x\to 0} x^2 = 0.$$

Both the functions, $\sqrt{x}$ and x^2, have the same limit as x tends to zero but this limit is approached at different rates. The function $\sqrt{x}$ tends to zero slower than does x, while the function x^2 tends to zero faster than x. The following table of values indicates this behavior.

x	0.1	0.01	0.001	0.0001
$\sqrt{x}$	0.316	0.1	0.0316	0.01
x^2	0.01	0.0001	0.000001	0.00000001.

Other functions which have the limit zero may approach zero at the same rate as some power of x. For example, $\lim_{x\to 0} \sin x = 0$, and if we tabulate the value of $\sin x$ to six decimal places as x tends to zero we have:

x	0.1	0.01	0.001	0.0001
$\sin x$	0.099833	0.009999	0.001000	0.000100.

This shows that $\sin x$ approaches zero at the same rate as does x and not say at the rate $\sqrt{x}$ or at the rate x^2. To indicate that x and $\sin x$ tend to zero at the same rate the 'big O' symbol is used, as follows:

$$\sin x = O(x) \quad \text{as } x \to 0.$$

Consider next the function $1 - \cos x$. We know that $\lim_{x\to 0} (1 - \cos x) = 0$ and the following table of values to six decimal places shows that $1 - \cos x$ tends to zero at the rate of $x^2/2$ as $x \to 0$.

x	0.5	0.1	0.05	0.01
$1 - \cos x$	0.122417	0.004996	0.001250	0.000050
$x^2/2$	0.125	0.005	0.00125	0.00005.

The coefficient 1/2 is omitted from the order relation in order to describe the rate that $1 - \cos x$ tends to zero. Thus we write the following:

$$1 - \cos x = O(x^2) \quad \text{as } x \to 0.$$

The symbol O implies that there is a constant of proportionality C such that as $x \to 0$ the function $1 - \cos x$ tends arbitrarily close to value Cx^2. The value of C is obtained from the limit of the ratio of $1 - \cos x$ and x^2 as x tends to zero,

$$\lim_{x \to 0} \frac{1 - \cos x}{x^2} = C.$$

In general if a function $f(x)$ approaches a limit value at the same rate as does another function $g(x)$ as $x \to x_0$ then we write $f(x) = O[g(x)]$ as $x \to x_0$. The functions $f(x)$ and $g(x)$ are said to be of the same order as $x \to x_0$. The test for this is the limit of the ratio. Thus if

$$\lim_{x \to x_0} \frac{f(x)}{g(x)} = C,$$

where C is finite then

$$f(x) = O(g(x)) \quad \text{as } x \to x_0.$$

Wherever possible the function $g(x)$ is chosen so that C is nonzero. However, it is sometimes helpful to allow C to be zero in the definition of the 'big O' symbol. It is essential that C is not infinite. Thus the following statements are true:

$$\sin x = O(x) \quad \text{as } x \to 0$$

$$\sin x = O(\sqrt{x}) \quad \text{as } x \to 0,$$

since in the first case we have

$$\lim_{x \to 0} \frac{\sin x}{x} = 1,$$

and in the second case

$$\lim_{x \to 0} \frac{\sin x}{\sqrt{x}} = 0.$$

However, it is not true to state that:

$$\sin x = O(x^2) \quad \text{as } x \to 0 \quad \text{(WRONG)}$$

because

$$\lim_{x \to 0} \frac{\sin x}{x^2} = \infty .$$

Although it is within the strict definition of the big O symbol that $\sin x = O(\sqrt{x})$ as $x \to 0$, it is less informative than stating that $\sin x = O(x)$ as $x \to 0$. In the latter case the nonzero limit of $\sin x/x$ as x tends to zero implies the ultimate proportionality as x tends to zero between $\sin x$ and x. Thus the rate at which $\sin x$ approaches zero is described by x and not by $\sqrt{x}$.

It is important to specify the domain of an order relation, i.e. x_0 must be stated. For example:

$$\frac{x}{1 + x^2} = O(x) \quad \text{as } x \to 0.$$

This is because the term x^2 in the denominator has a negligible effect as x tends to zero so that the function $x/(1 + x^2)$ approaches x asymptotically.

If we consider large x then the function $x/(1 + x^2)$ behaves very differently from x since now the x^2 term in the denominator is dominant. In fact

$$\frac{x}{1 + x^2} = O\left(\frac{1}{x}\right) \quad \text{as } x \to \infty .$$

Gauge functions

In the expression $f(x) = O(g(x))$ as $x \to x_0$ the function $g(x)$ is called a *gauge function.* Commonly used gauge functions are powers of x. They are chosen since their behavior is well known. However, we shall subsequently discover that powers of x provide neither a unique nor a sufficient set of gauge functions to describe the behavior of arbitrary functions.

The limit values of functions with which we are going to be most concerned are zero and infinity. In order to show that a function $f(x)$ approaches zero at the same rate as does a gauge function $g(x)$ as $x \to x_0$ we need to evaluate limits of ratios where the direct substitution $x = x_0$ would lead to an indeterminate expression. L'Hospital's rule allows such limits to be evaluated.

L'Hospital's rule

$$\lim_{x\to x_0}\frac{f(x)}{g(x)} = \lim_{x\to x_0}\frac{\dfrac{df}{dx}}{\dfrac{dg}{dx}} \quad \text{if } f(x_0) = g(x_0) = 0,$$

$$= \lim_{x\to x_0}\frac{\dfrac{d^2f}{dx^2}}{\dfrac{d^2g}{dx^2}} \quad \text{if } \frac{df}{dx}(x_0) = \frac{dg}{dx}(x_0) = 0, \quad \text{etc.}$$

The differentiation is stopped and the limit evaluated by setting $x = x_0$ when either or both derivatives are nonzero.

As an example of the use of l'Hospital's rule we prove the order relation.

$$1 - \cos x = O(x^2) \quad \text{as } x \to 0.$$

Let $f(x) = 1 - \cos x$ and $g(x) = x^2$. We have $f(0) = 1 - \cos 0 = 0$ and $g(0) = 0$, also

$$\frac{df}{dx}(0) = \sin 0 = 0 \quad \text{and} \quad \frac{dg}{dx}(0) = 0, \text{ while}$$

$$\frac{d^2f}{dx^2}(0) = \cos 0 = 1 \quad \text{and} \quad \frac{d^2g}{dx^2}(0) = 2, \quad \text{thus}$$

$$\lim_{x\to 0}\frac{1 - \cos x}{x^2} = \frac{d^2f}{dx^2}(0)\Big/\frac{d^2g}{dx^2}(0) = \frac{1}{2}.$$

This proves the order relation:

$$1 - \cos x = O(x^2) \quad \text{as } x \to 0.$$

Other examples of the use of l'Hospital's rule to establish order relations are:

(i) $e^x - 1 = O(x)$ as $x \to 0$.

Proof: $$\lim_{x\to 0}\frac{e^x - 1}{x} = \lim_{x\to 0}\frac{e^x}{1} = \frac{e^0}{1} = 1.$$

(ii) $\sin x - x = O(x^3)$ as $x \to 0$.

Proof: $$\lim_{x\to 0}\frac{\sin x - x}{x^3} = \lim_{x\to 0}\frac{\cos x - 1}{3x^2} = \lim_{x\to 0}\frac{-\sin x}{6x}$$

$$= \lim_{x\to 0}\frac{-\cos x}{6} = \frac{-\cos 0}{6} = -\frac{1}{6}.$$

(iii) $\cot x = O(1/x)$ as $x \to 0$.

$$\text{Proof:} \quad \lim_{x\to 0} \frac{\cot x}{1/x} = \lim_{x\to 0} \frac{x}{\tan x} = \lim_{x\to 0} \frac{1}{\sec^2 x}$$

$$= \frac{1}{\sec^2 0} = 1.$$

Quantities can be added to either side of an order relation. Thus since:

$$e^x - 1 = O(x) \quad \text{as } x \to 0,$$

we may add unity to both sides to obtain the expression

$$e^x = 1 + O(x) \quad \text{as } x \to 0.$$

Similarly we have the order relation

$$\cos x - 1 = O(x^2) \quad \text{as } x \to 0,$$

and on adding unity to both sides we obtain the expression

$$\cos x = 1 + O(x^2) \quad \text{as } x \to 0.$$

Order relations are most easily established using the Taylor expansion of a function about the point $x = x_0$. A truncated expansion is used and the behavior of the remainder is required.

Taylor's formula with remainder

$$f(x) = f(x_0) + (x - x_0)\frac{df}{dx}(x_0) + \frac{(x - x_0)^2}{2!}\frac{d^2f}{dx^2}(x_0) + \cdots$$

$$+ \frac{(x - x_0)^N}{N!}\frac{d^Nf}{dx^N}(x_0) + R_N, \qquad 2.1.1$$

where

$$R_N = \frac{(x - x_0)^{N+1}}{(N+1)!}\frac{d^{N+1}}{dx^{N+1}} f(z), \qquad 2.1.2$$

and z is a point lying between x_0 and x.

The validity of this formula depends on the existence of the first $N + 1$ derivatives of the function $f(x)$.

The special case when $x_0 = 0$ occurs often. Then 2.1.1 and 2.1.2 become Maclaurin's formula.

$$f(x) = f(0) + x\frac{df}{dx}(0) + \frac{x^2}{2!}\frac{d^2f}{dx^2}(0) + \cdots$$

$$+ \frac{x^N}{N!}\frac{d^Nf}{dx^N}(0) + R_N, \qquad 2.1.3$$

where

$$R_N = \frac{x^{N+1}}{(N+1)!}\frac{\mathrm{d}^{N+1}}{\mathrm{d}x^{N+1}} f(z), \tag{2.1.4}$$

and $0 < z < x$ if x is positive or $x < z < 0$ if x is negative.

The location of the point z and the value of the $(N + 1)$th derivative at $x = z$ is not important in our applications. All that is needed is that the derivative exists for z in the neighborhood of x_0; then we have

$$\lim_{x \to x_0} \frac{R_N}{(x - x_0)^{N+1}} = \frac{1}{(N+1)!}\frac{\mathrm{d}^{N+1}}{\mathrm{d}x^{N+1}} f(x_0).$$

This is the condition which shows that the remainder tends to zero in a manner which is asymptotically proportional to $(x - x_0)^{N+1}$. We may write

$$R_N = \mathrm{O}((x - x_0)^{N+1}) \quad \text{as } x \to x_0. \tag{2.1.5}$$

Reference has been made to an ultimate or asymptotic proportionality between R_N and $(x - x_0)^{N+1}$ and more generally between $f(x)$ and the gauge function $g(x)$. The proportionality constant is given by $\lim_{x \to x_0} [f(x)/g(x)]$. This proportionality is never usually achieved but rather approached arbitrarily closely as x tends to x_0. It is sometimes necessary to be more precise than simply to appeal to the notion of ultimate proportionality. We may instead assert that if

$$f(x) = \mathrm{O}(g(x)) \quad \text{as } x \to x_0,$$

then for any value of x in a neighborhood of x_0 there exists a positive constant K such that

$$|f(x)| < K|g(x)|.$$

To clarify this point consider the order relation:

$$\mathrm{e}^x - 1 = \mathrm{O}(x) \quad \text{as } x \to 0.$$

This is true because:

$$\lim_{x \to 0} \frac{\mathrm{e}^x - 1}{x} = \lim_{x \to 0} \frac{\mathrm{e}^x}{1} = 1,$$

and since this limit is unity the asymptotic constant of proportionality is unity. Thus $\mathrm{e}^x - 1$ tends to the value of x itself as x tends to zero. However, $\mathrm{e}^x - 1$ is always greater than x for positive values of x. We cannot bound $\mathrm{e}^x - 1$ by x itself but we certainly can by Kx where K exceeds unity. If we consider values of x less than 1.25 then we may choose $K = 2$, i.e. $|\mathrm{e}^x - 1| < 2|x|$ as is shown in Fig. 2.1. If we choose values of x less than 0.75 then K may be reduced to 1.5 i.e. $|\mathrm{e}^x - 1| < 1.5|x|$.

The value of K and the region of x values are unimportant. That for any neighbourhood of zero there exists a corresponding value of K such that $|\mathrm{e}^x - 1|$ is less than $K|x|$ is the essential implication of the order relation $\mathrm{e}^x - 1 = \mathrm{O}(x)$ as $x \to 0$.

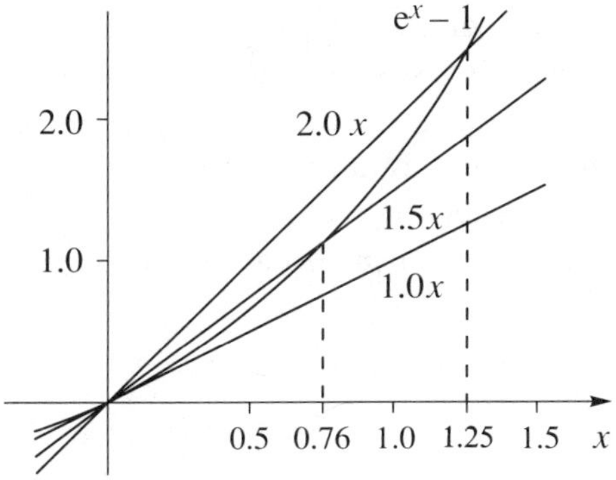

Fig. 2.1 Bounds for $e^x - 1$

Returning to Taylor's formula, equation 2.1.5 allows 2.1.1 to be written in the form

$$f(x) = f(x_0) + (x - x_0)\frac{df}{dx}(x_0) + \frac{(x - x_0)^2}{2!}\frac{d^2 f}{dx^2}(x_0) + \cdots$$

$$+ \frac{(x - x_0)^N}{N!}\frac{d^N f}{dx^N}(x_0) + O[(x - x_0)^{N+1}] \quad \text{as } x \to x_0. \qquad 2.1.6$$

The following well-known examples of Taylor's formula will be required in the subsequent examples:

$$e^x = 1 + x + \frac{x^2}{2!} + \frac{x^3}{3!} + O(x^4) \qquad \text{as } x \to 0$$

$$\ln(1 + x) = x - \frac{x^2}{2} + \frac{x^3}{3} + O(x^4) \qquad \text{as } x \to 0$$

$$\sin x = x - \frac{x^3}{3!} + \frac{x^5}{5!} + O(x^7) \qquad \text{as } x \to 0$$

$$\cos x = 1 - \frac{x^2}{2!} + \frac{x^4}{4!} + O(x^6) \qquad \text{as } x \to 0$$

$$\tan x = x + \frac{x^3}{3} + \frac{2}{15}x^5 + O(x^7) \qquad \text{as } x \to 0$$

$$(1 + x)^p = 1 + px + p\frac{(p - 1)}{2!}x^2 + p\frac{(p - 1)(p - 2)}{3!}x^3 + O(x^4) \quad \text{as } x \to 0.$$

The order of the remainder terms in the trigonometric expansions needs some clarification. If we consider $\sin x$ then Taylor's formula yields:

$$\sin x = x - \frac{x^3}{3!} + \frac{x^5}{5!} + R_5,$$

where $R_5 = \mathrm{O}(x^6)$ as $x \to 0$. This is an occasion where the big O symbol is used correctly but less informatively than we would like. It is used correctly since the limit of the ratio R_5/x^6 as x tends to zero is certainly noninfinite. However, the value of this limit is zero. Thus the direct use of Taylor's formula is not as informative as we would like because the remainder actually tends to zero faster than x^6. To prove that the remainder is in fact $\mathrm{O}(x^7)$ we must take one more term in the expansion, i.e.

$$\sin x = x - \frac{x^3}{3!} + \frac{x^5}{5!} - \frac{x^7}{7!} + \mathrm{O}(x^8) \quad \text{as } x \to 0,$$

and then use the result that the order of the combination $-x^7/7! + \mathrm{O}(x^8)$ is that of the dominant term x^7, i.e.

$$-\frac{x^7}{7!} + \mathrm{O}(x^8) = \mathrm{O}(x^7) \quad \text{as } x \to 0.$$

Thus

$$\sin x = x - \frac{x^3}{3!} + \frac{x^5}{5!} + \mathrm{O}(x^7) \quad \text{as } x \to 0.$$

The remainder behaves in this way for all series which consist either entirely of even or entirely of odd powers.

Properties of order symbols

The above discussion involved the use of one of the properties of order relations, namely that the sum of two quantities having different orders results in a quantity possessing the dominant order. Thus:

$$\mathrm{O}(x^2) + \mathrm{O}(x^3) = \mathrm{O}(x^2) \quad \text{as } x \to 0,$$

while

$$\mathrm{O}(x^2) + \mathrm{O}(x^3) = \mathrm{O}(x^3) \quad \text{as } x \to \infty.$$

The other rules for the manipulation of order symbols are as follows.

Multiplication by a number

Multiplication by a number does not change the order of a quantity, i.e.

$$C \cdot \mathrm{O}(x^n) = \mathrm{O}(x^n) \quad \text{as } x \to 0.$$

This is so regardless of the size of the constant C. Thus $10^6 \sin x = \mathrm{O}(x)$ as $x \to 0$ and $10^{-6} \sin x = \mathrm{O}(x)$ as $x \to 0$.

Multiplication and division of order symbols

Order symbols obey the usual rules of multiplication and division. Thus

$$\mathrm{O}(x^n).\,\mathrm{O}(x^m) = \mathrm{O}(x^{n+m})$$

$$\frac{\mathrm{O}(x^n)}{\mathrm{O}(x^m)} = \mathrm{O}(x^{n-m}).$$

Similarly powers of order symbols follow the usual rule, i.e.:

$$[\mathrm{O}(x^n)]^p = \mathrm{O}(x^{np}).$$

Functions of small quantities

Maclaurin's formula, 2.1.3, allows functions of small quantities to be expanded, e.g.

$$\exp[\mathrm{O}(x^n)] = 1 + \mathrm{O}(x^n) \quad \text{as } x \to 0 \quad (\text{for } n > 0)$$

$$[1 + \mathrm{O}(x^n)]^p = 1 + \mathrm{O}(x^n) \quad \text{as } x \to 0 \quad (\text{for } n > 0)$$

$$\sin[\mathrm{O}(x^n)] = \mathrm{O}(x^n) \quad \text{as } x \to 0 \quad (\text{for } n > 0)$$

$$\cos[\mathrm{O}(x^n)] = 1 + \mathrm{O}(x^{2n}) \quad \text{as } x \to 0 \quad (\text{for } n > 0).$$

Exercises

1 Prove the following order relations by evaluating the appropriate limits:

(i) $\dfrac{1 - \cos(x^2)}{\sin(x^3)} = \mathrm{O}(x) \quad \text{as } x \to 0.$

(ii) $\tan x - x = \mathrm{O}(x^3) \quad \text{as } x \to 0.$

(iii) $\ln(1 + \sqrt{x}) - \sqrt{x} = \mathrm{O}(x) \quad \text{as } x \to 0.$

(iv) $\mathrm{e}^{\sin\varepsilon} = \mathrm{O}(1) \quad \text{as } \varepsilon \to 0$

(v) $\mathrm{e}^{\sin\varepsilon} - 1 = \mathrm{O}(\varepsilon) \quad \text{as } \varepsilon \to 0.$

2 Repeat the above exercise using the appropriate Maclaurin expansions.

3 Use powers of x as gauge functions to describe the behavior of the following functions as $x \to 0$:

(i) $\sqrt{x(1-x)} - \sqrt{x}$ (ii) $\displaystyle\int_0^x \exp(-s^2)\,\mathrm{d}s$

(iii) $\displaystyle\int_0^x \exp(-s^2)\,\mathrm{d}s - x$ (iv) $\mathrm{e}^1 - \mathrm{e}^{1/(1+x)}$

(v) $\mathrm{e}^1 - \mathrm{e}^{\cos x}$ (vi) $\mathrm{e}^{\tan x} - \mathrm{e}^{\sin x}$.

4 Obtain the first three nonzero terms in the expansions in powers of ε of the following functions:

(i) $(1+\varepsilon+2\varepsilon^2)^{-1}$, (ii) $\sqrt{1-\varepsilon^3+\varepsilon^6}$, (iii) $\cos(\varepsilon-\varepsilon^2)$

(iv) $\ln(1+\sin\varepsilon)$, (v) $\ln(1+e^{\varepsilon})$, (vi) $e^{1-\cos\varepsilon}$.

The order symbol o (little o) is used to indicate that a function is smaller than a gauge function in the sense that it either tends to zero faster or to infinity slower than the gauge function. If

$$f(x) = o(g(x)) \quad \text{as } x \to x_0,$$

then

$$\lim_{x\to x_0} \frac{f(x)}{g(x)} = 0.$$

The definition of the big O order symbol includes the above case but whenever it is known that the limit of $f(x)/g(x)$ is zero as $x \to x_0$ the little o symbol should be used. If possible the big O symbol is reserved for cases when the above limit is nonzero.

Examples of little o order relations are:

(i) $\sin x = o(\sqrt{x}) \quad \text{as } x \to 0.$

To prove this we must show that:

$$\lim_{x\to 0} \frac{\sin x}{\sqrt{x}} = 0.$$

Either L'Hospital's rule or the Maclaurin expansion of $\sin x$ may be used to prove that the limit is zero. In this and the subsequent two examples the latter method will be used.

$$\lim_{x\to 0} \frac{\sin x}{\sqrt{x}} = \lim_{x\to 0} \left(\frac{x - O(x^3)}{\sqrt{x}}\right)$$

$$= \lim_{x\to 0} (\sqrt{x} - O(x^{5/2})) = 0.$$

Behavior of logarithmic and exponential functions

An important use of the little o symbol is in the order relation:

$$e^{-x} = o\left(\frac{1}{x^N}\right) \quad \text{as } x \to \infty.$$

This states that, as x tends to infinity, e^{-x} tends to zero faster than any power

$$\frac{1}{x}, \frac{1}{x^2}, \frac{1}{x^3}, \quad \text{etc.}$$

We must prove that:

$$\lim_{x \to \infty} \frac{e^{-x}}{1/x^N} = 0.$$

Therefore we must prove that:

$$\lim_{x \to \infty} \frac{x^N}{e^x} = 0.$$

The proof follows from bounding x^N/e^x from below by zero – this quotient is certainly nonnegative for all positive x – and bounding the quotient from above using a term from the Maclaurin expansion of e^x,

$$e^x \geqslant \frac{x^{N+1}}{(N+1)!} \quad \text{for} \quad x > 0,$$

thus

$$\frac{x^N}{e^x} \leqslant \frac{x^N}{x^{N+1}/(N+1)!} = \frac{(N+1)!}{x}.$$

Thus as $x \to \infty$ the bound from above tends to zero which is the bound from below and hence the limit is zero.

Although the above proof of the order relation is restricted to the case of integer N, the relation is true for all powers of x,

$$e^{-x} = o\left(\frac{1}{x^p}\right) \quad \text{as } x \to \infty \text{ for all } p. \qquad 2.1.7$$

An alternative form of this relation is obtained by replacing x with $1/y$ and letting $y \to 0^+$,

$$e^{-1/y} = o(y^p) \quad \text{as } y \to 0^+. \qquad 2.1.7'$$

The converse of relation 2.1.7 is that e^x tends to infinity faster than any power of x as $x \to \infty$, i.e.:

$$x^p = o(e^x) \quad \text{as } x \to \infty. \qquad 2.1.8$$

Exponentials are faster than powers and their inverse function, logarithms, are slower. So that

$$\ln(x) = o(x^p) \quad \text{as } x \to \infty \text{ for } p > 0. \qquad 2.1.9$$

In particular any small power such as $x^{1/2}$, $x^{1/4}$, $x^{1/10}$ etc. dominates $\ln(x)$ as $x \to \infty$. The relation 2.1.9 follows from 2.1.8 by setting $e^x = (y/p)^p$; then $x = p\ln(y/p)$ and if p is positive, $y \to \infty$ as $x \to \infty$ so that 2.1.8 becomes

$$p^p \ln(y) = o[(y/p)^p] \quad \text{as } y \to \infty \text{ for } p > 0.$$

The multiplicative coefficients p^p and $(1/p)^p$ may be omitted from this order relation to yield the expression 2.1.9.

The corresponding result as $x \to 0$ is obtained from 2.1.9 by setting $x = 1/y$ and considering the behavior as $y \to 0^+$. The expression 2.1.9 becomes

$$-\ln(y) = o(1/y^p) \quad \text{as } y \to 0^+ \text{ for } p > 0.$$

The minus sign may be omitted. Thus we have shown that $\ln(x)$ tends to infinity slower than any inverse power of x as $x \to 0$,

$$\ln(x) = o\left(\frac{1}{x^p}\right) \quad \text{as } x \to 0^+ \text{ for } p > 0.$$

Implied in these results is the description of the behavior of product functions as follows:

$$e^{-x}x^p \to 0 \quad \text{as } x \to \infty \text{ for all } p$$

$$e^x/x^p \to \infty \quad \text{as } x \to \infty \text{ for all } p$$

$$x^p \ln(x) \to 0 \quad \text{as } x \to 0^+ \text{ for all } p > 0$$

$$\frac{\ln(x)}{x^p} \to 0 \quad \text{as } x \to \infty \text{ for all } p > 0.$$

Exponential functions tend to zero or infinity faster than powers while the logarithm is slower to approach infinity than powers. Therefore in order to describe the behavior of arbitrary functions it is necessary to supplement the set of powers, x^p for all p, with functions such as $\ln(x)$, $\ln[\ln(x)]$, e^x, e^{e^x} and so on. Fortunately we will only rarely meet gauge functions other than the powers x^p.

2.2 Asymptotic sequences and expansions

In Chapter 1 we obtained expansions in inverse powers to describe the behavior of functions for large values of the independent variable. The example which we considered in some detail was

$$\begin{aligned} y &= e^x \int_x^\infty \frac{e^{-t}}{t}\,dt \\ &= \frac{1}{x} - \frac{1}{x^2} + \frac{2!}{x^3} + \cdots + (-1)^{N-1}\frac{(N-1)!}{x^N} + R_N, \end{aligned} \qquad 2.2.1$$

where

$$R_N = (-1)^N N!\, e^x \int_x^\infty \frac{e^{-t}}{t}\,dt.$$

We obtained the inequality

$$|R_N| \leqslant \frac{N!}{x^{N+1}}, \qquad 2.2.2$$

which shows that the remainder is bounded by the first neglected term of the truncated series.

The condition that the expansion 2.2.1 must satisfy to be an asymptotic expansion is not as strong as the bound on the remainder given by 2.2.2. The requirement is that the remainder should tend to zero faster than the previous terms as $x \to \infty$, i.e. $R_N = o(1/x^N)$ as $x \to \infty$.

The inverse powers of x all tend to zero as x tends to infinity. The dominant term is $1/x$ followed by $1/x^2$ and then $1/x^3$ etc. In terms of the little o order relation we have:

$$\frac{1}{x^2} = o\left(\frac{1}{x}\right) \quad \text{as } x \to \infty$$

$$\frac{1}{x^3} = o\left(\frac{1}{x^2}\right) = o\left(\frac{1}{x}\right) \quad \text{as } x \to \infty.$$

and so on.

The smallest term as $x \to \infty$ in the truncated expansion 2.2.1 is the term $1/x^N$. The remainder is required to be smaller than this term if the expansion is to be an asymptotic expansion. The condition

$$R_N = o(1/x^N) \quad \text{as } x \to \infty,$$

is to hold for all N and is equivalent to

$$R_N = O(1/x^{N+1}) \quad \text{as } x \to \infty,$$

since

$$R_N = (-1)^N \frac{N!}{x^{N+1}} + R_{N+1},$$

where $R_{N+1} = o(1/x^{N+1})$ so that

$$\begin{aligned} R_N &= O(1/x^{N+1}) + o(1/x^{N+1}) \\ &= O(1/x^{N+1}) \quad \text{as } x \to \infty. \end{aligned}$$

In general the expansion

$$f(x) = a_0 + \frac{a_1}{x} + \frac{a_2}{x^2} + \cdots + \frac{a_N}{x^N} + R_N, \qquad 2.2.3$$

is an asymptotic expansion as $x \to \infty$ if, for any N,

$$R_N = O\left(\frac{1}{x^{N+1}}\right) \quad \text{as } x \to \infty. \qquad 2.2.4$$

The following expression is used when 2.2.3 and 2.2.4 hold,

$$f(x) \sim \sum_{n=0}^{\infty} \frac{a_n}{x^n} \quad \text{as } x \to \infty . \qquad 2.2.5$$

The symbol $\sim$ (called tilde or, less formally, 'twiddle') is used rather than an 'equal to' sign because the infinite number of terms in the sum are never used. Indeed to do so may lead to a divergence. In fact we saw in Chapter 1 that the series 2.2.1 is divergent as $N \to \infty$, i.e.

$$\lim_{N \to \infty} R_N = \infty \quad \text{for any value of } x.$$

The expression 2.2.5 implies that

$$\lim_{x \to \infty} R_N = 0 \quad \text{for any value of } N,$$

and further that

$$\lim_{x \to \infty} \left(\frac{R_N}{1/x^N} \right) = 0 \quad \text{for any value of } N.$$

If a function, $f(x)$, has an asymptotic expansion in inverse powers of x then the coefficients a_n $(n = 0, 1, 2, \ldots)$ are given by the following limits.
Leading term

$$f(x) = a_0 + O\left(\frac{1}{x}\right) \quad \text{as } x \to \infty .$$

Thus,

$$\lim_{x \to \infty} f(x) = a_0 .$$

First order term

$$f(x) = a_0 + \frac{a_1}{x} + O\left(\frac{1}{x^2}\right) \quad \text{as } x \to \infty .$$

Divide through by $1/x$ to obtain

$$x(f(x) - a_0) = a_1 + O\left(\frac{1}{x}\right) \quad \text{as } x \to \infty .$$

Thus,

$$\lim_{x \to \infty} x[f(x) - a_0] = a_1 .$$

Second order term

$$f(x) = a_0 + \frac{a_1}{x} + \frac{a_2}{x^2} + O\left(\frac{1}{x^2}\right) \quad \text{as } x \to \infty .$$

Thus,

$$\lim_{x \to \infty} x^2 \left[f(x) - a_0 - \frac{a_1}{x} \right] = a_2.$$

The Nth order term

$$a_N = \lim_{x \to \infty} x^N \left(f(x) - \sum_{n=0}^{N-1} a_n/x^n \right). \qquad 2.2.6$$

The sequence $\{1, 1/x, 1/x^2, 1/x^3, \ldots\}$ is an example of an *asymptotic sequence* as $x \to \infty$. The characteristic feature of such sequences is that each member is dominated by the previous member. There are infinitely many asymptotic sequences. In constructing examples it is easier to deal with the limit zero than any other. Thus for the case $x \to \infty$ we let $\varepsilon = 1/x$ while for $x \to x_0$ we let $\varepsilon = x - x_0$ so that without loss of generality we may confine our attention to the limit $\varepsilon \to 0$. The standard asymptotic sequence is $\{1, \varepsilon, \varepsilon^2, \varepsilon^3, \ldots\}$ as $\varepsilon \to 0$. We have seen other examples in the first chapter, i.e. $\{1, \varepsilon^{1/2}, \varepsilon, \varepsilon^{3/2}, \ldots\}$. If we let $\delta_n(\varepsilon)$ represent members of an asymptotic sequence $\{\delta_0(\varepsilon), \delta_1(\varepsilon), \delta_2(\varepsilon), \ldots\}$ as $\varepsilon \to 0$ then the following condition must hold

$$\delta_{n+1}(\varepsilon) = o[\delta_n(\varepsilon)] \quad \text{as } \varepsilon \to 0.$$

Some examples of asymptotic sequences are:

(i) $\{1, \sin\varepsilon, (\sin\varepsilon)^2, (\sin\varepsilon)^3, \ldots\}$ i.e. $\delta_n = (\sin\varepsilon)^n$.

Proof:

$$\lim_{\varepsilon \to 0} \frac{\delta_{n+1}}{\delta_n} = \lim_{\varepsilon \to 0} \sin\varepsilon = 0.$$

(ii) $\{1, \ln(1+\varepsilon), \ln(1+\varepsilon^2), \ln(1+\varepsilon^3), \ldots\}$

i.e. $\delta_0 = 1$ and $\delta_n = \ln(1+\varepsilon^n)$ for $n \geqslant 1$.

Proof:

$$\lim_{\varepsilon \to 0} \frac{\delta_1}{\delta_0} = \lim_{\varepsilon \to 0} \ln(1+\varepsilon) = 0$$

$$\lim_{\varepsilon \to 0} \frac{\delta_{n+1}}{\delta_n} = \lim_{\varepsilon \to 0} \frac{\ln(1+\varepsilon^{n+1})}{\ln(1+\varepsilon^n)} \quad \text{for } n \geqslant 1.$$

Maclaurin's formula yields

$$\ln(1+\varepsilon^n) = \varepsilon^n + O(\varepsilon^{2n}) \quad \text{as } \varepsilon \to 0.$$

Hence

$$\lim \frac{\delta_{n+1}}{\delta_n} = \lim_{\varepsilon \to 0} \left(\frac{\varepsilon^{n+1} + O(\varepsilon^{2n+2})}{\varepsilon^n + O(\varepsilon^{2n})} \right) = 0.$$

(iii) $\{\tan(\varepsilon^{1/2}), \tan(\varepsilon), \tan(\varepsilon^{3/2}), \ldots\}$ i.e. $\delta_n = \tan(\varepsilon^{n/2})$.

(Notice here that unity is not a member of the sequence. There is no need for unity or any other particular function to be a member of the sequence.)
Proof:

$$\operatorname*{Lim}_{\varepsilon\to 0}\frac{\delta_{n+1}}{\delta_n} = \operatorname*{Lim}_{\varepsilon\to 0}\frac{\tan(\varepsilon^{(n+1)/2})}{\tan(\varepsilon^{n/2})}.$$

Maclaurin's formula yields

$$\operatorname*{Lim}_{\varepsilon\to 0}\frac{\delta_{n+1}}{\delta_n} = \operatorname*{Lim}_{\varepsilon\to 0}\left(\frac{\varepsilon^{(n+1)/2} + O(\varepsilon^{3(n+1)/2})}{\varepsilon^{n/2} + O(\varepsilon^{3n/2})}\right) = 0.$$

Products of functions may be used in asymptotic sequences e.g.

$$\left\{\frac{\ln(\varepsilon)}{\varepsilon}, \frac{1}{\varepsilon}, \ln(\varepsilon), 1, \varepsilon\ln(\varepsilon), \varepsilon, \varepsilon^2\ln(\varepsilon), \varepsilon^2, \ldots\right\}.$$

The general expression for an asymptotic expansion of a function, $f(\varepsilon)$, in terms of an asymptotic sequence, $\delta_n(\varepsilon)$, is

$$f(\varepsilon) \sim \sum_{n=0}^{\infty} a_n\delta_n(\varepsilon) \quad \text{as } \varepsilon \to 0, \tag{2.2.7}$$

where the coefficients a_n, are independent of ε. The expression 2.2.7, involving the tilde symbol $\sim$, means that for all N

$$f(\varepsilon) = \sum_{n=0}^{N} a_n\delta_n(\varepsilon) + R_N, \tag{2.2.8}$$

where

$$R_N = O[\delta_{N+1}(\varepsilon)] \quad \text{as } \varepsilon \to 0. \tag{2.2.9}$$

A straightforward generalization of the argument leading to equation 2.2.6 yields the following expression for the coefficients a_n in 2.2.7,

$$a_N = \operatorname*{Lim}_{\varepsilon\to 0}\left(\frac{f(\varepsilon) - \sum_{n=0}^{N-1} a_n\delta_n(\varepsilon)}{\delta_N(\varepsilon)}\right). \tag{2.2.10}$$

To illustrate the meaning of these expressions the function e^ε will be expanded using the asymptotic sequence $\{1, \ln(1+\varepsilon), \ln(1+\varepsilon^2), \ldots\}$.
Assume that

$$e^\varepsilon \sim a_0 + a_1\ln(1+\varepsilon) + a_2\ln(1+\varepsilon^2) + a_3\ln(1+\varepsilon^3) + \cdots. \tag{2.2.11}$$

Then

$$e^\varepsilon = a_0 + a_1\ln(1+\varepsilon) + o[\ln(1+\varepsilon)] \quad \text{as } \varepsilon \to 0,$$

so that taking the limit $\varepsilon \to 0$ yields

$$1 = a_0 + 0, \quad \text{i.e.} \quad a_0 = 1.$$

Thus

$$e^{\varepsilon} - 1 = a_1 \ln(1 + \varepsilon) + o[\ln(1 + \varepsilon)],$$

so that dividing by $\ln(1 + \varepsilon)$ and taking the limit as $\varepsilon \to 0$ leads to the expression

$$\operatorname*{Lim}_{\varepsilon \to 0} \left(\frac{e^{\varepsilon} - 1}{\ln(1 + \varepsilon)} \right) = a_1 + \operatorname*{Lim}_{\varepsilon \to 0} \left(\frac{o[\ln(1 + \varepsilon)]}{\ln(1 + \varepsilon)} \right).$$

The second member of the right-hand side is zero. Taylor's formula may be used to express the numerator and denominator of the left-hand side in powers of ε which then enables the limit to be evaluated. (L'Hospital's rule may be used instead but it becomes cumbersome for the subsequent terms.)

$$\operatorname*{Lim}_{\varepsilon \to 0} \left(\frac{e^{\varepsilon} - 1}{\ln(1 + \varepsilon)} \right) = \operatorname*{Lim}_{\varepsilon \to 0} \left(\frac{1 + \varepsilon + O(\varepsilon^2) - 1}{\varepsilon + O(\varepsilon^2)} \right) = 1.$$

Thus $a_1 = 1$ and

$$e^{\varepsilon} - 1 - \ln(1 + \varepsilon) = a_2 \ln(1 + \varepsilon^2) + o[\ln(1 + \varepsilon^2)] \quad \text{as } \varepsilon \to 0.$$

Dividing by $\ln(1 + \varepsilon^2)$ and taking the limit $\varepsilon \to 0$ yields

$$\begin{aligned} a_2 &= \operatorname*{Lim}_{\varepsilon \to 0} \left(\frac{e^{\varepsilon} - 1 - \ln(1 + \varepsilon)}{\ln(1 + \varepsilon^2)} \right) \\ &= \operatorname*{Lim}_{\varepsilon \to 0} \left(\frac{1 + \varepsilon + \varepsilon^2/2 + O(\varepsilon^3) - 1 - \varepsilon + \varepsilon^2/2 + O(\varepsilon^3)}{\varepsilon^2 + O(\varepsilon^3)} \right) = 1. \end{aligned}$$

Continuing the process leads to

$$e^{\varepsilon} - 1 - \ln(1 + \varepsilon) - \ln(1 + \varepsilon^2) = a_3 \ln(1 + \varepsilon^3) + o[\ln(1 + \varepsilon^3)] \quad \text{as } \varepsilon \to 0,$$

so that

$$\begin{aligned} a_3 &= \operatorname*{Lim}_{\varepsilon \to 0} \left(\frac{e^{\varepsilon} - 1 - \ln(1 + \varepsilon) - \ln(1 + \varepsilon^2)}{\ln(1 + \varepsilon^3)} \right) \\ &= \operatorname*{Lim}_{\varepsilon \to 0} \left(\frac{1+\varepsilon+\varepsilon^2/2+\varepsilon^3/6+O(\varepsilon^4)-1-\varepsilon+\varepsilon^2/2-\varepsilon^3/3+O(\varepsilon^4)-\varepsilon^2+O(\varepsilon^4)}{\varepsilon^3+O(\varepsilon^6)} \right) \\ &= -1/6. \end{aligned}$$

Substituting the values of a_0, a_1, a_2 and a_3 into the expansion 2.2.11 yields, $e^{\varepsilon} = 1 + \ln(1 + \varepsilon) + \ln(1 + \varepsilon^2) - \frac{1}{6}\ln(1 + \varepsilon^3) + O[\ln(1 + \varepsilon^4)]$ as $\varepsilon \to 0$.

It is not always possible to expand a function in terms of a particular asymptotic sequence. To demonstrate this let us attempt to expand the function $\cos(\varepsilon^{1/2} + \varepsilon)$ using the asymptotic sequence $\{1, \ln(1 + \varepsilon), \ln(1 + \varepsilon^2), \ldots\}$ as $\varepsilon \to 0$.

Assume that

$$\cos(\varepsilon^{1/2} + \varepsilon) \sim a_0 + a_1 \ln(1 + \varepsilon) + a_2 \ln(1 + \varepsilon^2) + \cdots \quad \text{as } \varepsilon \to 0.$$

Then

$$a_0 = \operatorname*{Lim}_{\varepsilon \to 0} \cos(\varepsilon^{1/2} + \varepsilon) = 1$$

$$a_1 = \operatorname*{Lim}_{\varepsilon \to 0} \left(\frac{\cos(\varepsilon^{1/2} + \varepsilon) - 1}{\ln(1 + \varepsilon)} \right) = \operatorname*{Lim}_{\varepsilon \to 0} \left(\frac{1 - (\varepsilon^{1/2} + \varepsilon)^2/2 + O(\varepsilon^2) - 1}{\varepsilon + O(\varepsilon^2)} \right) = -\frac{1}{2}$$

$$a_2 = \operatorname*{Lim}_{\varepsilon \to 0} \left(\frac{1 - \frac{1}{2}(\varepsilon + 2\varepsilon^{3/2} + \varepsilon^2) + \frac{1}{24}[\varepsilon^2 + O(\varepsilon^{5/2})] - 1 + \frac{1}{2}\varepsilon - \frac{1}{4}\varepsilon^2 + O(\varepsilon^3)}{\varepsilon^2 + O(\varepsilon^4)} \right)$$

$$= \operatorname*{Lim}_{\varepsilon \to 0} \left(-\frac{\varepsilon^{3/2} + O(\varepsilon^2)}{\varepsilon^2 + O(\varepsilon^4)} \right) = -\infty.$$

The occurrence of an infinite value for a coefficient in an assumed expansion invalidates the expansion. In this case the absence of a term of order $\varepsilon^{3/2}$ (and of subsequent fractional powers) in the asymptotic sequence $\{1, \ln(1 + \varepsilon), \ln(1 + \varepsilon^2) \ldots\}$ is the reason why an expansion does not exist for the function $\cos(\varepsilon^{1/2} + \varepsilon)$ in terms of the given asymptotic sequence.

Exercise

Obtain the first three nonzero coefficients in the expansions of

(i) $\ln(1+\varepsilon)$

(ii) e^{ε}

using the asymptotic sequence $\{1, \sin \varepsilon, (\sin \varepsilon)^2, (\sin \varepsilon)^3, \ldots\}$ as $\varepsilon \to 0$.

Uniqueness

If a function possesses an asymptotic expansion involving the sequence $\{\delta_0(\varepsilon), \delta_1(\varepsilon), \delta_2(\varepsilon), \ldots\}$ then the coefficients a_n, of the expansion 2.2.7, given by the expression 2.2.10 are unique. However, another function may share the same set of coefficients. Thus while functions have unique expansions, an expansion does not correspond to a unique function. An example of this is provided by the two functions $f_1 = \sin \varepsilon$ and $f_2 = \sin \varepsilon + \exp(-1/\varepsilon^2)$ and their expansions in terms of the asymptotic sequence $\{1, \varepsilon, \varepsilon^2, \varepsilon^3, \ldots\}$.

If we assume that $f_2 \sim a_0 + a_1 \varepsilon + a_2 \varepsilon^2 + \cdots$ as $\varepsilon \to 0$, then

$$a_0 = \operatorname*{Lim}_{\varepsilon \to 0} [\sin \varepsilon + \exp(-1/\varepsilon^2)] = 0$$

$$a_1 = \operatorname*{Lim}_{\varepsilon \to 0} \left(\frac{\sin \varepsilon + \exp(-1/\varepsilon^2)}{\varepsilon} \right) = \operatorname*{Lim}_{\varepsilon \to 0} \left(\frac{\sin \varepsilon}{\varepsilon} \right) + \operatorname*{Lim}_{\varepsilon \to 0} \left(\frac{\exp(-1/\varepsilon^2)}{\varepsilon} \right)$$

$$= \quad 1 \quad + \quad 0.$$

That the second limit is zero and that all subsequent limits of the form $\lim_{\varepsilon \to 0} [\exp(-1/\varepsilon^2)/\varepsilon^n]$ are zero can be seen from the order relation 2.1.7′. Thus

$$a_2 = \lim_{\varepsilon \to 0} \left(\frac{\sin \varepsilon + \exp(-1/\varepsilon^2) - \varepsilon}{\varepsilon^2} \right) = 0$$

and

$$a_3 = \lim_{\varepsilon \to 0} \left(\frac{\sin \varepsilon + \exp(-1/\varepsilon^2) - \varepsilon}{\varepsilon^3} \right) = -\frac{1}{6}.$$

The functions f_1 and f_2 share the same expansion in terms of the asymptotic sequence $\{\varepsilon^n\}$ because

$$\exp(-1/\varepsilon^2) \sim 0 \times 1 + 0 \times \varepsilon + 0 \times \varepsilon^2 + 0 \times \varepsilon^3 + \cdots \quad \text{as } \varepsilon \to 0.$$

We will see an example of the importance of the distinction between functions of the type f_1 and f_2 when we consider boundary layers. Then two functions typically differ by the quantity $\mathrm{e}^{-x/\varepsilon}$ with $0 < \varepsilon \ll 1$ where x is the independent variable. If x is of size unity then $\mathrm{e}^{-x/\varepsilon}$ is negligible in the sense that it tends to zero faster than any power of ε as $\varepsilon \to 0$. However, as x approaches zero the function $\mathrm{e}^{-x/\varepsilon}$ is of order unity.

Maclaurin series are asymptotic expansions

If a function possesses a Maclaurin series then

$$f(x) = \sum_{n=0}^{\infty} a_n x^n, \tag{2.2.12}$$

for $|x|$ less than some nonzero radius of convergence. This is also an asymptotic expansion as $x \to 0$, i.e.

$$f(x) \sim \sum_{n=0}^{\infty} a_n x^n \quad \text{as } x \to 0. \tag{2.2.13}$$

To prove this we must show that

$$f(x) = \sum_{n=0}^{N} a_n x^n + R_N, \tag{2.2.14}$$

where

$$R_N = \mathrm{O}(x^{N+1}) \quad \text{as } x \to 0.$$

We know from 2.1.4 that

$$R_N = \frac{x^{N+1}}{(N+1)!} \frac{\mathrm{d}^{N+1}}{\mathrm{d}x^{N+1}} f(z),$$

where $|z| < |x|$ and the existence of the $(N+1)$th derivative is guaranteed provided

$|x|$ is less than the radius of convergence of the series 2.2.12. Therefore

$$\lim_{x\to 0} \frac{R_N}{x^{N+1}} = \frac{1}{(N+1)!} \frac{\mathrm{d}^{N+1}}{\mathrm{d}x^{N+1}} f(0),$$

this is a finite number which proves that $R_N = O(x^{N+1})$ as $x \to 0$. Thus all Maclaurin series are asymptotic expansions.

2.3 Uniform and nonuniform expansions

Consider a function, $f(x; \varepsilon)$, which depends on both an independent variable, x, and a small parameter, ε. Suppose that $f(x; \varepsilon)$ is expanded using an asymptotic sequence $\{\delta_n(\varepsilon)\}$,

$$f(x; \varepsilon) = \sum_{n=0}^{N} a_n(x)\delta_n(\varepsilon) + R_N(x; \varepsilon). \qquad 2.3.1$$

The coefficients of the gauge functions $\delta_n(\varepsilon)$, are functions of x and the remainder after N terms is a function of both x and ε. For this to be an asymptotic expansion we require

$$R_N(x; \varepsilon) = O[\delta_{N+1}(\varepsilon)] \quad \text{as } \varepsilon \to 0. \qquad 2.3.2$$

For 2.3.1 to be a uniform asymptotic expansion the ultimate proportionality between R_N and δ_{N+1} must be bounded by a number independent of x, i.e.

$$|R_N(x; \varepsilon)| \leqslant K|\delta_{N+1}(\varepsilon)|, \qquad 2.3.3$$

for ε in the neighborhood near zero, where K is a fixed constant.

An example of a uniform asymptotic expansion is

$$f(x; \varepsilon) = \frac{1}{1 - \varepsilon \sin x} \sim 1 + \varepsilon \sin x + \varepsilon^2 (\sin x)^2 + \varepsilon^3 (\sin x)^3 + \cdots \quad \text{as } \varepsilon \to 0.$$

The remainder after the Nth term is uniformly of order ε^{N+1} since

$$R_N = f - \sum_{n=0}^{N} \varepsilon^n (\sin x)^n,$$

and

$$\lim_{\varepsilon \to 0} \left(\frac{R_N}{\varepsilon^{N+1}} \right) = (\sin x)^{N+1}.$$

Thus $R_N(x; \varepsilon)$ can be uniformly bounded,

$$|R_N(x; \varepsilon)| \leqslant K|\varepsilon^{N+1}|,$$

where K can be chosen to be any fixed value greater than $(\sin x)^{N+1}$, for example $K = 1.1$.

In contrast the expansion 2.3.1 is nonuniform if no fixed constant, K, exists such that the condition 2.3.3 is satisfied. An example of a nonuniform expansion is:

$$f(x, \varepsilon) \sim 1 + \varepsilon x + \varepsilon^2 x^2 + \varepsilon^3 x^3 + \cdots \quad \text{as } \varepsilon \to 0. \qquad 2.3.4$$

In this case

$$R_N = f - \sum_{n=0}^{N} \varepsilon^n x^n,$$

and

$$\operatorname*{Lim}_{\varepsilon \to 0} \left(\frac{R_N}{\varepsilon^{N+1}} \right) = x^{N+1}.$$

Thus no fixed constant K exists such that

$$|R_N| \leqslant K|\varepsilon^{N+1}|,$$

because for any choice of K, x can be chosen so that x^{N+1} exceeds this value.

Region of nonuniformity

The expansion 2.3.4 becomes nonuniform when subsequent terms are no longer small corrections to previous terms. This occurs when subsequent terms are of the same order or of dominant order than previous terms. Subsequent terms are of the same order in the expansion 2.3.4 when $x = O(1/\varepsilon)$ as $\varepsilon \to 0$. Subsequent terms dominate previous terms for larger x, e.g. $x = O(1/\varepsilon^2)$. The expansion is valid for $x = O(1)$ since then subsequent terms decrease by a factor of ε. The expansion remains valid for large x provided x is not as large as $1/\varepsilon$. For example, the expansion is valid for $x = O(1/\sqrt{\varepsilon})$ as $\varepsilon \to 0$.

The critical case is such that subsequent terms are of the same order. This determines the region of nonuniformity. In the example considered the region of nonuniformity occurs when $\varepsilon x = O(1)$, i.e. $x = O(1/\varepsilon)$ as $\varepsilon \to 0$.

Another example of a nonuniformity is provided by the expansion

$$1 + \varepsilon e^x + \varepsilon^2 e^{2x} + \varepsilon^3 e^{3x} + \cdots$$

The region of nonuniformity occurs when $\varepsilon e^x = O(1)$, i.e. $e^x = O(1/\varepsilon)$ so that $x = O[-\ln(\varepsilon)]$ as $\varepsilon \to 0$.

In the next example the nonuniformity occurs near $x = 0$, consider

$$1 + \frac{\varepsilon}{x} + \frac{\varepsilon^2}{x^2} + \frac{\varepsilon^3}{x^3} + \cdots.$$

The region of nonuniformity is given by $\varepsilon/x = O(1)$, i.e. $x = O(\varepsilon)$ as $\varepsilon \to 0$.

Consider next the expansions for small ε of the functions $\sin(x+\varepsilon)$ and $\sin[x(1+\varepsilon)]$.

$$\sin(x+\varepsilon) = \sin x \, . \cos\varepsilon + \cos x \, . \sin\varepsilon$$

$$= \sin x \, . \left(1 - \frac{\varepsilon^2}{2} + O(\varepsilon^4)\right) + \cos x \, . \left(\varepsilon - \frac{\varepsilon^3}{3!} + O(\varepsilon^5)\right)$$

$$= \sin x + \varepsilon \cos x - \frac{\varepsilon^2}{2}\sin x - \frac{\varepsilon^3}{6}\cos x + O(\varepsilon^4) \qquad 2.3.5$$

$$\sin[x(1+\varepsilon)] = \sin x \, . \cos\varepsilon x + \cos x \, . \sin\varepsilon x$$

$$= \sin x \, . \left(1 - \frac{\varepsilon^2 x^2}{2} + O(\varepsilon^4 x^4)\right)$$

$$+ \cos x \, . \left(\varepsilon x - \frac{\varepsilon^3 x^3}{3!} + O(\varepsilon^5 x^5)\right)$$

$$= \sin x + \varepsilon x \cos x - \frac{\varepsilon^2 x^2}{2}\sin x - \frac{\varepsilon^3 x^3}{6}\cos x + O(\varepsilon^4 x^4). \qquad 2.3.6$$

In the expansion 2.3.5 each coefficient of the ascending powers of ε is bounded by a fixed quantity so that subsequent terms in the expansion are small corrections to previous terms for all x. Thus the expansion is uniform for all x. In the expansion 2.3.6 the coefficients cannot be bounded by a fixed quantity and consequently it is possible for subsequent terms to have the same order as the previous terms. The region of nonuniformity is given by $x = O(1/\varepsilon)$ as $\varepsilon \to 0$. In establishing this result the trigonometric functions $\cos x$ and $\sin x$ are assigned order unity for all x.

The region of nonuniformity should not be confused with the radius of convergence. The series 2.3.6 converges for all values of ε and x and yet is nonuniform. In fact convergence is irrelevant since we shall always deal with a truncated series consisting of the first few terms. Our concern is over the behavior of the remainder associated with the truncated series. The remainder must tend to zero faster than the previous terms of an expansion to be a uniform expansion.

Nonuniformities do not occur in expansions

$$f(x;\varepsilon) \sim \sum_{n=0}^{\infty} f_n(x)\,\delta_n(\varepsilon) \quad \text{as } \varepsilon \to 0,$$

if the coefficient functions $f_n(x)$ are bounded. For a nonuniformity we require

$$f_{n+1}(x)\,\delta_{n+1}(\varepsilon) = O[f_n(x)\,\delta_n(\varepsilon)],$$

i.e.

$$f_{n+1}(x) = f_n(x)\,O\left(\frac{\delta_n(\varepsilon)}{\delta_{n+1}(\varepsilon)}\right). \qquad 2.3.7$$

The functions $\delta_n(\varepsilon)$ form an asymptotic sequence as $\varepsilon \to 0$ so that the ratio $\delta_n(\varepsilon)/\delta_{n+1}(\varepsilon)$ is singular. The nonuniformity condition, 2.3.7, requires not only that

$f_{n+1}(x)$ is singular but also that it is more singular than $f_n(x)$. Thus the expansion

$$x + \varepsilon x + \varepsilon^2 x + \varepsilon^3 x,$$

is uniform even though the coefficient functions are singular as $x \to \infty$. The singularity does not increase in strength with subsequent terms so that the increasing powers of ε ensure that subsequent terms are small corrections to previous terms regardless of how large x becomes.

Exercises

1 Obtain the region of nonuniformity as $\varepsilon \to 0$ of the following expansions:

(i) $1 + \dfrac{\varepsilon x^2}{2!} + \dfrac{\varepsilon^2 x^4}{4!} + \cdots$

(ii) $1 + \dfrac{\varepsilon^2}{2x} + \dfrac{\varepsilon^4}{3x^2} + \cdots$

(iii) $\sqrt{x} - \varepsilon + \dfrac{\varepsilon^2}{\sqrt{x}} - \dfrac{\varepsilon^3}{x} + \cdots$

(iv) $1 - \varepsilon e^{-x} + \varepsilon^2 e^{-2x} + \varepsilon^3 e^{-3x} + \cdots$.

2 Expand the following functions for small ε using the asymptotic sequence $\{1, \varepsilon, \varepsilon^2, \ldots\}$. Which of the expansions are nonuniform? What are their regions of nonuniformity?

(i) $e^{x+\varepsilon}$, (ii) $e^{x/\varepsilon}$, (iii) $1/(x^2+\varepsilon)$, (iv) $\ln\left(1+\dfrac{\varepsilon}{x}\right)$, (v) $\cos\left(\dfrac{\varepsilon}{1+x^2}\right)$, (vi) $\cos\left(\dfrac{\varepsilon}{x^2}\right)$.

2.4 Sources of nonuniformity

There are two common sources of nonuniformities in asymptotic expansions; they are:

1 Infinite domains which allow long-term effects of small perturbations to accumulate
2 Singularities in governing equations which lead to localized regions of rapid change.

Infinite domains

An example of a nonuniformity caused by an infinite domain is provided by the nonlinear oscillator equation

$$\frac{d^2 u}{dt^2} + u + \varepsilon u^3 = 0. \qquad 2.4.1$$

This is Duffing's equation. In the following chapter we will see how it arises in the description of two different mechanical systems. It also governs certain electrical systems. It is an example of a class of nonlinear oscillators which we will study in some detail. The variable u can represent a variety of quantities such as an angle of oscillation, the deformation of an elastic system, a current or a voltage. The independent variable, t, is time. We shall consider the behavior of the solution, u, for $t > 0$ subject to specified initial conditions, $u(0)$ and $\frac{du}{dt}(0)$.

Since the perturbation term, εu^3, in 2.4.1 is nonlinear, no exact solution in terms of elementary functions can be obtained.† Suppose the solution may be expanded using the standard asymptotic sequence $\delta_n(\varepsilon) = \varepsilon^n$,

$$u(t; \varepsilon) \sim u_0(t) + \varepsilon u_1(t) + \varepsilon^2 u_2(t) + \cdots. \qquad 2.4.2$$

On substituting this into 2.4.1 we obtain

$$\frac{d^2 u_0}{dt^2} + \varepsilon \frac{d^2 u_1}{dt^2} + \cdots + u_0 + \varepsilon u_1 + \cdots + \varepsilon u_0^3 + \cdots \sim 0.$$

The symbol ~ 0 rather than $= 0$ allows for the possibility of the asymptotic expansion being divergent. If we replace the expressions $+\cdots$ by $+O(\varepsilon^2)$ then the $\sim$ symbol may be replaced by the $=$ symbol, i.e.

$$\left(\frac{d^2 u_0}{dt^2} + u_0\right) + \varepsilon\left(\frac{d^2 u_1}{dt^2} + u_1 + u_0^3\right) + O(\varepsilon^2) = 0.$$

The right-hand side of this equation may be expressed in the form $0 = 0 + 0\varepsilon + 0\varepsilon^2 + \cdots$ and coefficients of powers of ε equated,

$$O(1): \quad \frac{d^2 u_0}{dt^2} + u_0 = 0 \qquad 2.4.3$$

$$O(\varepsilon): \quad \frac{d^2 u_1}{dt^2} + u_1 + u_0^3 = 0, \qquad 2.4.4$$

etc.

In order to obtain solutions for u_0, u_1 etc., initial conditions must be imposed. Suppose $u(0) = a$ and $\frac{du}{dt}(0) = b$. Then

$$u_0(0) + \varepsilon u_1(0) + \varepsilon^2 u_2(0) + \cdots \sim a + 0\varepsilon + 0\varepsilon^2 + \cdots \qquad 2.4.5$$

and

$$\frac{du_0}{dt}(0) + \varepsilon \frac{du_1}{dt}(0) + \varepsilon^2 \frac{du_2}{dt}(0) + \cdots \sim b + 0\varepsilon + 0\varepsilon^2 + \cdots \qquad 2.4.6$$

† Solutions can however be constructed using elliptic functions.

Equating coefficients of powers of ε in 2.4.5 and 2.4.6 leads to the following partitioning of the initial conditions:

$$u_0(0) = a, \quad u_1(0) = 0, \quad u_2(0) = 0, \text{ etc.}$$

$$\frac{du_0}{dt}(0) = b, \quad \frac{du_1}{dt}(0) = 0, \quad \frac{du_2}{dt}(0) = 0, \text{ etc.}$$

To simplify the subsequent analysis we will set $b = 0$. The solution of 2.4.3 is

$$u_0 = a \cos t.$$

Equation 2.4.4 becomes

$$\frac{d^2 u_1}{dt^2} + u_1 = -a^3 \cos^3 t.$$

The right-hand side can be expressed in terms of $\cos t$ and $\cos 3t$ using the multiple angle formulae. These are conveniently derived using Euler's identities

$$e^{\pm i\theta} = \cos\theta \pm i \sin\theta.$$

Thus

$$\cos^3 t = \tfrac{1}{8}(e^{it} + e^{-it})^3 = \tfrac{1}{8}(e^{3it} + e^{-3it} + 3e^{it} + 3e^{-it})$$
$$= \tfrac{1}{4}\cos 3t + \tfrac{3}{4}\cos t,$$

and

$$\frac{d^2 u_1}{dt^2} + u_1 = \frac{-a^3}{4}\cos 3t - \frac{3a^3}{4}\cos t.$$

The complementary function is $A\cos t + B\sin t$. The particular integral associated with the $\cos 3t$ member of the inhomogeneous term has the form $\alpha\cos 3t + \beta\sin 3t$ and the absence of the first derivative, du_1/dt, on the left-hand side causes $\beta = 0$. Thus we require

$$-9\alpha\cos 3t + \alpha\cos 3t = \frac{-a^3}{4}\cos 3t,$$

so that $\alpha = a^3/32$.

The last member of the inhomogeneous term is of the same form as the complementary function so that a trial particular integral of the form $\alpha t\cos t + \beta t\sin t$ should be used. The absence of du_1/dt causes $\alpha = 0$. Then

$$2\beta\cos t = -\tfrac{3}{4}a^3\cos t,$$

so that $\beta = -3a^3/8$.

Thus the general solution for u_1 is

$$u_1 = A\cos t + B\sin t + \frac{a^3}{32}\cos 3t - \frac{3a^3}{8} t\sin t.$$

The initial conditions $u_1(0) = \frac{du_1}{dt}(0) = 0$ lead to

$$A = -a^3/32, \quad B = 0,$$

so that

$$u_1 = \frac{a^3}{32}(\cos 3t - \cos t) - \frac{3a^3}{8} t \sin t,$$

and finally

$$u \sim a \cos t + \varepsilon \left[\frac{a^3}{32}(\cos 3t - \cos t) - \frac{3a^3}{8} t \sin t \right] + \cdots. \qquad 2.4.7$$

The term $t \sin t$ in the expansion 2.4.7 is called a *secular* term. It is an oscillatory term of growing amplitude. All the other terms are oscillatory of fixed amplitude. The secular term leads to a nonuniformity for large t. The region of nonuniformity is obtained by equating the order of the first and second terms,

$$\cos t = O(\varepsilon t \sin t) \quad \text{as } \varepsilon \to 0.$$

The trigonometric functions are treated as $O(1)$ terms. (We need not be concerned with those periodic occasions when the cosine or sine terms are zero.) Thus the region of nonuniformity is $t = O(1/\varepsilon)$ as $\varepsilon \to 0$.

In subsequent chapters we will consider techniques for overcoming this type of infinite domain nonuniformity.

Small parameter multiplying the highest derivative

The second common source of nonuniformities is associated with the presence of singularities. Consider the differential equation

$$\varepsilon \frac{dy}{dx} + y = e^{-x}, \qquad 2.4.8$$

where ε is a small positive parameter. The solution is to be obtained in the region $x > 0$ subject to the boundary condition $y(0) = 2$.

Suppose y has the expansion

$$y \sim y_0(x) + \varepsilon y_1(x) + \varepsilon^2 y_2(x) + \cdots.$$

The differential equation becomes

$$\varepsilon\left(\frac{dy_0}{dx} + \varepsilon \frac{dy_1}{dx} + \cdots\right) + y_0 + \varepsilon y_1 + \varepsilon^2 y_2 + \cdots \sim e^{-x}. \qquad 2.4.9$$

The boundary condition is partitioned as follows:

$$y_0(0) = 2, \quad y_1(0) = y_2(0) = \cdots = 0.$$

Equating coefficients of powers of ε on the left- and right-hand sides of 2.4.9 yields:

$O(1)$: $y_0 = e^{-x}$ with boundary condition $y_0(0) = 2$.

$O(\varepsilon)$: $y_1 = -\dfrac{dy_0}{dx} = e^{-x}$ with boundary condition $y_1(0) = 0$.

$O(\varepsilon^2)$: $y_2 = -\dfrac{dy_1}{dx} = e^{-x}$ with boundary condition $y_2(0) = 0$.

Clearly y_0 cannot satisfy the boundary condition $y_0(0) = 2$ as no constant of integration is available because the equation determining y_0 is an algebraic equation not a differential equation.

Similarly the equations for y_1 and y_2 are algebraic and the boundary conditions cannot be satisfied. Thus we have obtained the expansion

$$y \sim e^{-x} + \varepsilon e^{-x} + \varepsilon^2 e^{-x} + \cdots \qquad 2.4.10$$

but the boundary condition $y(0) = 2$ has not been satisfied.

The unperturbed problem, obtained by setting $\varepsilon = 0$, is not a differential equation but an algebraic equation $y = e^{-x}$. This, of course, cannot satisfy an arbitrarily imposed condition at $x = 0$. For any nonzero value of ε, equation 2.4.8 becomes a first order differential equation which can satisfy a boundary condition. This is an example of a singular perturbation where the behavior of the perturbed problem is very different from that of the unperturbed problem.

It turns out that the perturbation expansion 2.4.10 is a good approximation of the exact solution away from the region $x = 0$. To see this we must compare 2.4.10 with the exact solution. The general solution of 2.4.8 is found by standard techniques to be

$$y_{ex} = Ae^{-x/\varepsilon} + e^{-x}/(1 - \varepsilon).$$

The boundary condition requires

$$A = 2 - 1/(1 - \varepsilon) = (1 - 2\varepsilon)/(1 - \varepsilon).$$

Thus

$$y_{ex} = \frac{1 - 2\varepsilon}{1 - \varepsilon} \cdot e^{-x/\varepsilon} + \frac{e^{-x}}{1 - \varepsilon},$$

and on replacing $1/(1 - \varepsilon)$ by its Maclaurin expansion this becomes

$$y_{ex} = \underbrace{(1 - \varepsilon - \varepsilon^2 + \cdots)e^{-x/\varepsilon}}_{\text{I}} + \underbrace{(1 + \varepsilon + \varepsilon^2 + \cdots)e^{-x}}_{\text{II}} \qquad 2.4.11$$

The perturbation expansion 2.4.10 generates the second member, II, but fails to create the first member, I, of y_{ex}. The coefficient $e^{-x/\varepsilon}$ of I is a rapidly varying function which takes the value unity at $x = 0$ and rapidly decays to zero for positive x. The function value is negligible for x greater than about 3ε.

In Fig. 2.2 the exact solution and the leading term $y_0(x)$ of the perturbation expansion 2.4.10 are compared for $\varepsilon = 0.05$. Clearly $y_0(x)$ provides a good approximation away from the region $x = 0$. The region near $x = 0$ is called the *boundary layer*. These regions usually occur when the highest order derivative of a differential equation is multiplied by a small parameter. The unperturbed problem, obtained by setting the small parameter value to zero, is of lower order and consequently cannot satisfy all the boundary conditions. This leads to boundary layer regions where the solution varies rapidly in order to satisfy the boundary condition.

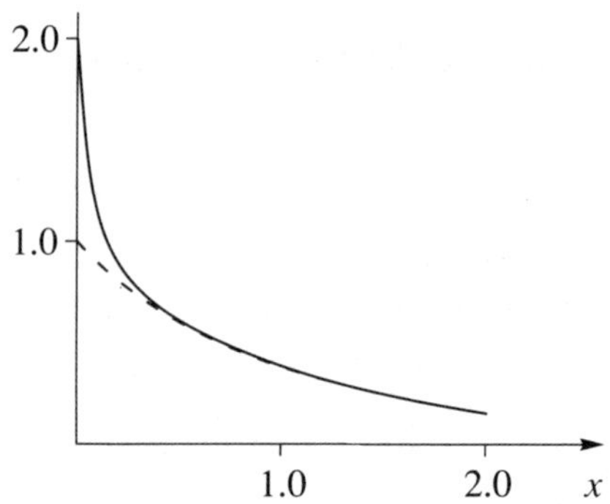

Fig. 2.2 y_{ex} —— and y_0 - - - for $\varepsilon = 0.05$

Boundary layers are regions of nonuniformity in perturbation expansions of the form 2.4.10. Consider the one-term expansion

$$y = e^{-x} + O(\varepsilon).$$

This is uniformly valid, i.e. the remainder is uniformly of order ε, provided the term I in the exact expression 2.4.11 is $O(\varepsilon)$ or less. Now the coefficient $e^{-x/\varepsilon}$ of the term I has the following behavior,

$$e^{-x/\varepsilon} = o(\varepsilon^n) \quad \text{as } \varepsilon \to 0 \text{ for all } n \text{ if } x = O(1)$$

$$e^{-x/\varepsilon} = O(1) \quad \text{as } \varepsilon \to 0 \text{ if } x = O(\varepsilon).$$

Thus omitting the term I is uniformly valid for $x = O(1)$ but introduces a nonuniformity for $x = O(\varepsilon)$. We may conclude that the region of nonuniformity of the expansion 2.4.10 is $x = O(\varepsilon)$ as $\varepsilon \to 0$.

Consider next a second order differential equation which exhibits boundary layer behavior,

$$\varepsilon \frac{d^2 y}{dx^2} + \frac{dy}{dx} + y = 0, \quad 0 < \varepsilon \ll 1. \qquad 2.4.12$$

A solution is to be obtained in the interval $0 < x < 1$ subject to the boundary conditions $y(0) = 0$, $y(1) = 1$.

The general solution of equation 2.4.12 is

$$y = A\exp(m_1 x) + B\exp(m_2 x),$$

where m_1 and m_2 are the solutions of the auxiliary equation

$$\varepsilon m^2 + m + 1 = 0,$$

i.e.

$$m_1 = \frac{-1 + \sqrt{1 - 4\varepsilon}}{2\varepsilon}, \quad m_2 = \frac{-1 - \sqrt{1 - 4\varepsilon}}{2\varepsilon}.$$

Using the binomial expansion

$$\begin{aligned}\sqrt{1-4\varepsilon} &= 1 + \tfrac{1}{2}(-4\varepsilon) + \tfrac{1}{2}(-\tfrac{1}{2})\tfrac{1}{2}(-4\varepsilon)^2 + \cdots \\ &= 1 - 2\varepsilon - 2\varepsilon^2 + \cdots\end{aligned}$$

leads to

$$m_1 = -1 - \varepsilon + \cdots, \quad m_2 = -\frac{1}{\varepsilon} + 1 + \varepsilon + \cdots$$

so that

$$y = Ae^{-x} + Be^{-x/\varepsilon}.e^{x} + O(\varepsilon).$$

The boundary conditions yield the following:

$$y(0) = 0 = A + B$$

$$y(1) = 1 = Ae^{-1} + Be^{-1/\varepsilon}.e^{1} + O(\varepsilon).$$

The term $e^{-1/\varepsilon}$ tends to zero faster than any power of ε and may therefore be neglected. Thus

$$A = e + O(\varepsilon) = -B$$

and

$$y = e^{1-x} - e^{-x/\varepsilon}.e^{1+x} + O(\varepsilon). \qquad 2.4.13$$

This solution behaves like the first term e^{1-x} away from the region $x = 0$ because the function $e^{-x/\varepsilon}$ is then negligible. On the other hand, in the region near $x = 0$ the

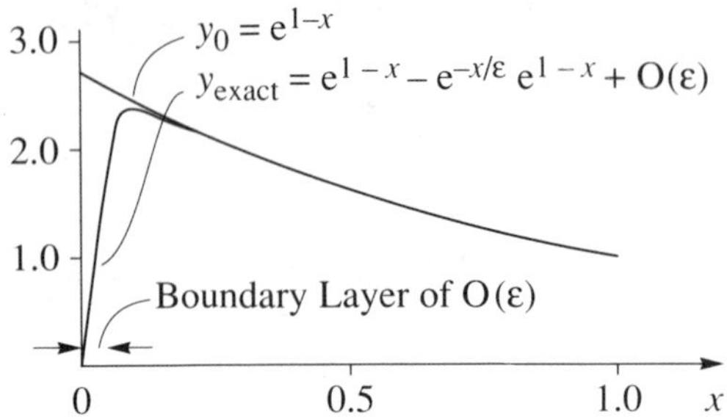

Fig. 2.3 Comparison of y_{exact} and y_0

function $e^{-x/\varepsilon}$ rapidly grows to take the value unity at $x = 0$. Thus there is a boundary layer at $x = 0$ (see Fig. 2.3).

The perturbation expansion

$$y \sim y_0(x) + \varepsilon y_1(x) + \cdots$$

leads in the usual way to the following order equations,

$$O(1): \quad \frac{dy_0}{dx} + y_0 = 0, \quad y_0(0) = 0, \quad y_0(1) = 1 \tag{2.4.14}$$

$$O(\varepsilon): \quad \frac{dy_1}{dx} + y_1 = -\frac{d^2 y_0}{dx^2}, \quad y_1(0) = 0, \quad y_1(1) = 0, \tag{2.4.15}$$

etc.

The equations 2.4.14 and 2.4.15 are first order and therefore cannot satisfy both boundary conditions. If the boundary condition at $x = 0$ is neglected then the solution of 2.4.14 ($y_0 = Ae^{-x}$) satisfying the boundary condition at $x = 1$ is $y_0 = e^{1-x}$. This closely approximates the exact solution away from the boundary layer at $x = 0$ (see Fig. 2.3). Had we chosen the alternative of neglecting the boundary condition at $x = 1$ and satisfying the $x = 0$ condition we obtain $y_0 = 0$ which fails to approximate the exact solution anywhere except at $x = 0$ itself.

The reason for choosing to neglect one boundary condition rather than another in order to obtain a useful approximation to the exact solution outside the boundary layer will be explained in Chapter 5. It will be shown how perturbation expansions are constructed both within and outside the boundary layers and matched in an intermediate region.

Boundary layers can occur in the solution of partial differential equations when a small parameter multiplies one of the highest derivatives. Consider for example the steady conduction and convection of heat in a two-dimensional rectangular region $0 < X < L$, $0 < Y < H$ with solid planes at $Y = 0$ and $Y = H$ held at the fixed temperature T_w. Suppose fluid with thermal diffusivity α flows between the planes with speed U_0 as shown in Fig. 2.4. Let the inlet temperature of the fluid be T_0 and the outlet temperature be T_w.

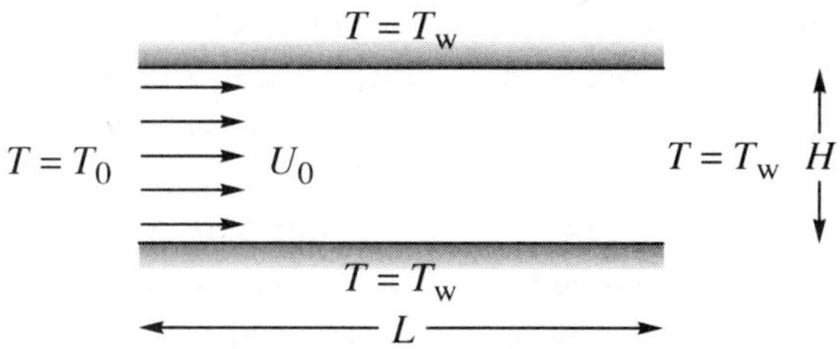

Fig. 2.4 Conduction and convection of heat with flow between parallel planes

The governing conduction convection equation for the fluid temperature is

$$U_0 \frac{\partial T}{\partial X} = \alpha \left(\frac{\partial^2 T}{\partial X^2} + \frac{\partial^2 T}{\partial Y^2} \right). \qquad 2.4.16$$

Introducing the following nondimensional variables

$$x = \frac{X}{L}, \quad y = \frac{Y}{H}, \quad \theta = \frac{T - T_w}{T_0 - T_w},$$

leads to the equation

$$\frac{U_0}{L} \frac{\partial \theta}{\partial x} = \alpha \left(\frac{1}{L^2} \frac{\partial^2 \theta}{\partial x^2} + \frac{1}{H^2} \frac{\partial^2 \theta}{\partial y^2} \right).$$

Multiply throughout by H^2/α and denote the nondimensional quantity $U_0 H^2/\alpha L$ by Pe (called the Péclet number) and H^2/L^2 by ε so that the equation becomes

$$\varepsilon \frac{\partial^2 \theta}{\partial x^2} + \frac{\partial^2 \theta}{\partial y^2} - Pe \frac{\partial \theta}{\partial x} = 0. \qquad 2.4.17$$

This is to be solved in the region $0 < x < 1$, $0 < y < 1$ with boundary conditions

$$\theta = 0 \text{ on } y = 0 \text{ and } y = 1$$

$$\theta = 0 \text{ on } x = 1 \text{ (outlet)}$$

$$\theta = 1 \text{ on } x = 0 \text{ (inlet)}. \qquad 2.4.18$$

We will suppose that the Péclet number is of order one and that $H \ll L$ so that ε is a small parameter. The full equation 2.4.17 is of elliptic form requiring boundary conditions to be specified on a closed region as indeed they are by equations 2.4.18.

A perturbation expansion

$$\theta(x, y; \varepsilon) \sim \theta_0(x_1 y) + \varepsilon \theta_1(x, y) + \cdots \qquad 2.4.19$$

will have coefficient functions θ_n $(n = 0, 1, \ldots)$ which satisfy differential equations of parabolic form, i.e.

$$\frac{\partial^2 \theta_0}{\partial y^2} - Pe \frac{\partial \theta_0}{\partial x} = 0.$$

The solutions of such equations cannot satisfy all of the boundary conditions. In fact there is a boundary layer at the outlet $x=1$ where the expansion 2.4.19 is invalid and it is the $x=1$ boundary condition which must be neglected in solving for the functions θ_n.

This concludes our review of some of the sources of nonuniformities in perturbation expansions. The nonuniformities associated with the accumulation of small effects over an infinite domain will be dealt with in Chapters 3 and 4 when we consider the method of strained coordinates and the multiple scale technique. Boundary layers will be studied in detail in Chapter 5 where a technique will be developed for obtaining perturbation expansions both outside and within the boundary layer. These expansions are then matched in an overlap region where both expansions are uniformly valid.

Chapter 3
Strained Coordinates

The method of strained coordinates is a technique for dealing with certain types of nonuniformities which occur in asymptotic expansions. In this chapter two related methods will be described, namely the Lindstedt–Poincaré[3] and the Lighthill[4] techniques. The former applies to systems which are periodic where the period of the motion is changed by a perturbation. It can be applied to various oscillators such as mechanical spring and mass systems, electrical systems and planetary motion.

Lighthill's method is a generalization of the Lindstedt–Poincaré method which enables strained coordinates to be applied to a far wider class of problems. The method has been found to be of particular value in the solution of the partial differential equations which occur in fluid dynamics.

3.1 The Lindstedt–Poincaré technique

The asymptotic expansions which were considered in the first two chapters are called straightforward Poincaré expansions. These are named in recognition of Henri Poincaré who, towards the end of the last century, provided the foundations of the theory of asymptotic analysis. A straightforward asymptotic expansion of a function, f, which depends on an independent variable, t, and a parameter, ε,

$$f(t;\varepsilon) \sim \sum_{n=0}^{\infty} f_n(t)\delta_n(\varepsilon) \quad \text{as } \varepsilon \to 0, \qquad 3.1.1$$

has coefficient functions $f_n(t)$ which are formally expressed as limits.

$$f_N(t) = \lim_{\varepsilon \to 0}\left[\frac{f(t;\varepsilon) - \sum_{n=0}^{N-1} f_n(t)\delta_n(\varepsilon)}{\delta_N(\varepsilon)}\right]. \qquad 3.1.2$$

Although not explicitly stated, it is implied that the limits are to be evaluated as $\varepsilon \to 0$ *while the independent variable, t, is kept fixed.*

Thus, for example, we can construct the expansion in powers of ε of the following functions,

$$\cos(t + \varepsilon) \sim \cos t - \varepsilon \sin t - \frac{\varepsilon^2}{2} \cos t + \cdots \qquad 3.1.3$$

and

$$\cos[t + \varepsilon(1 + t)] \sim \cos t - \varepsilon(1 + t) \sin t - \frac{\varepsilon^2(1 + t)^2}{2} \cos t + \cdots. \qquad 3.1.4$$

In practice these expansions are obtained from a knowledge of the Maclaurin series for $\cos\theta$ and $\sin\theta$ rather than by using the expression 3.1.2, i.e.

$$\cos[t + \varepsilon(1 + t)] = \cos t . \cos[\varepsilon(1 + t)] - \sin t . \sin[\varepsilon(1 + t)]$$

$$= \cos t . \left(1 - \frac{\varepsilon^2(1 + t)^2}{2} + \cdots\right) - \sin t . [\varepsilon(1 + t) + \cdots].$$

The same expansion is, of course, obtained if 3.1.2 is used with $\delta_n(\varepsilon) = \varepsilon^n$. The essential point is that the independent variable, t, is fixed at any arbitrary value while the expansions in powers ε or more generally in gauge functions $\delta_n(\varepsilon)$ are generated.

The expansion 3.1.3 is uniform for all t because the coefficients are bounded. The expansion 3.1.4 has coefficients which are unbounded as $t \to \infty$. The region of nonuniformity is such that $\varepsilon t = O(1)$, i.e.

$$t = O(1/\varepsilon) \quad \text{as } \varepsilon \to 0.$$

If we rewrite the function $\cos[t + \varepsilon(1 + t)]$ as $\cos[t(1 + \varepsilon) + \varepsilon]$ and introduce the new independent variable $\tau = t(1 + \varepsilon)$ then the function is now dependent on τ and ε.

$$f(t; \varepsilon) = \cos[t(1 + \varepsilon) + \varepsilon] = \cos(\tau + \varepsilon) = F(\tau; \varepsilon),$$

The expansion of $F(\tau; \varepsilon)$ for small ε is given by 3.1.3 with t replaced by τ,

$$\cos[t(1 + \varepsilon) + \varepsilon] \sim \cos\tau - \varepsilon \sin\tau - \frac{\varepsilon^2}{2} \cos\tau + \cdots. \qquad 3.1.5$$

This expansion is of the form

$$F(\tau; \varepsilon) \sim \sum_{n=0}^{\infty} F_n(\tau) \delta_n(\varepsilon) \quad \text{as } \varepsilon \to 0, \qquad 3.1.6$$

where

$$F_N(\tau) = \operatorname*{Lim}_{\varepsilon \to 0} \left[\frac{F(\tau; \varepsilon) - \sum_{n=0}^{N-1} F_n(\tau)\delta_n(\varepsilon)}{\delta_N(\varepsilon)} \right]. \qquad 3.1.7$$

The limit in 3.1.7 is to be evaluated *while the new independent variable, τ, is kept fixed.* This new variable, $\tau = t(1 + \varepsilon)$, is called a *strained coordinate.* The difference between t and τ is small in a relative sense, i.e.

$$\frac{\tau}{t} = 1 + \varepsilon.$$

However, the numerical difference between t and τ is

$$\tau - t = \varepsilon t,$$

which becomes large when $t = O(1/\varepsilon)$.

The introduction of a strained coordinate has led to a uniform expansion, 3.1.5, while the expansion using the original coordinate, 3.1.4, is nonuniform. The use of a strained coordinate to overcome a nonuniformity in a straightforward expansion is the idea behind the Lindstedt–Poincaré technique. This technique will now be demonstrated using a model problem for which the exact solution is available for the purpose of comparison.

Model problem

Consider the differential equation

$$\frac{d^2 u}{dt^2} + \omega_0^2 u = \varepsilon u, \tag{3.1.8}$$

with the initial conditions $u(0) = a$, $\dfrac{du}{dt}(0) = 0$.

We first construct a straightforward expansion assuming the form

$$u(t; \varepsilon) \sim u_0(t) + \varepsilon u_1(t) + \cdots.$$

Substituting this expansion into equation 3.1.8 yields

$$\frac{d^2 u_0}{dt^2} + \varepsilon \frac{d^2 u_1}{dt^2} + \cdots + \omega_0^2 (u_0 + \varepsilon u_1 + \cdots) \sim \varepsilon u_0 + \cdots.$$

The initial conditions are partitioned in the usual way so that

$$u_0(0) = a, \quad u_1(0) = 0, \ldots \quad \text{and} \quad \frac{du_0}{dt}(0) = 0, \quad \frac{du_1}{dt}(0) = 0, \text{ etc.}$$

The following order equations are obtained:

$$O(1): \quad \frac{d^2 u_0}{dt^2} + \omega_0^2 u_0 = 0, \quad u_0(0) = a, \quad \frac{du_0}{dt}(0) = 0 \tag{3.1.9}$$

$$O(\varepsilon): \quad \frac{d^2 u_1}{dt^2} + \omega_0^2 u_1 = u_0, \quad u_1(0) = 0, \quad \frac{du_1}{dt}(0) = 0. \tag{3.1.10}$$

The O(1) equation is the unperturbed problem obtained by setting $\varepsilon = 0$. It is the governing equation of a harmonic oscillator with angular frequency ω_0. The solution is $u_0 = a\cos\omega_0 t$.

The O(ε) system, 3.1.10, becomes

$$\frac{d^2 u_1}{dt^2} + \omega_0^2 u_1 = a\cos\omega_0 t. \qquad 3.1.11$$

The complementary function is $A\cos\omega_0 t + B\sin\omega_0 t$. The right-hand side of 3.1.11 is of this form so that a particular integral of the form $\alpha t\sin\omega_0 t$ must be sought. (The term $t\cos\omega_0 t$ is absent from the particular integral when the first derivative of u_1 is not present.) Substituting this trial particular integral into equation 3.1.11 leads to the condition $2\alpha\omega_0\cos\omega_0 t = a\cos\omega_0 t$. Thus $\alpha = a/2\omega_0$ and the general solution of 3.1.11 is

$$u_1 = A\cos\omega_0 t + B\sin\omega_0 t + \frac{at}{2\omega_0}\sin\omega_0 t.$$

The initial conditions require that $A = B = 0$.

The two-term straightforward expansion, u_{2T}, of the solution of equation 3.1.8 is

$$u_{2T} = a\cos\omega_0 t + \varepsilon\frac{at}{2\omega_0}\sin\omega_0 t. \qquad 3.1.12$$

The second member of the right-hand side is a secular term. It becomes arbitrarily large as t increases. Consequently the expansion is nonuniform for large t. The region of nonuniformity is

$$t = O(1/\varepsilon) \quad \text{as } \varepsilon \to 0.$$

The exact solution of equation 3.1.8 is easily found since it corresponds to a harmonic oscillator with angular frequency $\sqrt{\omega_0^2 - \varepsilon}$, thus

$$u = a\cos(\omega_0\sqrt{1 - \varepsilon/\omega_0^2}\,t). \qquad 3.1.13$$

This is periodic with an amplitude of oscillation a. The exact solution is therefore bounded, so we may conclude that the unbounded behavior associated with the secular term in the expansion 3.1.12 is spurious and does not correspond to a real feature of the full solution. The origin of the secular term can be seen by constructing a series in powers of ε (with t fixed) from 3.1.13. Using two terms in the expansion of the square root

$$\sqrt{1 - \varepsilon/\omega_0^2} = 1 + \frac{1}{2}\left(-\frac{\varepsilon}{\omega_0^2}\right) + \cdots$$

leads to

$$a\cos\left[\omega_0\left(1-\frac{\varepsilon}{2\omega_0^2}+\cdots\right)t\right]=a\cos\omega_0 t\,.\cos\left(\frac{\varepsilon t}{2\omega_0}+\cdots\right)$$

$$+\,a\sin\omega_0 t\,.\sin\left(\frac{\varepsilon t}{2\omega_0}+\cdots\right)$$

$$=a\cos\omega_0 t\,.\left(1-\frac{\varepsilon^2 t^2}{8\omega_0^2}+\cdots\right)$$

$$+\,a\sin\omega_0 t\,.\left(\frac{\varepsilon t}{2\omega_0}+\cdots\right)$$

$$=a\cos\omega_0 t+\varepsilon\frac{at}{2\omega_0}\sin\omega_0 t+\cdots. \qquad 3.1.14$$

The secular terms arise from the truncated expansions of $\cos(\varepsilon t/2\omega_0+\cdots)$ and $\sin(\varepsilon t/2\omega_0+\cdots)$.

The expansion 3.1.14 is an asymptotic expansion, i.e. the remainder has the required behavior, so we may write

$$u=a\cos\omega_0 t+R_{1T},$$

where the remainder of the one-term expansion is $O(\varepsilon)$. Similarly

$$u=a\cos\omega_0 t+\varepsilon\frac{at}{2\omega_0}\sin\omega_0 t+R_{2T},$$

where $R_{2T}=O(\varepsilon^2)$ as $\varepsilon\to 0$.

We can prove that R_{1T} is $O(\varepsilon)$ by evaluating the following limit,

$$L=\lim_{\varepsilon\to 0}\left[\frac{R_{1T}}{\varepsilon}\right]=\lim_{\varepsilon\to 0}\left[\frac{a\cos(\sqrt{\omega_0^2-\varepsilon}\,t)-a\cos\omega_0 t}{\varepsilon}\right].$$

L'Hospital's rule leads to

$$L=\left.\frac{\dfrac{d}{d\varepsilon}[a\cos(\sqrt{\omega_0^2-\varepsilon}\,t)]}{\dfrac{d}{d\varepsilon}(\varepsilon)}\right|_{\varepsilon=0}$$

$$=\left.\frac{\left(\dfrac{-at}{2\sqrt{\omega_0^2-\varepsilon}}\right)\cdot[-\sin(\sqrt{\omega_0^2-\varepsilon}\,t)]}{1}\right|_{\varepsilon=0}$$

$$=\frac{at}{2\omega_0}\sin\omega_0 t.$$

Thus R_{1T} is of order ε as $\varepsilon \to 0$. However, it cannot be bounded for all values of t in the form $|R_{1T}| < K|\varepsilon|$ where K is a constant independent of t, so R_{1T} is not uniformly of order ε. It is convenient to indicate this nonuniformity by writing

$$R_{1T} = \mathrm{O}(\varepsilon t) \quad \text{as } \varepsilon \to 0.$$

Similarly it can be shown that

$$R_{2T} = \mathrm{O}(\varepsilon^2 t^2) \quad \text{as } \varepsilon \to 0.$$

The exact solution shows that the effect of the perturbation is to modify the frequency from the unperturbed value ω_0 to the perturbed value $\omega_0\sqrt{1 - \varepsilon/\omega_0^2}$. The strained coordinate $\tau = \sqrt{1 - \varepsilon/\omega_0^2}\,t$ provides a uniform expansion

$$u = \cos \omega_0 \tau,$$

which happens to be exact. The remainder is precisely zero and certainly uniform.

Without knowledge of the exact solution, the form of the straining transformation $\tau = \sqrt{1 + \varepsilon/\omega_0^2}\,t$ is unknown. The Lindstedt–Poincaré technique provides a systematic approximation to the straining transformation by representing the ratio τ/t as a Maclaurin expansion

$$\tau/t = 1 + \varepsilon\omega_1 + \varepsilon^2\omega_2 + \cdots. \tag{3.1.15}$$

The leading term is unity since $\tau = t$ when $\varepsilon = 0$.

The operators $\mathrm{d}/\mathrm{d}t$ and $\mathrm{d}^2/\mathrm{d}t^2$ become

$$\frac{\mathrm{d}}{\mathrm{d}t} = \frac{\mathrm{d}\tau}{\mathrm{d}t}\frac{\mathrm{d}}{\mathrm{d}\tau} = (1 + \varepsilon\omega_1 + \varepsilon^2\omega_2 + \cdots)\frac{\mathrm{d}}{\mathrm{d}\tau}$$

$$\frac{\mathrm{d}^2}{\mathrm{d}t^2} = (1 + \varepsilon\omega_1 + \varepsilon^2\omega_2 + \cdots)^2\frac{\mathrm{d}^2}{\mathrm{d}\tau^2}.$$

Applying this transformation to the model problem 3.1.8 leads to

$$[1 + 2\varepsilon\omega_1 + \varepsilon^2(\omega_1^2 + 2\omega_2) + \cdots]\frac{\mathrm{d}^2 u}{\mathrm{d}\tau^2} + \omega_0^2 u = \varepsilon u, \tag{3.1.16}$$

where u is now to be regarded as a function of the strained variable τ. To avoid confusion we will subsequently use an overbar to denote the function of the original variable t, thus $u(\tau) = \bar{u}(t)$. Assume that $u(\tau; \varepsilon)$ possesses an expansion in terms of the asymptotic sequence ε^n,

$$\begin{aligned} u(\tau;\varepsilon) &\sim \sum_{n=0}^{\infty} \varepsilon^n u_n(\tau) \quad \text{as } \varepsilon \to 0 \\ &\sim u_0(\tau) + \varepsilon u_1(\tau) + \varepsilon^2 u_2(\tau) + \cdots. \end{aligned}$$

Substituting this into equation 3.1.16 yields the following order equations

$$O(1): \quad \frac{d^2u_0}{d\tau^2} + \omega_0^2 u_0 = 0$$

$$O(\varepsilon): \quad \frac{d^2u_1}{d\tau^2} + \omega_0^2 u_1 + 2\omega_1 \frac{d^2u_0}{d\tau^2} = u_0$$

$$O(\varepsilon^2): \quad \frac{d^2u_2}{d\tau^2} + \omega_0^2 u_2 + (\omega_1^2 + 2\omega_2)\frac{d^2u_0}{d\tau^2} + 2\omega_1 \frac{d^2u_1}{d\tau^2} = u_1.$$

The initial conditions require some care. If $t = 0$ then τ is also zero so that

$$u(\tau = 0) = \bar{u}(t = 0) = a$$

$$(1 + \varepsilon\omega_1 + \varepsilon^2\omega_2 + \cdots)\frac{du}{d\tau}(\tau = 0) = \frac{d\bar{u}}{dt}(0) = b.$$

These lead to

$$u_0(0) = a, \quad u_1(0) = 0, \quad u_2(0) = 0 \ldots$$

$$\frac{du_0}{d\tau}(0) = b, \quad \frac{du_1}{d\tau}(0) + \omega_1 \frac{du_0}{d\tau}(0) = 0$$

$$\frac{du_2}{d\tau}(0) + \omega_1 \frac{du_1}{d\tau}(0) + \omega_2 \frac{du_0}{d\tau}(0) = 0 \text{ etc.}$$

In the present example $b = 0$ which leads to the simplification $\frac{du_n}{d\tau}(0) = 0$ for all n.

The $O(1)$ system has the solution

$$u_0 = A\cos\tau + B\sin\tau,$$

and the initial conditions $u_0(0) = a$, $\frac{du_0}{d\tau}(0) = 0$ lead to $u_0 = a\cos\omega_0\tau$.

The $O(\varepsilon)$ equation becomes

$$\frac{d^2u_1}{d\tau^2} + \omega_0^2 u_1 = u_0 - 2\omega_1 \frac{d^2u_0}{d\tau^2} = (1 + 2\omega_1\omega_0^2)a\cos\omega_0\tau.$$

The complementary function is $A\cos\omega_0\tau + B\sin\omega_0\tau$. The right-hand side is of this form, so the particular integral will be of secular type, $\alpha\tau\sin\omega_0\tau$. The constant ω_1 is chosen to remove the source of the secular term, thus $\omega_1 = -1/2\omega_0^2$. Then the solution satisfying the zero initial conditions is $u_1(\tau) = 0$.

The $O(\varepsilon^2)$ equation becomes

$$\frac{d^2u_2}{d\tau^2} + \omega_0^2 u_2 = \left(\frac{1}{4\omega_0^4} + 2\omega_2\right) a\omega_0^2\cos\omega_0\tau.$$

The value of ω_2 is chosen to avoid secular terms occurring in the function u_2, thus $\omega_2 = -1/8\omega_0^4$. Again the solution satisfying the zero initial conditions is zero for all τ, $u_2(\tau) = 0$.

The strained coordinate is given by

$$\tau = t\left(1 - \frac{\varepsilon}{2\omega_0^2} - \frac{\varepsilon^2}{8\omega_0^4} + \cdots\right),$$

and the expansion of the solution consists of the single term

$$u = a\cos\omega_0\tau.$$

The exact solution is $a\cos(\omega_0\sqrt{1 - \varepsilon/\omega_0^2}t)$. Expanding the square root leads to

$$a\cos\left[\omega_0\left(1 - \frac{\varepsilon}{2\omega_0^2} - \frac{\varepsilon^2}{8\omega_0^4} + \cdots\right)t\right].$$

Clearly the Lindstedt–Poincaré technique generates this expansion by the process of determining the constants ω_n in the straining transformation 3.1.15 from the requirement that secular terms are absent from the expansion.

3.2 Duffing's equation

Duffing's equation was considered in Section 2.4 where it provided an example of a nonuniformity arising from the occurrence of secular terms. We will see how the asymptotic expansion of the solution of Duffing's equation can be rendered uniform by the Lindstedt–Poincaré technique. First we consider two mechanical systems which are governed by Duffing's equation.

The pendulum

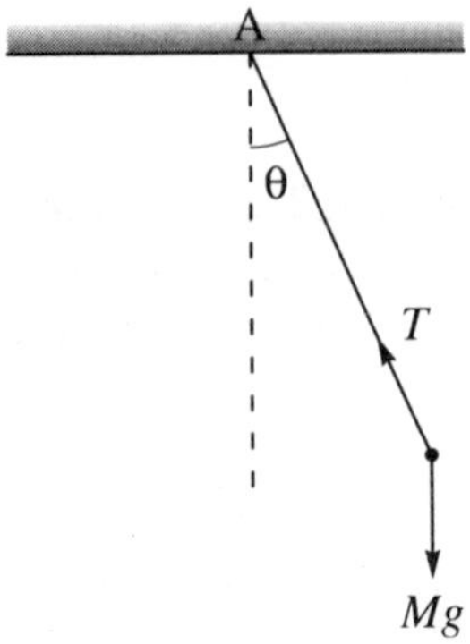

A point mass M is connected by a rod of length L to a hinge at A. The mass swings in a vertical plane under the action of gravity. The length of the rod is fixed and its mass is negligible. The two forces acting on the mass are gravity and the tension in the rod. The component of motion tangential to the sector of the circle on which the mass moves is driven by the force $-Mg\sin\theta$. The tangential acceleration of a particle moving on a circle of fixed radius is $L(\mathrm{d}^2\theta/\mathrm{d}T^2)$ and Newton's second law yields the governing equation,

$$ML\frac{\mathrm{d}^2\theta}{\mathrm{d}T^2} = -Mg\sin\theta.$$

Suppose the pendulum is released from rest when $\theta = \theta_0$ at $T = 0$. After introducing the scaled angle of oscillation $u = \theta/\theta_0$ and the frequency related parameter $\Omega = \sqrt{g/L}$ the governing equation becomes

$$\frac{\mathrm{d}^2u}{\mathrm{d}T^2} = -\Omega^2\frac{\sin(\theta_0 u)}{\theta_0}$$

with initial conditions $u = 1$ and $\mathrm{d}u/\mathrm{d}T = 0$ at $T = 0$.

If θ_0 is small the sine term can be approximated by a truncated Maclaurin expansion. On keeping two terms in the expansion we obtain

$$\frac{\mathrm{d}^2u}{\mathrm{d}T^2} = -\Omega^2\left(u - \frac{\theta_0^2u^3}{6}\right).$$

Introducing nondimensional time $t = \Omega T$ and the parameter $\varepsilon = \theta_0^2/6$ leads to Duffing's equation

$$\frac{\mathrm{d}^2u}{\mathrm{d}t^2} + u - \varepsilon u^3 = 0. \qquad 3.2.1$$

A mass and spring oscillator

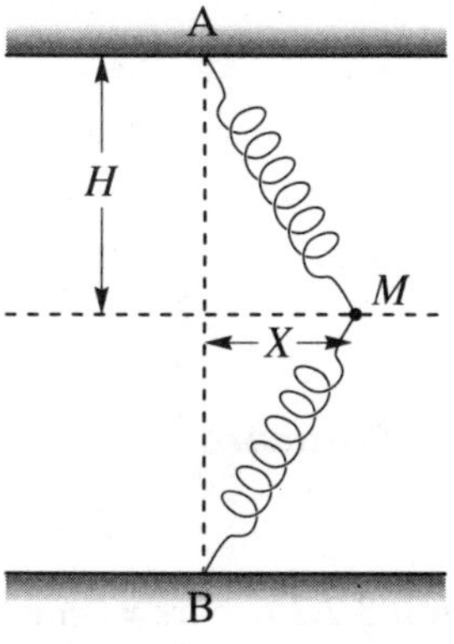

A point mass, M, is connected to the fixed points A and B by identical springs of negligible mass. The natural length of each spring is L and the spring constant is Λ. The separation between A and B is $2H$ where $H > L$. The mass is displaced a perpendicular distance X_0 from AB and released. The mass oscillates on a path perpendicular to AB. Gravity is neglected and the spring forces, F, vary linearly with the extension.

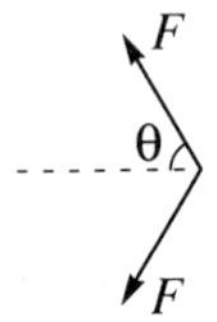

$$F = \Lambda\left(\frac{\sqrt{H^2 + X^2} - L}{L}\right).$$

Newton's second law yields

$$M\frac{\mathrm{d}^2 X}{\mathrm{d}T^2} = -2F\cos\theta,$$

and $\cos\theta = X/\sqrt{H^2 + X^2}$ so that

$$M\frac{\mathrm{d}^2 X}{\mathrm{d}T^2} = -\frac{2\Lambda}{L}X\left(1 - \frac{L}{\sqrt{H^2 + X^2}}\right).$$

When the displacement X is small compared with H the inverse square root term may be replaced by a truncated Maclaurin expansion. On keeping two terms in the expansion we obtain

$$\frac{\mathrm{d}^2 X}{\mathrm{d}T^2} = -\frac{2\Lambda}{ML}X\left[1 - \frac{L}{H}\left(1 - \frac{1}{2}\frac{X^2}{H^2}\right)\right].$$

Introducing the nondimensional displacement $u = X/X_0$ and the frequency related parameter $\Omega = \sqrt{2\Lambda(H - L)/MLH}$ the governing equation becomes

$$\frac{\mathrm{d}^2 u}{\mathrm{d}T^2} = -\Omega^2\left(u + \frac{L}{2(H - L)}\frac{X_0^2}{H^2}u^3\right).$$

Then on introducing nondimensional time $t = \Omega T$ and the parameter

$\varepsilon = LX_0^2/2(H - L)H_0^2$, we obtain Duffing's equation,

$$\frac{\mathrm{d}^2 u}{\mathrm{d}t^2} + u + \varepsilon u^3 = 0. \qquad 3.2.2$$

Solutions of Duffing's equation using the Lindstedt–Poincaré technique

In the above mass spring system the nonlinear term in equation 3.2.2 occurs with a coefficient ε which is positive while in equation 3.2.1 describing the pendulum, this coefficient is negative. For sufficiently small values of $|\varepsilon|$ the solutions of Duffing's equation are bounded. This can easily be shown for positive values of ε in equation 3.2.2. First multiply throughout by $\mathrm{d}u/\mathrm{d}t$ and then express the resulting equation in the form

$$\frac{\mathrm{d}}{\mathrm{d}t}\left[\frac{1}{2}\left(\frac{\mathrm{d}u}{\mathrm{d}t}\right)^2+\frac{1}{2}u^2+\frac{\varepsilon}{4}u^4\right]=0.$$

The term in the square brackets can be interpreted as the total energy, ϕ, where $\frac{1}{2}\left(\frac{\mathrm{d}u}{\mathrm{d}t}\right)^2$ represents kinetic energy and the remaining terms represent the potential energy.

The total energy is a constant determined by the initial values of u and $\mathrm{d}u/\mathrm{d}t$. In the case of positive ε the energy expression is positive definite. This allows a bound, $u_{\max}$, to be obtained for u. The bound is obtained by setting $\mathrm{d}u/\mathrm{d}t = 0$,

$$\frac{1}{2}u_{\max}^2+\frac{\varepsilon}{4}u_{\max}^4=\phi.$$

Thus

$$u_{\max}=\sqrt{-\frac{1}{\varepsilon}+\frac{1}{\varepsilon}\sqrt{1+4\phi\varepsilon}},$$

and $|u| \leqslant u_{\max}$.

When ε is negative the energy expression is no longer positive definite. However, in this case it can be shown that solutions whose energy ϕ is less than $1/4|\varepsilon|$ are bounded (see Grimshaw[2a]). We shall be concerned with solutions of Duffing's equation for small values of $|\varepsilon|$ with initial conditions which are O(1). These will therefore be bounded solutions.

We saw in Section 2.4 that a straightforward perturbation expansion for the solution of Duffing's equation led to secular terms. These terms cannot represent an actual unboundedness in the solution. They arise, just as in the previous model problem, from a truncated expansion of the full solution. We seek a strained coordinate which will provide an expansion in which secular terms are absent. This is provided by the Lindstedt–Poincaré technique as follows.

Consider Duffing's equation 3.2.2 subject to the initial conditions $u(0) = 1$, $\frac{\mathrm{d}u}{\mathrm{d}t}(0) = 0$. Introduce the strained coordinate τ where

$$\tau = t(1+\varepsilon\omega_1+\varepsilon^2\omega_2+\cdots),$$

so that Duffing's equation becomes

$$(1+\varepsilon\omega_1+\varepsilon^2\omega_2+\cdots)^2\frac{\mathrm{d}^2u}{\mathrm{d}\tau^2}+u+\varepsilon u^3=0. \qquad 3.2.3$$

Assume that u possesses an expansion of the form

$$u(\tau;\varepsilon) \sim u_0(\tau) + \varepsilon u_1(\tau) + \varepsilon^2 u_2(\tau) + \cdots$$

and substitute into 3.2.3 to obtain the following order equations:

$$\mathrm{O}(1): \quad \frac{d^2u_0}{d\tau^2} + u_0 = 0, \quad u_0(0) = 1, \quad \frac{du_0}{d\tau}(0) = 0$$

$$\mathrm{O}(\varepsilon): \quad \frac{d^2u_1}{d\tau^2} + u_1 = -u_0^3 - 2\omega_1\frac{d^2u_0}{d\tau^2}, \quad u_1(0) = \frac{du_1}{d\tau}(0) = 0$$

$$\mathrm{O}(\varepsilon^2): \quad \frac{d^2u_2}{d\tau^2} + u_2 = -3u_0^2u_1 - 2\omega_1\frac{d^2u_1}{d\tau^2} - (\omega_1^2 + 2\omega_2)\frac{d^2u_0}{d\tau^2},$$

$$u_2(0) = \frac{du_2}{d\tau}(0) = 0.$$

The solution of the O(1) equation is $u_0 = \cos t$. Substituting this function into the $\mathrm{O}(\varepsilon)$ equation leads to

$$\frac{d^2u_1}{d\tau^2} + u_1 = -\cos^3\tau + 2\omega_1\cos\tau. \qquad 3.2.4$$

The first member of the right-hand side of 3.2.4 can be expressed in terms of multiple angles. From Euler's identity, $\cos\tau = (e^{i\tau} + e^{-i\tau})/2$, we have

$$\cos^3\tau = \tfrac{1}{8}(e^{i\tau} + e^{-i\tau})^3 = \tfrac{1}{8}(e^{3i\tau} + e^{-3i\tau} + 3e^{i\tau} + 3e^{-i\tau})$$
$$= \tfrac{1}{4}\cos 3\tau + \tfrac{3}{4}\cos\tau.$$

The complementary function associated with 3.2.4 has the form $A\cos\tau + B\sin\tau$. The $\cos\tau$ component of the multiple angle decomposition (Fourier series) of the right-hand side of 3.2.4 will generate a secular term of the form $\tau\sin\tau$ in the particular integral. This is removed by choosing $\omega_1 = 3/8$. Then 3.2.4 becomes

$$\frac{d^2u_1}{d\tau^2} + u_1 = -\frac{1}{4}\cos 3\tau.$$

A particular integral will be of the form $\alpha\cos 3\tau$ where $-9\alpha + \alpha = -1/4$, so $\alpha = 1/32$. The general solution is

$$u_1 = A\cos\tau + B\sin\tau + \tfrac{1}{32}\cos 3\tau,$$

and the initial conditions yield

$$u_1(0) = A + \tfrac{1}{32} = 0$$

$$\frac{du_1}{d\tau}(0) = B = 0.$$

Thus we have obtained the $O(\varepsilon)$ coefficient function

$$u_1 = -\tfrac{1}{32}(\cos\tau - \cos 3\tau).$$

The $O(\varepsilon^2)$ equation has a complementary function of the form $A\cos\tau + B\sin\tau$. To avoid secular terms the right-hand side of the $O(\varepsilon^2)$ equation must not contain $\cos\tau$ or $\sin\tau$ terms in its multiple angle decomposition. The right-hand side is

$$\frac{3}{32}\cos^2\tau(\cos\tau - \cos 3\tau) - 2\cdot\frac{3}{8}\cdot\frac{1}{32}(\cos\tau - 9\cos 3\tau) + \left(\frac{9}{64} + 2\omega_2\right)\cos\tau.$$

This contains multiple angles and the product terms $\cos^3\tau$ and $\cos^2\tau\cos 3\tau$. The former has already been decomposed, for the latter we have

$$\begin{aligned}\cos^2\tau\,.\cos 3\tau &= \tfrac{1}{8}(e^{i\tau} + e^{-i\tau})^2(e^{3i\tau} + e^{-3i\tau})\\ &= \tfrac{1}{8}(e^{2i\tau} + 2 + e^{-2i\tau})(e^{3i\tau} + e^{-3i\tau})\\ &= \tfrac{1}{8}[e^{5i\tau} + e^{-5i\tau} + 2(e^{3i\tau} + e^{-3i\tau}) + e^{i\tau} + e^{-i\tau}]\\ &= \tfrac{1}{4}\cos 5\tau + \tfrac{1}{2}\cos 3\tau + \tfrac{1}{4}\cos\tau.\end{aligned}$$

The overall coefficient of $\cos\tau$ is

$$\frac{3}{32}\cdot\left(\frac{3}{4} - \frac{1}{4}\right) - \frac{3}{4}\cdot\frac{1}{32} + \frac{9}{64} + 2\omega_2.$$

This must be equal to zero to avoid secular terms, thus $\omega_2 = -21/256$.

The two-term, uniformly valid expansion of the solution of Duffing's equation is

$$u = \cos\tau + \frac{\varepsilon}{32}(\cos 3\tau - \cos\tau) + O(\varepsilon^2), \tag{3.2.5}$$

where

$$\tau = \left(1 + \frac{3}{8}\varepsilon - \frac{21}{256}\varepsilon^2 + O(\varepsilon^3)\right)t \quad \text{as } \varepsilon \to 0. \tag{3.2.6}$$

It is important to notice that the straining transformation is given to a higher order than the expansion of the solution. We did not complete the solution for the term u_2 but only used the $O(\varepsilon^2)$ system to determine ω_2. The point is that it would be futile to determine u_2 and use an expansion of the form

$$u = \cos\tau + \frac{\varepsilon}{32}(\cos 3\tau - \cos\tau) + \varepsilon^2 u_2 + O(\varepsilon^3), \tag{3.2.7}$$

along with the strained coordinate 3.2.6 because the remainder term in 3.2.7 will contain secular terms and is therefore of order $\varepsilon^3 t$. Of course this term can be made uniformly of order ε^3 by the correct choice of ω_3 but at the stage given by 3.2.6 only ω_2 is known, so 3.2.7 has a remainder $O(\varepsilon^3 t)$. Therefore when $t = O(1/\varepsilon)$ the unknown remainder term has the same order as the previous term, $\varepsilon^2 u_2$, and it is pointless to use the term $\varepsilon^2 u_2$ because the expansion has an error of $O(\varepsilon^2)$.

Similarly if a one-term expansion is required for the solution then a two-term expansion must be obtained for the straining transformation,

$$u = \cos\tau + O(\varepsilon),$$

where

$$\tau = \left(1 + \frac{3}{8}\varepsilon + O(\varepsilon^2)\right)t \quad \text{as } \varepsilon \to 0.$$

Exercise

Use the Lindstedt–Poincaré technique to show that the two-term uniformly valid expansion of the solution of

$$\frac{d^2u}{dt^2} + u = \varepsilon u\left(\frac{du}{dt}\right)^2, \quad u(0) = 1, \quad \frac{du}{dt}(0) = 0,$$

is

$$u = \cos\tau + \frac{\varepsilon}{32}(\cos 3\tau - \cos\tau) + O(\varepsilon^2)$$

where

$$\tau = \left(1 - \frac{\varepsilon}{8} + \frac{3}{256}\varepsilon^2 + O(\varepsilon^3)\right)t \quad \text{as } \varepsilon \to 0.$$

3.3 Lighthill's technique

The method of strained coordinates has been extended by Lighthill[4] to allow a broader class of straining transformation than that used in the Lindstedt–Poincaré technique. Lighthill's method has been applied to various branches of continuum mechanics including flow past aerofoils and wave propagation in solids and fluids.

Lighthill's method can be applied to partial differential equations where one or more of the independent variables is strained. The straining may be applied to space and/or time variables or to the combinations known as characteristic variables in the theory of hyperbolic partial differential equations. (See for example Lighthill[5], Nayfeh[6a], Nayfeh and Kluwick[7].)

In this section Lighthill's technique will be introduced and applied to various ordinary differential equations. In the following section the technique will be applied to the study of flow past thin aerofoils.

We denote the original independent variable by x and the strained coordinate by s. Lighthill uses the straining transformation

$$x \sim s + \varepsilon f_1(s) + \varepsilon^2 f_2(s) + \cdots \qquad 3.3.1$$

where the coefficients of ε^n are functions of the strained coordinate. The Lindstedt–Poincaré transformation is of the form

$$\frac{s}{x} \sim 1 + \varepsilon\omega_1 + \varepsilon^2\omega_2 + \cdots$$

thus

$$x \sim \frac{s}{1 + \varepsilon\omega_1 + \varepsilon^2\omega_2 + \cdots} = s(1 - \varepsilon\omega_1 - \varepsilon^2\omega_2 - \cdots + \varepsilon^2\omega_1 + \cdots).$$

This shows that the Lindstedt–Poincaré transformation is a special case of Lighthill's with $f_1 = -\omega_1 s$, $f_2 = (\omega_1 - \omega_2)s$ and in general $f_n(s) = \alpha_n s$ where the α_n are constants.

The standard procedure with Lighthill's technique is to introduce the new variable into the governing equation and boundary or initial conditions using the relation

$$\frac{\mathrm{d}}{\mathrm{d}x} = \frac{\mathrm{d}s}{\mathrm{d}x}\frac{\mathrm{d}}{\mathrm{d}s} = \left(\frac{\mathrm{d}x}{\mathrm{d}s}\right)^{-1}\cdot\frac{\mathrm{d}}{\mathrm{d}s} \sim \frac{1}{1 + \varepsilon\dfrac{\mathrm{d}f_1}{\mathrm{d}s} + \varepsilon^2\dfrac{\mathrm{d}f_2}{\mathrm{d}s} + \cdots}\cdot\frac{\mathrm{d}}{\mathrm{d}s}$$

$$\sim \left\{1 - \varepsilon\frac{\mathrm{d}f_1}{\mathrm{d}s} - \varepsilon^2\left[\frac{\mathrm{d}f_2}{\mathrm{d}s} - \left(\frac{\mathrm{d}f_1}{\mathrm{d}s}\right)^2\right] + \cdots\right\}\frac{\mathrm{d}}{\mathrm{d}s}.$$

The procedure is analogous to the Lindstedt–Poincaré technique in that an expansion of the form

$$u(s;\varepsilon) \sim u_0(s) + \varepsilon u_1(s) + \cdots \qquad 3.3.2$$

is assumed for the dependent variable. This is substituted into the transformed governing equation and order equations generated for u_0, u_1, etc, with associated boundary/initial conditions. It is sometimes necessary to use augmented asymptotic sequences involving fractional powers of ε or logarithmic terms.

Renormalization

A procedure which is equivalent to the above but which is manipulatively less complicated is called *renormalization* (see Pritulo[8]). The method starts from a straightforward expansion of the original governing equation

$$\bar{u}(x;\varepsilon) \sim \bar{u}_0(x) + \varepsilon\bar{u}_1(x) + \cdots \qquad 3.3.3$$

This will in general be nonuniform. The transformation 3.3.1 is then introduced and u is re-expressed in an asymptotic expansion for fixed s to obtain the form 3.3.2. (Bars have been used in 3.3.3 because in general the functions of x are different from the functions of s.)

Lighthill's condition for determining the functions $f_n(s)$ in the transformation 3.3.1 is that the expansion 3.3.2 should be uniform. We saw in Section 2.3 that for a nonuniformity to occur, the coefficient functions must be singular and that the singularity must grow. Lighthill's condition for rendering the expansion uniform is:

Subsequent coefficient functions should be no more singular than previous functions	Lighthill's Condition

This must be applied to both the function expansion 3.3.2 and the straining transformation 3.3.1, i.e. $u_{n+1}(s)$ must be no more singular than $u_n(s)$ and $f_{n+1}(s)$ must be no more singular than $f_n(s)$. (We shall see that for the case of s near zero there is an exception to this rule in that f_0 is not singular while $f_1, f_2, \ldots$ may be singular. The point is that this singularity must not grow.)

Shift in the singularity of a differential equation

Lighthill described the use of his technique by considering a model problem which is similar to, but more complicated than, the following example which we will consider

$$(x + \varepsilon y)\frac{\mathrm{d}y}{\mathrm{d}x} + y = 0, \quad 0 < \varepsilon \ll 1, \quad y(1) = 1. \qquad 3.3.4$$

We will construct a two-term uniformly valid expansion using renormalization. The straightforward expansion

$$y(x;\varepsilon) \sim y_0(x) + \varepsilon y_1(x) + \varepsilon^2 y_2(x) + \cdots$$

leads to the following order equations:

$$\mathrm{O}(1): \quad x\frac{\mathrm{d}y_0}{\mathrm{d}x} + y_0 = 0, \quad y_0(1) = 1 \qquad 3.3.5\text{a}$$

$$\mathrm{O}(\varepsilon): \quad x\frac{\mathrm{d}y_1}{\mathrm{d}x} + y_1 = -y_0\frac{\mathrm{d}y_0}{\mathrm{d}x}, \quad y_1(1) = 0 \qquad 3.3.5\text{b}$$

$$\mathrm{O}(\varepsilon^2): \quad x\frac{\mathrm{d}y_2}{\mathrm{d}x} + y_2 = -y_0\frac{\mathrm{d}y_1}{\mathrm{d}x} - y_1\frac{\mathrm{d}y_0}{\mathrm{d}x}, \quad y_2(1) = 0, \qquad 3.3.5\text{c}$$

etc.

The O(1) equation may be written in the form $(\mathrm{d}/\mathrm{d}x)(xy_0) = 0$, with solution $y_0 = A/x$. The boundary condition $y_0(1) = 1$ requires that $A = 1$, so that $y_0 = 1/x$.

The $\mathrm{O}(\varepsilon)$ equation becomes

$$\frac{\mathrm{d}}{\mathrm{d}x}(xy_1) = \frac{1}{x^3}.$$

Thus $xy_1 = -1/2x^2 + A$ and the boundary condition requires $A = 1/2$, so that

$$y_1 = \frac{1}{2x} - \frac{1}{2x^3}.$$

The $O(\varepsilon^2)$ equation becomes

$$\frac{d}{dx}(xy_2) = \frac{1}{x}\left(\frac{1}{2x^2} - \frac{3}{2x^4}\right) + \left(\frac{1}{2x} - \frac{1}{2x^3}\right)\frac{1}{x^2}$$

and the solution satisfying the boundary condition $y_2(1) = 0$ is

$$y_2 = -\frac{1}{2x^3} + \frac{1}{2x^5}.$$

The two-term straightforward expansion is

$$y \sim \frac{1}{x} + \varepsilon\left(\frac{1}{2x} - \frac{1}{2x^3}\right) + \varepsilon^2\left(-\frac{1}{2x^3} + \frac{1}{2x^5}\right) + \cdots. \tag{3.3.6}$$

The coefficient functions are singular as $x \to 0$ and the singularity increases, i.e.

$$y_0 = O\left(\frac{1}{x}\right), \quad y_1 = O\left(\frac{1}{x^3}\right) \quad \text{and} \quad y_2 = O\left(\frac{1}{x^5}\right) \quad \text{as } x \to 0.$$

The region of nonuniformity is such that $\varepsilon/x^2 = O(1)$, i.e.

$$x = O(\sqrt{\varepsilon}) \quad \text{as } \varepsilon \to 0.$$

Each of the equations 3.3.5 is singular when $x = 0$ while the original equation is singular when $x + \varepsilon y = 0$. The perturbation expansion has shifted the location of the singularity in the x–y plane from the line $x + \varepsilon y = 0$ to the line $x = 0$. In fact the exact solution of 3.3.4 can be constructed because it is of homogeneous form in x and y; it is

$$y_{ex} = \frac{-x + \sqrt{x^2 + \varepsilon(2 + \varepsilon)}}{\varepsilon}. \tag{3.3.7}$$

This is not singular at $x = 0$, indeed $y(0) = \sqrt{1 + 2/\varepsilon}$. For large negative x the exact solution behaves as follows:

$$\begin{aligned} y_{ex} &= \frac{-x + |x|\sqrt{1 + \varepsilon(2 + \varepsilon)/x^2}}{\varepsilon} \\ &= \frac{-x - x[1 + \varepsilon(2 + \varepsilon)/2x^2 + \cdots]}{\varepsilon} \\ &= -\frac{2x}{\varepsilon} - \frac{(2 + \varepsilon)}{2x}. \end{aligned}$$

Thus it asymptotically approaches the line $y = -2x/\varepsilon$.

In Fig. 3.1 the exact solution and the leading term $y_0(x)$ of the straightforward expansion are shown along with location of the singularity of the differential equation.

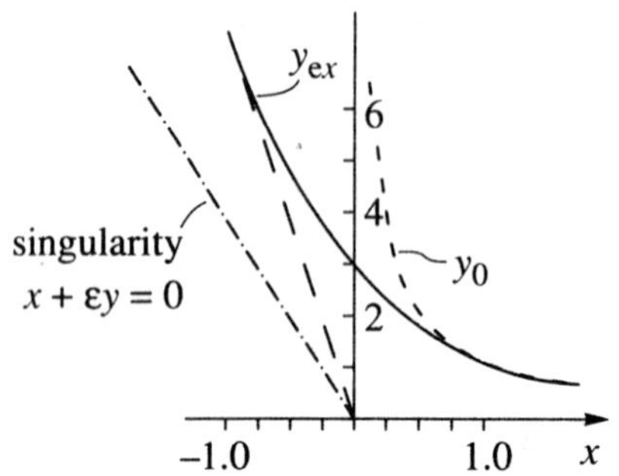

Fig. 3.1 The functions y_{ex} —— and $y_0(x)$ - - - for the case $\varepsilon = 1/4$

The singularity of the equation is not encountered by the exact solution corresponding to the boundary condition $y(1) = 1$. However the straightforward perturbation expansion does encounter the shifted singularity at $x = 0$. This is not a feature of the exact solution, so we are justified in seeking an expansion which is uniform in the region $x = 0$. This we achieve by renormalizing the straightforward expansion 3.3.6 using the strained coordinate transformation 3.3.1 and Lighthill's condition that subsequent terms be no more singular than previous terms.

The expansion 3.3.6 expressed in terms of the strained coordinate, s, is

$$y \sim \frac{1}{s+\varepsilon f_1+\varepsilon^2 f_2+\cdots} + \varepsilon\left(\frac{1}{2(s+\varepsilon f_1+\cdots)} - \frac{1}{2(s+\varepsilon f_1+\cdots)^3}\right)$$
$$+ \varepsilon^2\left(-\frac{1}{2(s+\cdots)^3} + \frac{1}{2(s+\cdots)^5}\right) + \cdots.$$

Working to $O(\varepsilon^2)$ this becomes

$$y \sim \frac{1}{s}\left(1+\frac{\varepsilon f_1}{s}+\frac{\varepsilon^2 f_2}{s}+\cdots\right)^{-1} + \varepsilon\left[\frac{1}{2s}\left(1+\frac{\varepsilon f_1}{s}+\cdots\right)^{-1}\right.$$
$$\left.-\frac{1}{2s^3}\left(1+\frac{\varepsilon f_1}{s}+\cdots\right)^{-3}\right]$$
$$+\varepsilon^2\left(-\frac{1}{2s^3}+\cdots+\frac{1}{2s^5}+\cdots\right)$$
$$\sim \frac{1}{s}\left(1-\frac{\varepsilon f_1}{s}-\frac{\varepsilon^2 f_2}{s}+\frac{\varepsilon^2 f_1^2}{s^2}+\cdots\right)$$
$$+\varepsilon\left[\frac{1}{2s}\left(1-\frac{\varepsilon f_1}{s}\right)-\frac{1}{2s^3}\left(1-3\frac{\varepsilon f_1}{s}\right)+\cdots\right]$$
$$+\varepsilon^2\left(-\frac{1}{2s^3}+\frac{1}{2s^5}+\cdots\right).$$

Collecting coefficients of powers of ε yields the expansion of y *with s fixed*

$$y(s;\varepsilon) \sim \frac{1}{s} + \varepsilon\left(-\frac{f_1}{s^2} + \frac{1}{2s} - \frac{1}{2s^3}\right)$$
$$+ \varepsilon^2\left(-\frac{f_2}{s^2} + \frac{f_1^2}{s^3} - \frac{f_1}{2s^2} + \frac{3}{2}\frac{f_1}{s^4} - \frac{1}{2s^3} + \frac{1}{2s^5}\right) + \cdots .$$

This process has generated the fixed s expansion

$$y \sim y_0(s) + \varepsilon y_1(s) + \varepsilon^2 y_2(s) + \cdots$$

from the fixed x expansion

$$y \sim \bar{y}_0(x) + \varepsilon \bar{y}_1(x) + \varepsilon^2 \bar{y}_2(x) + \cdots .$$

The bars indicate that the functions of x are in general different from the functions of s although in fact the leading order terms are the same.

We have the following coefficient functions:

$$y_0(s) = \frac{1}{s}$$

$$y_1(s) = \frac{1}{2s} - \frac{f_1}{s^2} - \frac{1}{2s^3}$$

$$y_2(s) = -\frac{f_2}{s^2} + \frac{f_1^2}{s^3} - \frac{f_1}{2s^2} + \frac{3}{2}\frac{f_1}{s^4} - \frac{1}{2s^3} + \frac{1}{2s^5}.$$

The leading order term, y_0, is singular as $s \to 0$. Lighthill's condition requires that y_1 be no more singular than this. A possible choice for f_1 is $f_1 = -1/2s$, then $y_1 = 1/2s$ and

$$y_2 = -\frac{f_2}{s^2} + \cancel{\frac{1}{4s^5}} + \frac{1}{4s^3} - \cancel{\frac{3}{4s^5}} - \frac{1}{2s^3} + \cancel{\frac{1}{2s^5}}.$$

A possible choice for f_2 is $f_2 = -1/4s$, then $y_2 = 0$.

The expansion generated by renormalization is

$$y \sim \frac{1}{s} + \frac{\varepsilon}{2s} + \varepsilon^2 0 + \cdots$$

It is accidental that y_2 happens to be zero. The actual function is not important. What is required is that y_2 does not introduce a nonuniformity. Then, just as with the Lindstedt–Poincaré technique, we work to a higher order in the straining transformation than we do in the function expansion. Thus we write

$$y = \frac{1}{s} + \frac{\varepsilon}{2s} + O(\varepsilon^2), \tag{3.3.8}$$

where

$$x = s - \frac{\varepsilon}{2s} - \frac{\varepsilon^2}{4s} + O(\varepsilon^3) \quad \text{as } \varepsilon \to 0. \qquad 3.3.9$$

The region of nonuniformity of the straightforward expansion 3.3.6 is $x = O(\sqrt{\varepsilon})$ as $\varepsilon \to 0$. Away from this region the straining transformation has little effect. If for example $x = O(1)$ then $s = O(1)$ and 3.3.9 may be solved for s to give to leading order, $s = x + O(\varepsilon)$. Then 3.3.8 yields $y = 1/x + O(\varepsilon)$. However for $x = O(\sqrt{\varepsilon})$ the straining transformation has a significant effect. From 3.3.9 we see that $s = O(\sqrt{\varepsilon})$ and solving for s to leading order involves the quadratic expression

$$x = s - \frac{\varepsilon}{2s}, \quad \text{i.e.} \quad s^2 - sx - \varepsilon/2 = 0.$$

Thus

$$s = \frac{x + \sqrt{x^2 + 2\varepsilon}}{2},$$

where the positive sign is chosen for the square root to select the branch associated with the relation $s = x + O(\varepsilon)$ for $x = O(1)$.

Then to leading order,

$$y \simeq \frac{1}{s} = \frac{2}{x + \sqrt{x^2 + 2\varepsilon}}. \qquad 3.3.10$$

In particular, when $x = 0$ we obtain the approximation $y \simeq \sqrt{2/\varepsilon}$.

The exact solution 3.3.7 yields for $x = 0$ the following value of y,

$$y = \frac{\sqrt{\varepsilon(2 + \varepsilon)}}{\varepsilon} = \sqrt{\frac{2}{\varepsilon} + 1} = \sqrt{\frac{2}{\varepsilon}} \cdot \sqrt{1 + \frac{\varepsilon}{2}}$$

$$= \sqrt{\frac{2}{\varepsilon}} \cdot [1 + O(\varepsilon)].$$

Thus the exact solution yields the same leading order approximation for $y(x = 0)$ as the renormalized expansion.

The choice of the functions $f_n(s)$ is not unique. We may choose any function provided y_n is no more singular than y_{n-1}. To demonstrate this consider the one-term expansion

$$y = \frac{1}{s} + O(\varepsilon),$$

where

$$x = s + \varepsilon f_1(s) + O(\varepsilon^2),$$

and $f_1(s)$ is to be chosen to ensure that $y_1(s)$ is no more singular than $1/s$. We have

$$y_1(s) = \frac{1}{2s} - \frac{f_1}{s^2} - \frac{1}{2s^3}, \qquad 3.3.11$$

and our first choice for f_1 was $-1/2s$. Then $x = s - \varepsilon/2s + O(\varepsilon^2)$ with the leading order expansion given by 3.3.10.

If we make an alternative choice of f_1 in 3.3.11 which also ensures y_1 is no more singular than $O(1/s)$ then to leading order 3.3.10 still holds. For example, let us choose $f_1 = -1/2s + s/2$. (This happens to make y_1 equal zero but this is irrelevant.) Then

$$x = s + \varepsilon\left(\frac{s}{2} - \frac{1}{2s}\right) + O(\varepsilon^2),$$

and again the region of nonuniformity of the original straightforward expansion $x = O(\sqrt{\varepsilon})$ corresponds to $s = O(\sqrt{\varepsilon})$. The leading order quadratic relation between s and x is

$$s^2\left(1 + \frac{\varepsilon}{2}\right) - sx - \frac{\varepsilon}{2} = 0,$$

with the solution

$$s = \frac{x + \sqrt{x^2 + \varepsilon(2 + \varepsilon)}}{2 + \varepsilon}, \qquad 3.3.12$$

where the positive square root must be chosen so that when $x = O(1)$ we have $s \simeq x$.

The term 2ε in the square root member of 3.3.12 cannot be neglected when $x = O(\sqrt{\varepsilon})$. However, the term ε^2 in the square root is a small term and we may expand as follows:

$$\begin{aligned}\sqrt{x^2 + 2\varepsilon + \varepsilon^2} &= \sqrt{x^2 + 2\varepsilon}\,.\sqrt{1 + \varepsilon^2/(x^2 + 2\varepsilon)}\\ &= \sqrt{x^2 + 2\varepsilon}\,.\left[1 + O\left(\frac{\varepsilon^2}{x^2 + 2\varepsilon}\right)\right]\\ &= \sqrt{x^2 + 2\varepsilon}\,.[1 + O(\varepsilon)].\end{aligned}$$

The denominator of equation 3.3.12 has the usual expansion

$$\frac{1}{2 + \varepsilon} = \frac{1}{2}[1 + O(\varepsilon)].$$

Thus 3.3.12 may be expressed in the form

$$s = \left(\frac{x + \sqrt{x^2 + 2\varepsilon}}{2}\right).[1 + O(\varepsilon)].$$

hen the one-term expansion $y = 1/s$ becomes

$$y = \frac{2}{x + \sqrt{x^2 + 2\varepsilon}} \cdot [1 + \mathrm{O}(\varepsilon)],$$

which is the same to leading order as the expression 3.3.10.

Although the choice of $f_1(s)$ is not unique, the singular component of $f_1(s)$ namely $-1/2s$ is uniquely determined by the need to remove the $\mathrm{O}(1/s^3)$ singularity from the expression 3.3.11 for $y_1(s)$. We may conclude that the function $f_1(s)$ is arbitrary up to any additive regular function of s. The arbitrary regular function will not change the leading order term in the renormalized expansion. This property extends to higher order expansions. For example, a two-term renormalized expansion is independent to the second order of accuracy of the choice of additive regular functions for $f_1(s)$ and $f_2(s)$.

Worked example

Obtain a two-term straightforward expansion for the solution of

$$(x + \varepsilon y)\frac{dy}{dx} + 2y = 2, \quad 0 < \varepsilon \ll 1, \quad y(1) = 2.$$

Determine the region of nonuniformity. Obtain a one-term uniformly valid expansion using the method of renormalization. Obtain a leading order approximation for the value of y when $x = 0$.

Solution

Let $y \sim y_0(x) + \varepsilon y_1(x) + \cdots$
The order equations are

$$\mathrm{O}(1)\colon x\frac{dy}{dx} + 2y_0 = 2, \quad y_0(1) = 2$$

$$\mathrm{O}(\varepsilon)\colon x\frac{dy_1}{dx} + 2y_1 = -y_0\frac{dy_0}{dx}, \quad y_1(1) = 0.$$

The integrating factor for the O(1) equation is x, leading to

$$x^2\frac{dy_0}{dx} + 2xy_0 = \frac{d}{dx}(x^2 y_0) = 2x.$$

Thus $x^2 y_0 = x^2 + C$ and the boundary condition gives $C = 1$ so that

$$y_0 = 1 + 1/x^2.$$

The integrating factor for the $\mathrm{O}(\varepsilon)$ equation is x and we obtain

$$\frac{d}{dx}(x^2 y_1) = -xy_0\frac{dy_0}{dx} = -x\left(1 + \frac{1}{x^2}\right)\left(-\frac{2}{x^3}\right) = \frac{2}{x^2} + \frac{2}{x^4}.$$

After integrating we have

$$x^2 y_1 = -\frac{2}{x} - \frac{2}{3x^3} + C$$

and the boundary condition gives $C = 8/3$. Thus

$$y_1 = \frac{8}{3x^2} - \frac{2}{x^3} - \frac{2}{3x^5},$$

and the two-term straightforward expansion is

$$y_{2T} = 1 + \frac{1}{x^2} + \varepsilon\left(\frac{8}{3x^2} - \frac{2}{x^3} - \frac{2}{3x^5}\right).$$

The region of nonuniformity is obtained by equating the order of y_0 and εy_1 as $\varepsilon \to 0$,

$$1/x^2 = O(\varepsilon/x^5).$$

Hence the region of nonuniformity is $x = O(\varepsilon^{1/3})$ as $\varepsilon \to 0$.

To renormalize the straightforward expansion we introduce the strained coordinate s where

$$x \sim s + \varepsilon f_1(s) + \cdots$$

Then

$$y_{2T} = 1 + \frac{1}{(s + \varepsilon f_1 + \cdots)^2} + \varepsilon\left(\frac{8}{3}\frac{1}{(s + \cdots)^2} - \frac{2}{(s + \cdots)^3} - \frac{2}{3(s + \cdots)^5}\right)$$

$$= 1 + \frac{1}{s^2}\left(1 - 2\frac{\varepsilon f_1}{s} + \cdots\right) + \varepsilon\left(\frac{8}{3s^2} - \frac{2}{s^3} - \frac{2}{3s^5}\right) + \cdots$$

$$= 1 + \frac{1}{s^2} + \varepsilon\left(-2\frac{f_1}{s^3} + \frac{8}{3s^2} - \frac{2}{s^3} - \frac{2}{3s^5}\right) + \cdots$$

The leading order term has a singularity of $O(1/s^2)$. The singularities of $O(1/s^3)$ and $O(1/s^5)$ in the second term must be removed by an appropriate choice of $f_1(s)$. The simplest choice is to set

$$-2\frac{f_1}{s^3} - \frac{2}{s^3} - \frac{2}{3s^5} = 0.$$

Thus $f_1 = -1 - 1/3s^2$ and the one-term uniformly valid expansion is

$$y = 1 + \frac{1}{s^2} + O(\varepsilon)$$

where

$$x = s - \varepsilon\left(1 + \frac{1}{3s^2}\right) + O(\varepsilon^2) \quad \text{as } \varepsilon \to 0.$$

When $x = 0$ the dominant approximation for s is the equation

$$0 = s - \varepsilon\frac{1}{3s^2} \quad \text{i.e.} \quad s \doteq \left(\frac{\varepsilon}{3}\right)^{1/3},$$

and the leading order approximation for y is

$$y \triangleq \left(\frac{3}{\varepsilon}\right)^{2/3}.$$

Exercise

Obtain the two-term straightforward expansion of the solution of

$$(x + \varepsilon y)\frac{dy}{dx} + 4y = 1, \quad 0 < \varepsilon \ll 1, \quad y(1) = 1/2.$$

Determine the region of nonuniformity.

Obtain, using the method of renormalization, the uniformly valid one-term expansion

$$y = \frac{1}{4}\left(1 + \frac{1}{s^4}\right) + O(\varepsilon)$$

where

$$x = s - \frac{\varepsilon}{20}\left(5 + \frac{1}{s^4}\right) + O(\varepsilon^2) \quad \text{as } \varepsilon \to 0.$$

Application to nonlinear oscillators

Renormalization can be used to obtain uniformly valid expansions for second order oscillators such as Duffing's and related equations. Consider for example the equation

$$\frac{d^2u}{dt^2} + u = \varepsilon u\left[1 - \left(\frac{du}{dt}\right)^2\right] \qquad 3.3.13$$

with initial conditions $u(0) = a$, $(du/dt)(0) = 0$.

We will obtain a uniformly valid one-term expansion by applying the renormalization technique to the two-term straightforward expansion

$$u_0(t) + \varepsilon u_1(t).$$

The order equations are

$$O(1): \frac{d^2u_0}{dt^2} + u_0 = 0, \quad u_0(0) = a, \quad \frac{du_0}{dt}(0) = 0$$

$$O(\varepsilon): \frac{d^2u_1}{dt^2} + u_1 = u_0\left[1 - \left(\frac{du_0}{dt}\right)^2\right], \quad u_1(0) = \frac{du_1}{dt}(0) = 0.$$

The O(1) solution is $u_0 = a\cos t$. The $O(\varepsilon)$ equation becomes

$$\frac{d^2u_1}{dt^2} + u_1 = a\cos t - a^3\cos t\,.\sin^2 t.$$

The combination $\cos t . \sin^2 t$ is expressed in terms of multiple angles using $\cos t = (e^{it} + e^{-it})/2$ and $\sin t = (e^{it} - e^{-it})/2i$, thus

$$\cos t . \sin^2 t = (e^{it} + e^{-it}).(e^{it} - e^{-it})^2/(-8)$$
$$= (e^{3it} - e^{it} + e^{-3it} - e^{-it})/(-8) = -\tfrac{1}{4}\cos 3t + \tfrac{1}{4}\cos t.$$

So the $O(\varepsilon)$ equation becomes

$$\frac{d^2u_1}{dt^2} + u_1 = -\frac{a^3}{4}\cos 3t + \left(a + \frac{a^3}{4}\right)\cos t.$$

The complementary function has the form $A\cos t + B\sin t$. A particular integral is $\alpha\cos 3t + \beta t\sin t$ where

$$-9\alpha + \alpha = -a^3/4 \qquad \text{and} \qquad 2\beta = a + a^3/4.$$

Thus the general solution for u_1 is

$$u_1 = A\cos t + B\sin t + \frac{a^3}{32}\cos 3t + \left(\frac{a}{2} + \frac{a^3}{8}\right)t\sin t.$$

The initial conditions yield $A = -a^3/32$ and $B = 0$. The two-term straightforward expansion is therefore

$$u = a\cos t + \varepsilon\left[\frac{a^3}{32}(\cos 3t - \cos t) + \left(\frac{a}{2} + \frac{a^3}{8}\right)t\sin t\right]. \qquad 3.3.14$$

The secular term $t\sin t$ leads to a nonuniformity. The region of nonuniformity is $t = O(1/\varepsilon)$. We introduce the strained coordinate s where,

$$t \sim s + \varepsilon f_1(s) + \cdots$$

then

$$u \sim a\cos s - a\varepsilon f_1\sin s + \cdots + \varepsilon\left[\frac{a^3}{32}(\cos 3s - \cos s) + \left(\frac{a}{2} + \frac{a^3}{8}\right)s\sin s + \cdots\right]$$

and secular terms are removed in the $O(\varepsilon)$ term if $f_1 = s\left(\frac{1}{2} + \frac{a^2}{8}\right)$.

Thus the one-term uniformly valid expansion is

$$u = a\cos s + O(\varepsilon)$$

where

$$t = s + \varepsilon s\left(\frac{1}{2} + \frac{a^2}{8}\right) + O(\varepsilon^2)$$

i.e.

$$s = t\left[1 - \varepsilon\left(\frac{1}{2} + \frac{a^2}{8}\right) + O(\varepsilon^2)\right]$$

so that

$$u = a\cos t\left[1 - \varepsilon\left(\frac{1}{2} + \frac{a^2}{8}\right) + \cdots\right] + O(\varepsilon) \quad \text{as } \varepsilon \to 0.$$

The same expansion would have been obtained using the Lindstedt–Poincaré straining transformation because Lighthill's straining transformation reduces to that form for second order oscillator problems.

An example of the failure of renormalization

Renormalization does not always succeed in rendering the expansions of oscillator problems uniform. An example of its failure is provided by the governing equation for the van der Pol oscillator,

$$\frac{d^2u}{dt^2} + u = \varepsilon(1 - u^2)\frac{du}{dt}, \quad u(0) = 1, \quad \frac{du}{dt}(0) = 0.$$

The straightforward two-term expansion is easily found to be

$$u = \cos t + \varepsilon\left(\frac{3}{8}t\cos t - \frac{9}{32}\sin t - \frac{1}{32}\sin 3t\right).$$

The region of nonuniformity is again associated with the secular term and is given by $t = O(1/\varepsilon)$. If we attempt to renormalize using $t = s + \varepsilon f_1$, we obtain

$$u = \cos s - \varepsilon f_1 \sin s + \varepsilon\left(\frac{3}{8}s\cos s - \frac{9}{32}\sin s - \frac{1}{32}\sin 3s\right) + \cdots$$

To remove secular terms in the $O(\varepsilon)$ term we require,

$$-f\sin s + \frac{3}{8}s\cos s = 0, \quad \text{i.e.} \quad f = \frac{3}{8}s\cot s.$$

Then

$$u = \cos s + O(\varepsilon)$$

where

$$t = s + \frac{3}{8}\varepsilon s\cot s + O(\varepsilon^2).$$

This is invalid because the cotangent function is singular when

$$s = 0, \pi, 2\pi, \ldots .$$

The failure of renormalization when applied to oscillator problems will be considered further in the following chapter. It will be shown that renormalization cannot deal with a time varying amplitude of oscillation. The multiple scale technique is able to deal with this effect and a uniformly valid expansion for the solution o. der Pol's equation will be obtained in Chapter 4.

Exercises

1 Obtain a two-term straightforward expansion for the solution of each of the following equations, determine the region of nonuniformity, obtain a one-term uniformly valid expansion by renormalization and obtain an approximation for the value of y when $x = 0$,

(i) $(x + \varepsilon y)\dfrac{dy}{dx} + 3y = x^2, 0 < \varepsilon \ll 1, y(1) = 1$

(ii) $(x + \varepsilon y)\dfrac{dy}{dx} + 2y = 1, 0 < \varepsilon \ll 1, y(1) = 2$

(iii) $(x + \varepsilon y)\dfrac{dy}{dx} + 3y = 2, 0 < \varepsilon \ll 1, y(1) = 1$

(iv) $(x + \varepsilon y)\dfrac{dy}{dx} + 4y = 0, 0 < \varepsilon \ll 1, y(1) = 1.$

2 Use the renormalization technique to obtain a one-term uniformly valid expansion for the solutions of

(i) $\dfrac{d^2u}{dt^2} + u = \varepsilon u^5, \quad u(0) = 1, \quad \dfrac{du}{dt}(0) = 0.$

(ii) $\dfrac{d^2u}{dt^2} + u = \varepsilon(u - u^3), \; u(0) = 2, \quad \dfrac{du}{dt}(0) = 0.$

(iii) $\dfrac{d^2u}{dt^2} + u = \varepsilon u\left(\dfrac{du}{dt}\right)^4, \; u(0) = 0, \quad \dfrac{du}{dt}(0) = 1.$

3 Obtain, using the renormalization technique, the two-term uniformly valid expansion for the solution of Duffing's equation and verify that this is the same as that obtained in Section 3.2 using the Lindstedt–Poincaré technique (equations 3.2.5 and 3.2.6).

4 Show that renormalization fails to generate a uniformly valid one-term expansion for the solution of the equation

$$\frac{d^2u}{dt^2} + u + \varepsilon\left(\frac{du}{dt}\right)^3 = 0, \quad u(0) = a, \quad \frac{du}{dt}(0) = 0.$$

3.4 Flow past an aerofoil

Lighthill[9] has applied his technique to the study of fluid flow past a thin aerofoil. The thickness of the aerofoil provides a small parameter allowing the construction of a perturbation expansion. The boundary condition on the aerofoil surface is transformed to a condition applying on the centre plane. It is far more convenient to solve the governing partial differential equation in a region with a plane boundary than in a region with a boundary coinciding with the aerofoil profile.

The straightforward perturbation expansion for the fluid velocity is nonuniform due to the occurrence of a singularity in the flow on the aerofoil surface. This singularity grows in severity with subsequent terms in the expansion. Renormalization shifts the singularity to its correct location within the aerofoil shape and consequently yields a uniformly valid expansion for the fluid velocity. Van Dyke[10] has presented the solution of the thin aerofoil problem using a number of techniques and considers a range of profile shapes. We confine our study to two simple aerofoil profiles. Exact solutions are available for both profiles and allow the technique to be verified.

It is helpful to perform the analysis using the powerful technique of the complex potential for inviscid two-dimensional fluid flow. An account of the technique can be found in most texts on fluid dynamics. (A very thorough description of the application of complex variable theory to hydrodynamics can be found in the text book written by Milne-Thomson[11].) The key results will be briefly described here.

Consider the steady flow of an ideal, incompressible fluid past a symmetric aerofoil (Fig. 3.2).

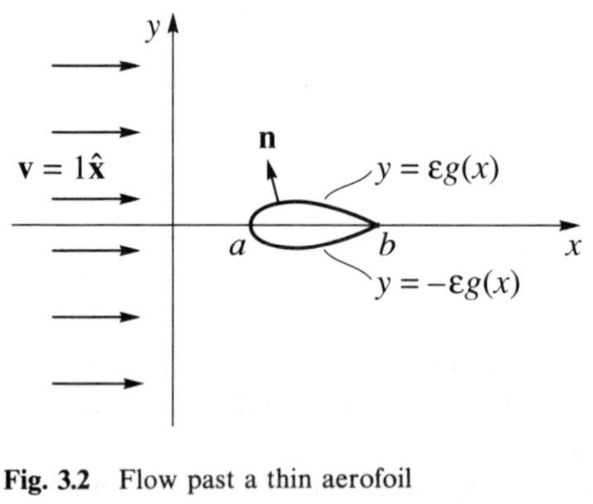

Fig. 3.2 Flow past a thin aerofoil

The nondimensional velocity field has components u and v,

$$\mathbf{v} = u\hat{\mathbf{x}} + v\hat{\mathbf{y}}$$

where $\hat{\mathbf{x}}$ and $\hat{\mathbf{y}}$ denote unit vectors. The incoming stream is given by $\mathbf{v} = 1\hat{\mathbf{x}}$. The nondimensional equation describing the profile shape is $y = \pm\varepsilon g(x)$ for $a < x < b$ where the small parameter, ε, is the width to length ratio of the aerofoil.

The restriction to ideal fluids means that viscous forces are neglected. There will in fact be a narrow region called a boundary layer, near the aerofoil surface, in which viscous effects are not negligible. The viscous forces in the boundary layer decelerate the tangential fluid velocity component to zero so that no slip occurs on the aerofoil surface. In this study we will neglect the boundary layer and consider only the flow outside this region. We may imagine the aerofoil to be slightly thickened to take into account the boundary layer. The no slip boundary condition does not apply to the inviscid flow solution. There is only the no penetration boundary condition to be satisfied at the aerofoil surface. In practice the boundary layer may separate from the aerofoil. This often occurs near the trailing edge. Our analysis remains valid until the region of boundary layer separation whose effects are neglected in this study.

We will assume that the flow is irrotational ($\text{curl}\,\mathbf{v} = \mathbf{0}$). This is justified since any tendency for the fluid to spin (i.e. to have nonzero curl) is both caused by the decelerating effects at the aerofoil surface and dissipated by the viscous forces associated with shearing. Therefore outside the boundary layer it is usually valid to assume irrotational flow. This implies the existence of a velocity potential ϕ such that $\mathbf{v} = \text{grad}\,\phi$. In two dimensions we have

$$u = \frac{\partial \phi}{\partial x} \quad \text{and} \quad v = \frac{\partial \phi}{\partial y}.$$

The complex potential

The condition of incompressible flow ($\text{div}\;\mathbf{v} = 0$) implies the existence of a stream function, ψ, such that

$$u = \frac{\partial \psi}{\partial y} \quad \text{and} \quad v = -\frac{\partial \psi}{\partial x}.$$

Thus the fluid velocity may be expressed in terms of derivatives of either the ϕ or ψ fields,

$$u = \frac{\partial \phi}{\partial x} = \frac{\partial \psi}{\partial y}$$

$$v = \frac{\partial \phi}{\partial y} = -\frac{\partial \psi}{\partial x}.$$

These relationships between the partial derivatives of ϕ and ψ are the Cauchy–Riemann equations and imply the existence of an analytic complex function w of the complex variable z ($= x + \mathrm{i}y$), called the complex potential,

$$w(z) = \phi(x, y) + \mathrm{i}\psi(x, y).$$

The derivative, $\mathrm{d}w/\mathrm{d}z$, can be evaluated in any direction in the complex z plane. In particular

$$\frac{\partial w}{\partial x} = \frac{\partial z}{\partial x}\frac{\mathrm{d}w}{\mathrm{d}z} = \frac{\mathrm{d}w}{\mathrm{d}z}, \quad \text{so} \quad \frac{\mathrm{d}w}{\mathrm{d}z} = \frac{\partial \phi}{\partial x} + \mathrm{i}\frac{\partial \psi}{\partial x} = u - \mathrm{i}v, \qquad 3.4.1$$

and

$$\frac{\partial w}{\partial y} = \frac{\partial z}{\partial y}\frac{dw}{dz} = i\frac{dw}{dz}, \quad \text{so} \quad \frac{dw}{dz} = -i\frac{\partial \phi}{\partial y} + \frac{\partial \psi}{\partial y} = u - iv. \tag{3.4.2}$$

These relationships will be helpful in our subsequent study.

A particular property of the ψ field which is of great help in analyzing flows, is that curves of constant ψ values correspond to fluid particle paths in steady flow. This results from the following consideration. Suppose $f(x, y) = 0$ is a curve of constant ψ, then on this curve

$$d\psi = \frac{\partial \psi}{\partial x}dx + \frac{\partial \psi}{\partial y}dy = 0 \quad \text{and} \quad df = \frac{\partial f}{\partial x}dx + \frac{\partial f}{\partial y}dy = 0.$$

Eliminating dx and dy yields the condition

$$\frac{\partial \psi}{\partial y}\frac{\partial f}{\partial x} - \frac{\partial \psi}{\partial x}\frac{\partial f}{\partial y} = 0.$$

In terms of the velocity field this becomes

$$u\frac{\partial f}{\partial x} + v\frac{\partial f}{\partial y} = 0.$$

Thus the gradient of the function f is perpendicular to the velocity field $\mathbf{v}$. Since the normal to the curve $f = 0$ is given by the gradient of f, this proves that the fluid velocity is tangent to curves of constant values of the stream function. These curves are called streamlines. Any streamline can be replaced by a solid boundary since the no penetration boundary condition is satisfied on a streamline.

All analytic functions $w(z)$ represent a possible fluid flow. The actual function $w(z)$ representing a particular flow is determined by the boundary conditions associated with the flow. In the examples which we will consider it is possible to guess the form of the function $w(z)$ from the nature of the boundary conditions.

The undisturbed stream $\mathbf{v} = 1\hat{\mathbf{x}}$ has the corresponding complex potential $w_0 = z$ (any constant may be added to w without changing the flow fields u and v). The effect of the aerofoil will vanish at great distances from it. Thus at large distances $w \to w_0$.

The flow is assumed to be symmetric about the x axis so that only the upper half-plane need be considered. Referring to Fig. 3.2 the boundary conditions are

$$v = 0 \quad \text{for} \quad x < a \text{ and } x > b \text{ on } y = 0,$$

$$\mathbf{v}.\mathbf{n} = 0 \text{ for } \quad a < x < b \text{ on } y = \varepsilon g(x).$$

The gradient to the surface $y - \varepsilon g(x) = 0$ has components $\left(-\varepsilon\frac{dg}{dx}, 1\right)$, i.e.

$$\mathbf{n} = \frac{-\hat{\mathbf{x}}\varepsilon\dfrac{dg}{dx} + \hat{\mathbf{y}}}{\sqrt{1 + \varepsilon^2\left(\dfrac{dg}{dx}\right)^2}}.$$

The condition $\mathbf{v}.\mathbf{n} = 0$ becomes

$$-\varepsilon u \frac{dg}{dx} + v = 0 \quad \text{on} \quad y = \varepsilon g, \quad a < x < b. \tag{3.4.3}$$

This boundary condition is difficult to impose on the aerofoil profile $y = \varepsilon g$. To overcome this the small parameter, ε, is used to develop a perturbation expansion for the complex potential with boundary conditions which are transferred to the $y = 0$ plane.

Let us assume

$$w(z;\varepsilon) \sim w_0(z) + \varepsilon w_1(z) + \varepsilon^2 w_2(z) + \varepsilon^3 w_3(z) + \cdots.$$

It turns out that a nonuniformity first occurs in this expansion in the third order term, so it is necessary to carry out the subsequent analysis to this order.

The leading approximation, $w_0(z)$, corresponds to the case of unperturbed flow ($\varepsilon = 0$) thus $w_0 = z$. The remaining potentials $w_1(z)$, $w_2(z)$, etc. represent the effect of the aerofoil. The corresponding velocity fields, obtained from the real and imaginary parts of dw_1/dz, dw_2/dz, etc., should vanish at infinite distances from the aerofoil.

On the aerofoil surface the boundary condition 3.4.3 becomes, when expressed in terms of the velocity potential,

$$-\varepsilon \frac{dg}{dx} \cdot \frac{\partial \phi}{\partial x}(x, y = \varepsilon g) + \frac{\partial \phi}{\partial y}(x, y = \varepsilon g) = 0 \quad a < x < b. \tag{3.4.4}$$

This condition can be transferred to the plane $y = 0$ by the use of the following Maclaurin expansions,

$$\frac{\partial \phi}{\partial x}(x, \varepsilon g) = \frac{\partial \phi}{\partial x}(x, 0) + \varepsilon g \frac{\partial^2 \phi}{\partial x \partial y}(x, 0) + \frac{\varepsilon^2 g^2}{2} \frac{\partial^3 \phi}{\partial x \partial y^2}(x, 0) + \cdots$$

and

$$\frac{\partial \phi}{\partial y}(x, \varepsilon g) = \frac{\partial \phi}{\partial y}(x, 0) + \varepsilon g \frac{\partial^2 \phi}{\partial y^2}(x, 0) + \frac{\varepsilon^2 g^2}{2} \frac{\partial^3 \phi}{\partial y^3}(x, 0) + \cdots.$$

Expanding $\phi(x, y; \varepsilon)$,

$$\phi(x, y; \varepsilon) \sim \phi_0(x, y) + \varepsilon \phi_1(x, y) + \varepsilon^2 \phi_2(x, y) + \varepsilon^3 \phi_3(x, y) + \cdots$$

and substituting the above Maclaurin expansions into the boundary condition 3.4.4 leads to the following expression,

$$\begin{aligned}
&\varepsilon\left[\frac{dg}{dx}\frac{d\phi_0}{dx}\right] + \varepsilon^2\left[\frac{dg}{dx}\left(\frac{\partial \phi_1}{\partial x} + g\frac{\partial^2 \phi_0}{\partial x \partial y}\right)\right] + \varepsilon^3\left[\frac{dg}{dx}\left(\frac{\partial \phi_2}{\partial x} + g\frac{\partial^2 \phi_1}{\partial x \partial y} + \frac{g^2}{2}\frac{\partial^3 \phi_0}{\partial x \partial y^2}\right)\right] + \cdots \\
&\sim \frac{\partial \phi_0}{\partial y} + \varepsilon\left[\frac{\partial \phi_1}{\partial y} + g\frac{\partial^2 \phi_0}{\partial y^2}\right] + \varepsilon^2\left[\frac{\partial \phi_2}{\partial y} + g\frac{\partial^2 \phi_1}{\partial y^2} + \frac{g^2}{2}\frac{\partial^3 \phi_0}{\partial y^3}\right] \\
&\quad + \varepsilon^4\left[\frac{\partial \phi_3}{\partial y} + g\frac{\partial^2 \phi_2}{\partial y^2} + \frac{g^2}{2}\frac{\partial^3 \phi_1}{\partial y^3} + \frac{g^3}{6}\frac{\partial^4 \phi_0}{\partial y^4}\right] + \cdots
\end{aligned}$$

where all ϕ derivatives are evaluated at $y = 0$ and x lies in the range $a < x < b$.

The undisturbed flow has $\phi_0 = \text{Re}(w_0) = x$ so that all y derivatives of ϕ_0 are zero. The O(1) boundary condition, $\partial\phi_0/\partial y = 0$, is automatically satisfied.

The remaining order equations are:

$$\text{O}(\varepsilon): \quad \frac{\partial\phi_1}{\partial y} = \frac{dg}{dx}\frac{\partial\phi_0}{\partial x} \tag{3.4.5}$$

$$\text{O}(\varepsilon^2): \quad \frac{\partial\phi_2}{\partial y} = -g\frac{\partial^2\phi_1}{\partial y^2} + \frac{dg}{dx}\frac{\partial\phi_1}{\partial x} \tag{3.4.6}$$

$$\text{O}(\varepsilon^3): \quad \frac{\partial\phi_3}{\partial y} = -g\frac{\partial^2\phi_2}{\partial y^2} - \frac{g^2}{2}\frac{\partial^3\phi_1}{\partial y^3} + \frac{dg}{dx}\left(\frac{\partial\phi_2}{\partial x} + g\frac{\partial^2\phi_1}{\partial x\partial y}\right) \tag{3.4.7}$$

where, in each equation, $y = 0$ and $a < x < b$.

From the equation 3.4.2 we see that

$$\frac{\partial\phi}{\partial y} = -\text{Im}\left(\frac{dw}{dz}\right).$$

Thus the left-hand sides of the above three order equations yield minus the imaginary part of dw_n/dz for $n = 1, 2, 3$ at $y = 0$ for $a < x < b$.

The parabolic aerofoil

To proceed further we must choose an aerofoil shape, $g(x)$. The simplest case to analyze is the parabola $y = \pm\varepsilon\sqrt{x}$ for $x > 0$. Then $g = \sqrt{x}$ and $dg/dx = 1/2\sqrt{x}$. Equation 3.4.5 becomes

$$\text{Im}\left(\frac{dw_1}{dz}\right)\bigg|_{y=0} = -\frac{1}{2\sqrt{x}} \quad \text{for } x > 0$$

while for $x < 0$, $v(x, 0) = 0$ so $\text{Im}\left(\frac{dw_n}{dz}\right)\bigg|_{y-0} = 0.$

The boundary condition suggests the trial form

$w_1 = K\sqrt{z}$, then

$$\frac{dw_1}{dz} = \frac{K}{2\sqrt{z}} \quad \text{and} \quad \frac{dw_1}{dz}\bigg|_{y=0} = \frac{K}{2\sqrt{x}}.$$

Thus $K = -i$ so that

$$\text{Im}\left(\frac{dw_1}{dz}\right)\bigg|_{y=0} = \begin{cases} -\dfrac{1}{2\sqrt{x}} & \text{for } x > 0 \\ 0 & \text{for } x < 0 \end{cases}.$$

We may add Cz to w_1 without affecting this boundary condition (provided C is real). However the condition that $\mathrm{d}w_1/\mathrm{d}z$ should vanish far from the aerofoil forces C to be zero. The above remark applies to all the potentials w_n ($n \geqslant 1$) and serves to eliminate the addition of Cz terms.

Equation 3.4.6 becomes

$$-\operatorname{Im}\left(\frac{\mathrm{d}w_2}{\mathrm{d}z}\right)\Big|_{y=0} = -\sqrt{x}\frac{\partial^2\phi_1}{\partial y^2}(x, 0) + \frac{1}{2\sqrt{x}}\frac{\partial\phi_1}{\partial x}(x, 0),\ x > 0. \qquad 3.4.8$$

From equations 3.4.1 and 3.4.2 we have

$$\frac{\partial w}{\partial x} = \frac{\mathrm{d}w}{\mathrm{d}z} \quad \text{and} \quad \frac{\partial w}{\partial y} = \mathrm{i}\frac{\mathrm{d}w}{\mathrm{d}z}$$

so
$$\frac{\partial\phi_1}{\partial x} + \mathrm{i}\frac{\partial\psi_1}{\partial x} = \frac{\mathrm{d}w_1}{\mathrm{d}z}, \quad \text{i.e.} \quad \frac{\partial\phi_1}{\partial x} = \operatorname{Re}\left(\frac{\mathrm{d}w_1}{\mathrm{d}z}\right)$$

and
$$\frac{\partial\phi_1}{\partial y} + \mathrm{i}\frac{\partial\psi_1}{\partial y} = \mathrm{i}\frac{\mathrm{d}w_1}{\mathrm{d}z} \quad \text{thus} \quad \frac{\partial^2\phi_1}{\partial y^2} + \mathrm{i}\frac{\partial^2\psi_1}{\partial y^2} = -\frac{\mathrm{d}^2w_1}{\mathrm{d}z^2},$$

i.e.
$$\frac{\partial^2\phi_1}{\partial y^2} = \operatorname{Re}\left(-\frac{\mathrm{d}^2w_1}{\mathrm{d}z^2}\right).$$

Therefore

$$\frac{\partial\phi_1}{\partial x} = \operatorname{Re}\left(\frac{-\mathrm{i}}{2\sqrt{z}}\right), \quad \frac{\partial^2\phi_1}{\partial y^2} = \operatorname{Re}\left(\frac{-\mathrm{i}}{4z^{3/2}}\right)$$

and for $y = 0$, $x > 0$ these yield

$$\frac{\partial\phi_1}{\partial x}(x, 0) = \operatorname{Re}\left(\frac{-\mathrm{i}}{2\sqrt{x}}\right) = 0$$

$$\frac{\partial^2\phi_1}{\partial y^2}(x, 0) = \operatorname{Re}\left(\frac{-\mathrm{i}}{4x^{3/2}}\right) = 0.$$

Equation 3.4.8 becomes

$$\operatorname{Im}\left(\frac{\mathrm{d}w_2}{\mathrm{d}z}\right)\Big|_{y=0} = 0 \quad x > 0.$$

The only solution with the correct behavior for large z is $w_2 = 0$ (constants are, of course, irrelevant).

Equation 3.4.7 determines w_3,

$$-\operatorname{Im}\left(\frac{\mathrm{d}w_3}{\mathrm{d}z}\right)\Big|_{y=0} = -\sqrt{x}\frac{\partial^2\phi_2}{\partial y^2} - \frac{x}{2}\frac{\partial^3\phi_1}{\partial y^3}$$
$$+\frac{1}{2\sqrt{x}}\left(\frac{\partial\phi_2}{\partial x} + \sqrt{x}\frac{\partial^2\phi_1}{\partial x\partial y}\right) \quad \text{for } y = 0 \quad \text{and} \quad x > 0. \qquad 3.4.9$$

We have $w_2 = 0$ so ϕ_2 is zero. We need $\dfrac{\partial^3\phi_1}{\partial y^3}(x, 0)$ and $\dfrac{\partial^2\phi_1}{\partial x\partial y}(x, 0)$. These are obtained as follows:

$$\frac{\partial^3\phi_1}{\partial y^3} = \mathrm{Re}\left(\frac{\partial^3 w_1}{\partial y^3}\right) = \mathrm{Re}\left(\mathrm{i}^3\frac{\mathrm{d}^3 w_1}{\mathrm{d}z^3}\right) = \mathrm{Re}\left(-\frac{\mathrm{d}^3\sqrt{z}}{\mathrm{d}z^3}\right) = \mathrm{Re}\left(-\frac{3}{8}\frac{1}{z^{5/2}}\right);$$

therefore

$$\frac{\partial^3\phi_1}{\partial y^3}(x, 0) = -\frac{3}{8x^{5/2}} \quad \text{for } x > 0.$$

$$\frac{\partial^2\phi_1}{\partial x\partial y} = \mathrm{Re}\left(\frac{\partial^2 w_1}{\partial x\partial y}\right) = \mathrm{Re}\left(\mathrm{i}\frac{\mathrm{d}^2 w_1}{\mathrm{d}z^2}\right) = \mathrm{Re}\left(\frac{\mathrm{d}^2\sqrt{z}}{\mathrm{d}z^2}\right) = \mathrm{Re}\left(-\frac{1}{4z^{3/2}}\right),$$

so that

$$\frac{\partial^2\phi_1}{\partial x\partial y}(x, 0) = -\frac{1}{4x^{3/2}} \quad \text{for } x > 0.$$

Equation 3.4.9 becomes

$$-\mathrm{Im}\left(\frac{\mathrm{d}w_3}{\mathrm{d}z}\right)\bigg|_{y=0} = \frac{3}{16x^{3/2}} - \frac{1}{8x^{3/2}} = \frac{1}{16x^{3/2}} \quad \text{for } x > 0.$$

This suggests the trial form $w_3 = K/\sqrt{z}$. Substituting into the above condition yields

$$\mathrm{Im}\left(\frac{K}{2x^{3/2}}\right) = \frac{1}{16x^{3/2}} \quad \text{for } x > 0,$$

i.e. $K = \mathrm{i}/8$.

Thus we have obtained the expansion to $O(\varepsilon^3)$ for the complex potential

$$w \sim z - \mathrm{i}\varepsilon\sqrt{z} + \frac{\mathrm{i}\varepsilon^3}{8\sqrt{z}} + \cdots$$

The velocity field is obtained from the real and imaginary parts of the following expression,

$$\frac{\mathrm{d}w}{\mathrm{d}z} = u - \mathrm{i}v \sim 1 - \frac{\mathrm{i}\varepsilon}{2\sqrt{z}} - \frac{\mathrm{i}\varepsilon^3}{16z^{3/2}} + \cdots \qquad 3.4.10$$

The velocity expansion is singular at $z = 0$ and the singularity grows with higher order terms. The region of nonuniformity of the expansion 3.4.10 is such that $z = O(\varepsilon^2)$ as $\varepsilon \to 0$. To renormalize this expansion we introduce the strained coordinate s where

$$z \sim s + \varepsilon f_1(s) + \varepsilon^2 f_2(s) + \cdots.$$

Then

$$\frac{1}{\sqrt{z}}=\frac{1}{\sqrt{s}}\left(1+\frac{\varepsilon f_1}{s}+\frac{\varepsilon^2 f_2}{s}+\cdots\right)^{-1/2}$$

$$=\frac{1}{\sqrt{s}}\left(1-\frac{\varepsilon f_1}{2s}-\frac{\varepsilon^2 f_2}{2s}+\cdots+\frac{3}{8}\frac{\varepsilon^2 f_1^2}{s^2}+\cdots\right)$$

and

$$\frac{1}{z^{3/2}}=\frac{1}{s^{3/2}}(1+\cdots)$$

so, working to $O(\varepsilon^3)$,

$$\frac{dw}{dz}\sim 1-\frac{i\varepsilon}{2\sqrt{s}}+\frac{i}{2\sqrt{s}}\varepsilon^2\frac{f_1}{2s}+\frac{i}{2\sqrt{s}}\varepsilon^3\left(\frac{f_2}{2s}-\frac{3}{8}\frac{f_1^2}{s^2}-\frac{1}{8s}\right)+\cdots.$$

The simplest choice for f_1 to ensure that the $O(\varepsilon^2)$ term is no more singular than the $O(\varepsilon)$ term is to set $f_1=0$. Similarly the simplest choice for f_2 is $f_2=1/4$.

Then the first order uniformly valid expansion is

$$\frac{dw}{dz}=1-\frac{i\varepsilon}{2\sqrt{s}}+O(\varepsilon^2)\quad \text{as } \varepsilon\to 0,$$

where

$$z=s+\frac{\varepsilon^2}{4}+O(\varepsilon^3)\quad \text{as } \varepsilon\to 0.$$

We may eliminate s to obtain the approximation

$$\frac{dw}{dz}\simeq 1-\frac{i\varepsilon}{2\sqrt{z-\varepsilon^2/4}}. \tag{3.4.11}$$

Lighthill's technique has rendered the expansion 3.4.10 uniformly valid by shifting the singularity at $z=0$ of the straightforward expansion to the point $z=\varepsilon^2/4$. The point $z=0$ lies on the surface of the aerofoil while the point $z=\varepsilon^2/4$ lies within the aerofoil and consequently the singularity of the renormalized expansion lies outside the flow region.

Comparison with the exact solution

It so happens that, in the case of flow past a parabola, the approximation 3.4.11 is in fact the exact solution. To verify this we must show that the curve $y=\pm\varepsilon\sqrt{x}$

corresponds to a streamline. Integrating equation 3.4.11 yields

$$w = z - i\varepsilon\sqrt{z - \varepsilon^2/4}, \qquad 3.4.12$$

where no arbitrary constant is added since we choose $w \to z$ as $z \to \infty$.

The parabola $y = \pm\varepsilon\sqrt{x}$ passes through the origin and the value of the complex potential there is $-i\varepsilon\sqrt{-\varepsilon^2/4}$. Thus w has zero imaginary part at the origin. Now $w = \phi + i\psi$ so the streamline passing through the origin has the value $\psi = 0$. We have, in general, the equation

$$w = \phi + i\psi = z - i\varepsilon\sqrt{z - \varepsilon^2/4}.$$

Setting $\psi = 0$ and rearranging this expression yields

$$(\phi - z)^2 = -\varepsilon^2(z - \varepsilon^2/4),$$

with real part

$$(\phi - x)^2 - y^2 = -\varepsilon^2(x - \varepsilon^2/4),$$

and imaginary part

$$-2y(\phi - x) = -\varepsilon^2 y.$$

Thus the streamline $\psi = 0$ either corresponds to the line $y = 0$, $x < \varepsilon^2/4$ or yields the constraint $\phi - x = \varepsilon^2/2$ with a corresponding curve

$$\frac{\varepsilon^4}{4} - y^2 = -\varepsilon^2 x + \varepsilon^4/4,$$

i.e. $y = \pm\varepsilon\sqrt{x}$ (see Fig. 3.3).

Thus, in this particular case, renormalization has yielded the exact solution, 3.4.11, for flow past a parabolic aerofoil.

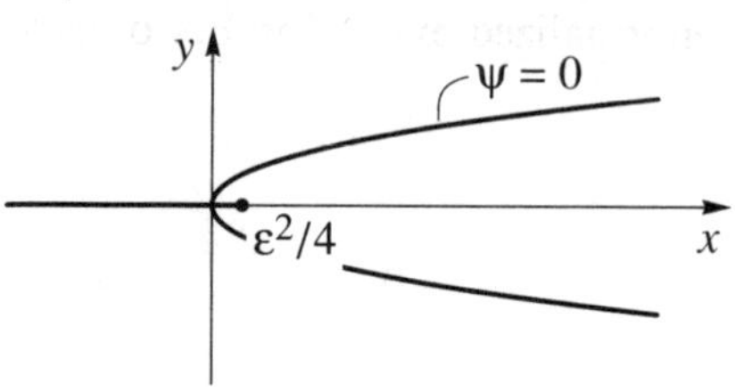

Fig. 3.3 The $\psi = 0$ streamline for the complex potential 3.4.12

The elliptic aerofoil

Next we briefly consider flow past an elliptic aerofoil as shown in Fig. 3.4.

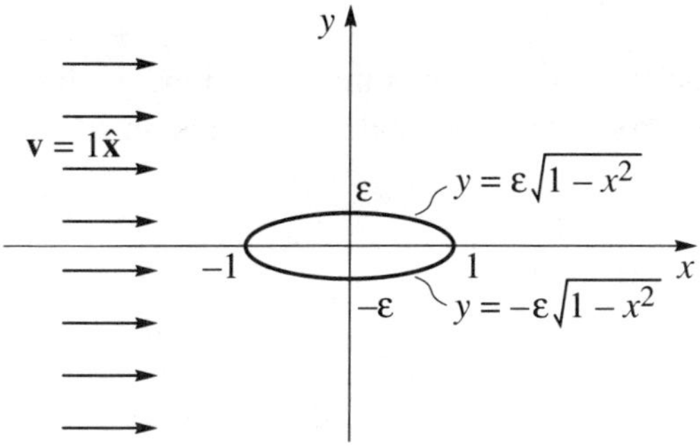

Fig. 3.4 Flow past an elliptic aerofoil

Repeating the previous procedure for case $g = \sqrt{1 - x^2}$ leads eventually to the following straightforward expansion for the complex potential:

$$\frac{dw}{dz} \sim 1 + (\varepsilon + \varepsilon^2)\left(1 - \frac{z}{\sqrt{z^2 - 1}}\right) + \varepsilon^3\left(1 - \frac{z}{\sqrt{z^2 - 1}} + \frac{z}{2(z^2 - 1)^{3/2}}\right) + \cdots. \tag{3.4.13}$$

This expansion is singular at $z = \pm 1$ and the singularity grows with higher order terms.

Renormalization can be performed with a strained coordinate, s, where

$$z \sim s + \varepsilon^2 f_2(s) + \cdots.$$

(The term $\varepsilon f_1(s)$ can be omitted since, just as in the previous example, f_1 may be chosen to be zero.) The term $z/\sqrt{z^2 - 1}$ has the expansion

$$\begin{aligned}\frac{s + \varepsilon^2 f_2}{\sqrt{s^2 + 2\varepsilon^2 s f_2 + \varepsilon^4 f_2^2 - 1}} &= \frac{s + \varepsilon^2 f_2}{\sqrt{s^2 - 1}}\left(1 - \frac{\varepsilon^2 s f_2}{s^2 - 1} + \cdots\right) \\ &= \frac{s}{\sqrt{s^2 - 1}} + \varepsilon^2\left(\frac{f_2}{\sqrt{s^2 - 1}} - \frac{s^2 f_2}{(s^2 - 1)^{3/2}}\right) \\ &= \frac{s}{\sqrt{s^2 - 1}} - \frac{\varepsilon^2 f_2}{(s^2 - 1)^{3/2}}.\end{aligned}$$

So on substituting into 3.4.13 we obtain

$$\frac{dw}{dz} \sim 1 + (\varepsilon + \varepsilon^2)\left(1 - \frac{s}{\sqrt{s^2 - 1}}\right)$$

$$+ \varepsilon^3\left(\frac{f_2}{(s^2 - 1)^{3/2}} + 1 - \frac{s}{\sqrt{s^2 - 1}} + \frac{s}{2(s^2 - 1)^{3/2}}\right) + \cdots.$$

The simplest choice for f_2 to ensure that the $O(\varepsilon^3)$ term is no more singular than the previous terms is $f_2 = -s/2$. Then the first order renormalized expansion is

$$\frac{dw}{dz} \sim 1 + \varepsilon\left(1 - \frac{s}{\sqrt{s^2 - 1}}\right) + \cdots$$

where

$$z \sim s - \varepsilon^2 s/2 + \cdots.$$

Eliminating s yields

$$\frac{dw}{dz} \sim 1 + \varepsilon\left(1 - \frac{z}{\sqrt{z^2 - (1 - \varepsilon^2/2)^2}}\right) + \cdots$$

$$\sim 1 + \varepsilon\left(1 - \frac{z}{\sqrt{z^2 - 1 + \varepsilon^2}}\right) + \cdots. \qquad 3.4.14$$

The singularities of the straightforward expansion occur at $z = \pm 1$. These are the leading and trailing edges of the aerofoil. The renormalized expansion has the singularities shifted to $z = \pm\sqrt{1 - \varepsilon^2}$ (to leading order). Thus the singularities occur within the aerofoil and out of the flow region.

Comparison with the exact solution

The exact solution for flow past an ellipse can be obtained from the solution for flow past a circle using a conformal transformation (see Milne-Thomson[11]). The exact velocity potential is

$$w = z + \left(\frac{\varepsilon}{1 - \varepsilon}\right)(z - \sqrt{z^2 - 1 + \varepsilon^2}). \qquad 3.4.15$$

The $\psi = 0$ streamline again corresponds to the aerofoil shape. To verify this we set $w = \phi + i0$ and rearrange 3.4.15 to obtain

$$(1 - \varepsilon)\phi - z = -\varepsilon\sqrt{z^2 - 1 + \varepsilon^2}.$$

Squaring and equating real and imaginary parts on either side of the equation yields

$$[(1 - \varepsilon)\phi - x]^2 - y^2 = \varepsilon^2(x^2 - y^2 - 1 + \varepsilon^2),$$

and

$$-[(1-\varepsilon)\phi - x]y = \varepsilon^2 xy.$$

Thus the streamline $\psi = 0$ corresponds either to the lines $y = 0$, $|x| > \sqrt{1-\varepsilon^2}$ or yields the constraint $(1-\varepsilon)\phi - x = -\varepsilon^2 x$ so that

$$\varepsilon^4 x^2 - y^2 = \varepsilon^2(x^2 - y^2 - 1 + \varepsilon^2).$$

This latter streamline equation may be rearranged to yield

$$\varepsilon^2(1-\varepsilon^2)x^2 + (1-\varepsilon^2)y^2 = \varepsilon^2(1-\varepsilon^2),$$

i.e. the ellipse

$$x^2 + \frac{y^2}{\varepsilon^2} = 1 \quad \text{(see Fig. 3.5).}$$

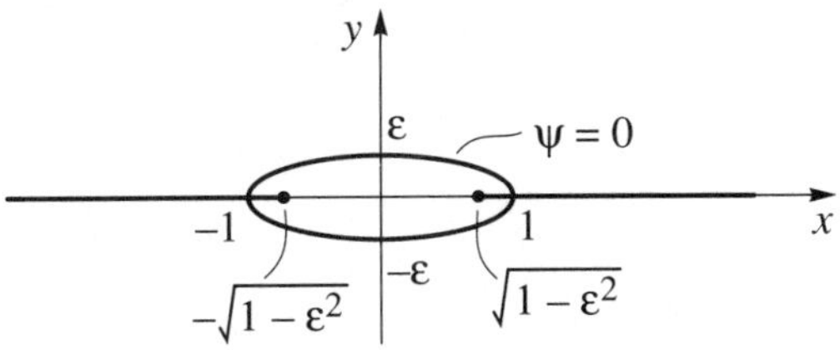

Fig. 3.5 The $\psi = 0$ streamline for the complex potential 3.4.15

Integrating the first order, renormalized expansion 3.4.14 yields the potential

$$w \sim z + \varepsilon(z - \sqrt{z^2 - 1 + \varepsilon^2}) + \cdots.$$

This is the same as the first order expansion of the exact potential 3.4.15 with the form $\sqrt{z^2 - 1 + \varepsilon^2}$ preserved.

Differentiating the exact potential, 3.4.15, and expanding for small ε yields the straightforward expansion 3.4.13. This illustrates the source of the nonuniformity, namely the expansion of the square root $\sqrt{z^2 - 1 + \varepsilon^2}$ when $z^2 - 1 = O(\varepsilon^2)$.

The application of renormalization to the thin aerofoil problem graphically demonstrates Lighthill's technique of shifting the singularity of the straightforward expansion by the use of strained coordinates. Unfortunately, difficulties arise when attempting to obtain higher order expansions for the aerofoil flow (see Fox[12]). The technique is not applicable to blunt or sharp aerofoils but restricted to round nosed aerofoils (see Van Dyke[10]). Despite these restrictions, when applicable, the method is very effective in that it provides a useful perturbation approximation with relatively little mathematical manipulation.

Lighthill's technique has proved to be of great value in applications involving wave propagation. It appears that the above-mentioned difficulties encountered with the aerofoil expansions are a consequence of the fact that the governing partial differential equation for the fluid potential (Laplace's equation) is elliptic. The hyperbolic equations which govern wave propagation are more suited to the application of Lighthill's technique (see Lighthill[5]).

Chapter 4
Multiple Scales

Some processes have more than one characteristic length or time scale associated with them, for example

mechanical vibrations with a slowly varying amplitude
the slow precession of planetary orbits
turbulence – there are various length scales of the turbulent eddies along with the length scale of objects over which fluid flows
surface roughness – the length scale of the surface finish is far smaller than the overall dimensions of the surface
noise in electrical signals – this is a high frequency (short time scale) effect superimposed on the signal information.

The failure to recognize a dependence on more than one coordinate scale is a common source of nonuniformity in perturbation expansions. The multiple scale technique can provide a method of rendering these expansions uniform. It is particularly effective in dealing with weakly nonlinear oscillators where the nonlinearity is a small perturbation.

The strong nonlinearity in the governing equations of fluid dynamics accounts for some of the difficulties in the mathematical description of turbulence. Currently there is no completely satisfactory theory of turbulence. The subject will not be considered in this book. However, a problem which has some similarities with turbulence is that of describing the effect of surface roughness in contact phenomena.

Surface roughness causes the pressure distribution in a lubricated bearing to have random fluctuations and it is the average pressure which is required. We will consider an application of the multiple scale technique to the derivation of an average Reynolds equation which determines the average pressure.

The chapter concludes with a description of the Krylov – Bogoliubov technique for dealing with different scales.

4.1 Second order systems

We will begin our study of the multiple scale technique with a consideration of its application to weakly nonlinear oscillators with governing equation

$$\frac{d^2u}{dt^2} + \omega_0^2 u = \varepsilon F\left(u, \frac{du}{dt}\right). \qquad 4.1.1$$

In the previous chapter we saw how renormalization could overcome nonuniformities in the perturbation expansion of the solutions of 4.1.1 for some forms of the function F (e.g. Duffing's equation with $F = -u^3$) but failed for other forms of F (e.g. van der Pol's equation with $F = (1 - u^2)(\mathrm{d}u/\mathrm{d}t)$). Multiple scales can successfully deal with all forms of the function $F(u, \mathrm{d}u/\mathrm{d}t)$. Before introducing the method it is helpful to briefly review the types of solutions of second order differential equations which can occur.

Consider first the differential equation

$$\frac{\mathrm{d}^2u}{\mathrm{d}t^2} + \omega_0^2 u = 0. \tag{4.1.2}$$

The general solution is $u = A\cos\omega_0 t + B\sin\omega_0 t$. The initial conditions provide the particular values of the constants A and B,

$$A = u(0), \quad B = \frac{1}{\omega_0}\frac{\mathrm{d}u}{\mathrm{d}t}(0).$$

There are infinitely many solutions corresponding to different values of A and B. All the solutions are periodic, i.e. u cycles through a fixed set of values for a given choice of A and B. The period of the cycle is $2\pi/\omega_0$.

The phase plane

It is helpful to represent the solution of second order systems using the phase plane. This consists of rectangular axes x, y where $x = u$ and $y = \mathrm{d}u/\mathrm{d}t$. The solution of 4.1.2 has the following x, y functions

$$x = A\cos\omega_0 t + B\sin\omega_0 t$$

$$y = -\omega_0 A\sin\omega_0 t + \omega_0 B\cos\omega_0 t;$$

eliminating t yields the solution curve (sometimes called a trajectory) in the phase plane

$$x^2 + \frac{y^2}{\omega_0^2} = A^2 + B^2.$$

Thus the solution curves are ellipses with semi-axes $\sqrt{A^2 + B^2}$ and $\omega_0\sqrt{A^2 + B^2}$.

Two solution curves are shown in Fig. 4.1 along with their starting points. The initial conditions are $u(0) = 1$, $(\mathrm{d}u/\mathrm{d}t)(0) = 0$ and $u(0) = 1$, $(\mathrm{d}u/\mathrm{d}t)(0) = 3$ respectively. In both cases $\omega_0 = 2$. The semi-axes are respectively,

$$\sqrt{1+0}, \quad 2\sqrt{1+0} \qquad \text{and} \qquad \sqrt{1+(3/2)^2}, \quad 2\sqrt{1+(3/2)^2}.$$

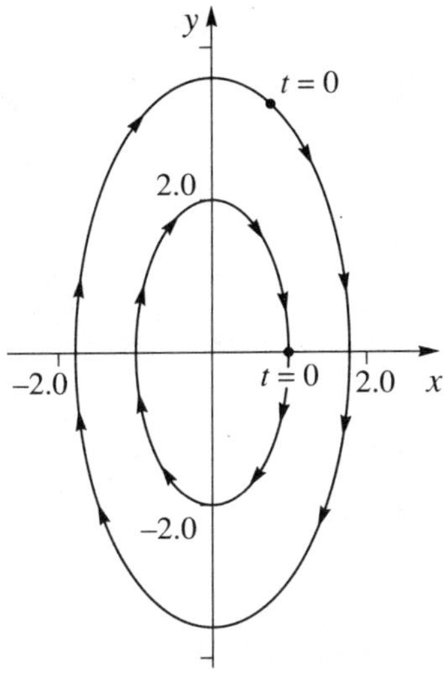

Fig. 4.1 Phase plane solution curves of equation 4.1.2

The direction of the trajectories is determined by the sign of y. If y is positive then, since $y = \mathrm{d}x/\mathrm{d}t$, x will increase with time and conversely, if y is negative, x will decrease with time.

Consider next the differential equation

$$\frac{\mathrm{d}^2u}{\mathrm{d}t^2} + \frac{\mathrm{d}u}{\mathrm{d}t} + 2u = 0. \qquad 4.1.3$$

The general solution is

$$u = \exp(-\tfrac{1}{2}t)(A\cos\sqrt{7}t/2 + B\sin\sqrt{7}t/2),$$

where

$$A = u(0) \quad \text{and} \quad -A/2 + \sqrt{7}B/2 = \frac{\mathrm{d}u}{\mathrm{d}t}(0).$$

Both u and $\mathrm{d}u/\mathrm{d}t$ oscillate with a decaying amplitude. All the solution trajectories will spiral into the origin of the phase plane. Two solution trajectories are shown in Fig. 4.2 for the initial conditions $u(0) = 0$, $(\mathrm{d}u/\mathrm{d}t)(0) = 1$ and $u(0) = 1$, $(\mathrm{d}u/\mathrm{d}t)(0) = 1$ respectively.

We will confine our study to *autonomous* second order differential equations. By this we mean that the coefficients of the terms in the differential equation are independent of time and that there are no time dependent forcing functions. There are basically three types of behavior of solutions of autonomous linear differential equations, they either

orbit the origin of the phase plane, i.e. they are periodic
decay towards the origin, or
diverge to infinity.

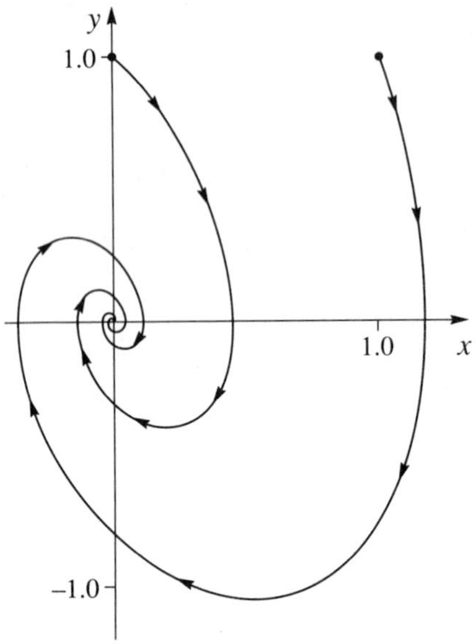

Fig. 4.2 Phase plane solution curves of equation 4.1.3

The latter two solution types can occur with or without a spiral motion. The solution type is dependent on the coefficients of the differential equation.

Nonlinear autonomous systems and the limit cycle

Nonlinear second order differential equations provide a rich variety of solution types. Both the form of the differential equation and the values of the initial conditions determine the solution type. A particular form of solution behavior for which there is no counterpart amongst linear equations is that associated with the occurrence of a *limit cycle.* As an example consider the differential equation

$$\frac{d^2u}{dt^2} = -u + \left[1 - u^2 - \left(\frac{du}{dt}\right)^2\right]\frac{du}{dt}. \tag{4.1.4}$$

In terms of the phase plane variables

$x = u,\ y = \dfrac{dx}{dt} = \dfrac{du}{dt}$, this becomes

$$\frac{dy}{dt} = -x + (1 - x^2 - y^2)y \tag{4.1.5}$$

along with

$$\frac{dx}{dt} = y. \tag{4.1.6}$$

Multiply 4.1.5 by y, 4.1.6 by x and add to obtain

$$\frac{\mathrm{d}}{\mathrm{d}t}(x^2 + y^2) = 2(1 - x^2 - y^2)y^2.$$

In terms of polar coordinates r, θ, where $x = r\cos\theta$, and $y = r\sin\theta$, this becomes

$$\frac{\mathrm{d}}{\mathrm{d}t}(r^2) = 2(1 - r^2)r^2\sin^2\theta. \qquad 4.1.7$$

This shows that there is a periodic solution corresponding to $r = 1$ since then r is constant. Furthermore if r is greater than unity the right-hand side of 4.1.7 is negative, showing that r will decrease, while if r is less than unity the right-hand side is positive, so r will increase. Thus the periodic solution is approached by all solutions. It is called a stable *limit cycle*. Some solution trajectories are shown in Fig. 4.3.

These solutions have been obtained using the improved Euler numerical solution scheme. The FORTRAN program overleaf solves for the variables u and $\mathrm{d}u/\mathrm{d}t$ (denoted by XO and YO) at time intervals DT. The first estimates for the values of X and Y after a time step DT are obtained from the equations

$$\frac{\mathrm{X1} - \mathrm{XO}}{\mathrm{DT}} = \left.\frac{\mathrm{d}u}{\mathrm{d}t}\right|_{\substack{\text{evaluated}\\ \text{with Y=YO}}} = \mathrm{YO}$$

$$\frac{\mathrm{Y1} - \mathrm{YO}}{\mathrm{DT}} = \left.\frac{\mathrm{d}^2u}{\mathrm{d}t^2}\right|_{\substack{\text{evaluated with}\\ \text{X=XO, Y=YO}}} = -\mathrm{XO} + (1.0 - \mathrm{XO.XO} - \mathrm{YO.YO}).\mathrm{YO}.$$

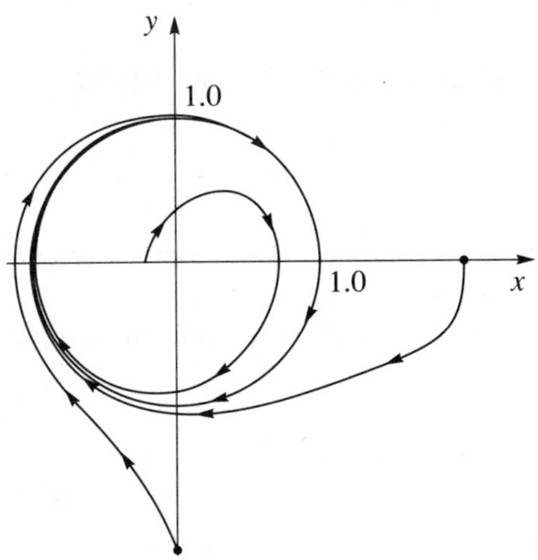

Fig. 4.3 Trajectories approaching the limit cycle of equation 4.1.4

```
CCC     TO SOLVE D2U/DT2 +U=(1-U*U-(DU/DT)*(DU/DT))*(DU/DT)
CCC     SETTING THE TIME STEP,INITIAL VALUES,NUMBER OF STEPS
CCC     AND THE PRINT FREQUENCY-----------------------------------
        PRINT*,'DT= U= DUDT= NSTEPS= IPF='
        READ*,DT,X0,Y0,NSTEPS,IPF
CCC---------------------------------------------------------------
        T=0.0
        IP=IPF
        DO 100 I=0,NSTEPS
CCC     PRINTING THE OUTPUT---------------------------------------
        IF(IP.EQ.IPF)THEN
        WRITE(1,1000)T,X0,Y0
        WRITE(20,*)T,X0
        WRITE(21,*)X0,Y0
        IP=0
        ENDIF
CCC---------------------------------------------------------------
        IP=IP+1
CCC     OBTAINING THE NEW VALUES OF X0 AND Y0---------------------
        X1=X0+DT*Y0
        Y1=Y0+DT*(-X0+(1.0-X0*X0-Y0*Y0)*Y0)
        X2=X0+DT*Y1
        Y2=Y0+DT*(-X1+(1.0-X1*X1-Y1*Y1)*Y1)
        X0=(X1+X2)/2.0
        Y0=(Y1+Y2)/2.0
        T=T+DT
CCC---------------------------------------------------------------
100     CONTINUE
1000    FORMAT(3F10.5)
        END
```

FORTRAN program to solve $\dfrac{d^2u}{dt^2} + u = \left[1 - u^2 - \left(\dfrac{du}{dt}\right)^2\right]\dfrac{du}{dt}$

A second estimate is obtained using the latest estimates for u and du/dt,

$$\frac{X2 - XO}{DT} = \left.\frac{du}{dt}\right|_{\substack{\text{evaluated}\\ \text{with } Y=Y1}} = Y1$$

$$\frac{Y2 - YO}{DT} = \left.\frac{d^2u}{dt^2}\right|_{\substack{\text{evaluated with}\\ X=X1,\ Y=Y1}} = -X1 + (1.0 - X1.X1 - Y1.Y1).Y1.$$

The new values of X and Y are obtained by averaging X1, X2 and Y1, Y2

$$XO = (X1 + X2)/2$$

$$YO = (Y1 + Y2)/2.$$

A full discussion of both linear and nonlinear second order autonomous systems and systems with forcing functions may be found in Grimshaw.[2]

The solution behavior of nonlinear autonomous second order differential equations can be grouped into the following four broad classes:

periodic solutions
solutions tend towards a periodic solution (limit cycle)
solutions tend towards a fixed value
solutions tend to infinity.

In the context of the weakly nonlinear oscillator (equation 4.1.1) the nonuniformity which occurs in the straightforward expansion can only be dealt with by renormalization if the solutions of the perturbed system are periodic. We will see in the next section that renormalization fails if the solutions of the perturbed system are of the latter three types in the above classification.

4.2 Limitation of renormalization

An understanding of the limitation of renormalization when applied to second order systems can be gained by comparing the following two linear problems,

$$\frac{d^2u}{dt^2} + u = \varepsilon u, \quad u(0) = 1, \quad \frac{du}{dt}(0) = 0 \tag{4.2.1}$$

$$\frac{d^2u}{dt^2} + u = -\varepsilon\frac{du}{dt}, \quad u(0) = 1, \quad \frac{du}{dt}(0) = 0. \tag{4.2.2}$$

We will attempt to obtain a uniformly valid one-term expansion by renormalization. The technique requires that we start with a two-term straightforward expansion, $u_0(t) + \varepsilon u_1(t)$. Both 4.2.1 and 4.2.2 share the leading order term $u_0 = \cos t$.

The $O(\varepsilon)$ equation associated with 4.2.1 is

$$\frac{d^2u_1}{dt^2} + u_1 = u_0(=\cos t), \quad u_1(0) = \frac{du_1}{dt}(0) = 0.$$

The solution is $u_1 = \frac{1}{2}t\sin t$.

The $O(\varepsilon)$ equation associated with 4.2.2 is

$$\frac{d^2u_1}{dt^2} + u_1 = -\frac{du_0}{dt}(=\sin t), \quad u_1(0) = \frac{du_1}{dt}(0) = 0.$$

This has solution $u_1 = \frac{1}{2}(-t\cos t + \sin t)$.

On introducing the strained coordinate, s, where

$$t = s + \varepsilon f_1(s) + O(\varepsilon^2),$$

the two-term expansion for the solution of 4.2.1 becomes

$$\begin{aligned}\cos t + \tfrac{1}{2}\varepsilon t\sin t &= \cos(s + \varepsilon f_1 + \cdots) + \tfrac{1}{2}\varepsilon s\sin s + \cdots \\ &= \cos s - \varepsilon f_1 \sin s + \tfrac{1}{2}\varepsilon s\sin s + \cdots.\end{aligned}$$

Secular terms in the $O(\varepsilon)$ term are removed by choosing $f_1 = s/2$. Then we have

$$u = \cos s + O(\varepsilon),$$

where

$$t = s + \varepsilon s/2 + O(\varepsilon^2),$$

$$\text{i.e. } s = t\left(1 + \frac{\varepsilon}{2} + O(\varepsilon^2)\right)^{-1} = t\left(1 - \frac{\varepsilon}{2} + O(\varepsilon^2)\right).$$

Thus

$$u = \cos t\left(1 - \frac{\varepsilon}{2} + \cdots\right).$$

The exact solution of 4.2.1 is

$$u = \cos\sqrt{1 - \varepsilon t} = \cos t\left(1 - \frac{\varepsilon}{2} + \cdots\right),$$

which corresponds to that generated by renormalization. This example demonstrates the general result, namely that renormalization provides a means of evaluating the frequency change associated with perturbed systems whose solutions are periodic.

We next attempt to renormalize the two-term straightforward expansion of the solution of 4.2.2. We have

$$\begin{aligned}\cos t - \tfrac{1}{2}\varepsilon(t\cos t - \sin t) &= \cos(s + \varepsilon f_1 + \cdots) - \tfrac{1}{2}\varepsilon(s\cos s - \sin s) + \cdots \\ &= \cos s - \varepsilon f_1 \sin s - \tfrac{1}{2}\varepsilon s\cos s + \tfrac{1}{2}\varepsilon\sin s + \cdots.\end{aligned}$$

To remove secular terms in the $O(\varepsilon)$ term we require $f_1 = -\frac{1}{2}s\cot s$. This is not acceptable because f_1 becomes singular at $s = \pi, 2\pi, 3\pi \ldots$ The difficulty is that in order to render uniform the expansion for the solution, the straining transformation has become nonuniform and thus the technique has failed.

Perturbations which cause a time-changing amplitude

The exact solution of 4.2.2 is

$$u = e^{-\varepsilon t/2}\left(\cos\sqrt{1 - (\varepsilon/2)^2}\,t + \frac{\varepsilon}{2\sqrt{1 - (\varepsilon/2)^2}}\sin\sqrt{1 - (\varepsilon/2)^2}\,t\right).$$

This is not a periodic function. The perturbation has introduced both a change in the frequency of oscillation (from 1 to $\sqrt{1 - (\varepsilon/2)^2}$) and a decaying amplitude of oscillation. It is generally the case that renormalization fails to deal with perturbations which cause the solution to become nonperiodic.

Nayfeh[6b] has provided the following method of determining when renormalization will succeed. Without loss of generality we may absorb the unperturbed frequency, ω_0, of equation 4.1.1 into the definition of nondimensional time. Further, we may choose the zero of time to occur when du/dt is zero so that the initial conditions $u(0) = a$ and $(du/dt)(0) = 0$ may be used for the weakly nonlinear oscillator

$$\frac{d^2u}{dt^2} + u = \varepsilon F\left(u, \frac{du}{dt}\right). \qquad 4.2.3$$

The two-term straightforward expansion, $u_0(t) + \varepsilon u_1(t)$, has the leading order member $u_0 = a\cos t$ and therefore u_1 satisfies

$$\frac{d^2 u_1}{dt^2} + u_1 = \varepsilon F(a\cos t, -a\sin t). \tag{4.2.4}$$

Since F is a function of the periodic quantities $\cos t$ and $\sin t$, it too will have period 2π and may therefore be Fourier decomposed as follows:

$$F(a\cos t, -a\sin t) = \frac{a_0}{2} + \sum_{n=1}^{\infty} \{a_n \cos nt + b_n \sin nt\}.$$

Usually the average value of F is zero so that $a_0 = 0$ but this is not crucial to the analysis.

The complementary function of the solution of equation 4.2.4 has the form $A\cos t + B\sin t$. The Fourier component $a_1 \cos t + b_1 \sin t$ of F will generate, as part of a particular integral of 4.2.4, the secular term

$$\frac{t}{2}(a_1 \sin t - b_1 \cos t).$$

Thus the two-term straightforward expansion of 4.2.4 is

$$a\cos t + \varepsilon\left[\frac{t}{2}(a_1 \sin t - b_1 \cos t) + \text{nonsecular terms}\right].$$

We attempt to renormalize using the strained coordinate s where

$$t = s + \varepsilon f_1 + \cdots.$$

The two-term expansion becomes

$$a\cos s - \varepsilon a f_1 \sin s + \varepsilon\frac{s}{2}(a_1 \sin s - b_1 \sin s) + \text{nonsecular terms}.$$

In order to remove secular terms from the $O(\varepsilon)$ term we require $f_1 = a_1 s/2a - b_1 s \cot s/2a$. The cotangent is singular and thus renders the straining transformation nonuniform. Consequently renormalization fails unless $b_1 = 0$. The condition for renormalization to succeed is therefore that $F(a\cos t, -a\sin t)$ when Fourier decomposed should not contain $\sin t$. Furthermore the form of the renormalized expansion is

$$u = a\cos s + O(\varepsilon),$$

where

$$t = s + \varepsilon\frac{a_1}{2a}s + O(\varepsilon^2),$$

i.e.

$$u = a\cos\left(1 - \varepsilon\frac{a_1}{2a} + \cdots\right)t.$$

Thus the renormalized expansion can only yield periodic solutions. It is therefore applicable to equations of the form 4.2.3 if the effect of the perturbation is to modify the frequency of the unperturbed periodic solution $a\cos t$. It is not applicable to those cases where the perturbation introduces a time-changing amplitude.

The Fourier decomposition test when applied to Duffing's equation with $F = -u^3$ yields

$$F(a\cos t,\ -a\sin t) = -a^3\cos^3 t = \frac{a^3}{4}(\cos 3t + 3\cos t),$$

and since $\sin t$ is absent we may conclude that renormalization will succeed. We saw in Section 3.3 that this indeed is the case.

Consider next the van der Pol equation with $F = (1 - u^2)(\mathrm{d}u/\mathrm{d}t)$.

$$F(a\cos t,\ -a\sin t) = (1 - a^2\cos^2 t).(-a\sin t)$$

and $\cos^2 t\sin t = (\sin t + \sin 3t)/4$ so that

$$F = \sin t\left(-a + \frac{a^3}{4}\right) + \frac{a^3}{4}\sin 3t.$$

Thus renormalization fails for the van der Pol equation unless $a = 2$. We shall see in the next section that $a = 2$ corresponds to the limit cycle of the van der Pol oscillator. It is only this special solution which is periodic.

4.3 The method of multiple scales

The multiple scale technique will be introduced by considering the problem 4.2.2 of the previous section which we found could not be successfully dealt with by renormalization . The exact solution of

$$\frac{\mathrm{d}^2u}{\mathrm{d}t^2} + u = -\varepsilon\frac{\mathrm{d}u}{\mathrm{d}t}, \quad u(0) = 1, \quad \frac{\mathrm{d}u}{\mathrm{d}t}(0) = 0, \qquad 4.3.1$$

is

$$u_{\mathrm{ex}} = \mathrm{e}^{-\varepsilon t/2}\left(\cos\sqrt{1-(\varepsilon/2)^2}\,t + \frac{\varepsilon}{2\sqrt{1-(\varepsilon/2)^2}}\sin\sqrt{1-(\varepsilon/2)^2}\,t\right). \qquad 4.3.2$$

The straightforward two-term expansion was obtained in the previous section, it is

$$u_{2T} = \cos t - \frac{\varepsilon}{2}(t\cos t - \sin t). \qquad 4.3.3$$

This can be constructed from the exact solution by expanding the exponential, square root and trigonometric functions. Nonuniformities are generated in forming the expansions of the exponential and trigonometric functions. While it is uniformly true to state that

$$u = \cos t + O(\varepsilon) \quad \text{for } t = O(1),$$

it is not uniformly valid for $t = O(1/\varepsilon)$. If we are interested in values of t which are $O(1/\varepsilon)$ then the combination εt must be preserved in the exponential function. Then it is uniformly valid to state that

$$u = e^{-\varepsilon t/2} \cos t + O(\varepsilon) \quad \text{for } t = O(1/\varepsilon). \qquad 4.3.4$$

If we are interested in times which are $O(1/\varepsilon^2)$ then 4.3.4 is no longer valid. In this case terms of the form $\varepsilon^2 t$ must be preserved in the cosine function appearing in 4.3.2,

$$\cos\sqrt{1-(\varepsilon/2)^2}\,t = \cos\left[1 - \frac{1}{2}\left(\frac{\varepsilon}{2}\right)^2 + O(\varepsilon^4)\right]t = \cos\left(1 - \frac{\varepsilon^2}{8}\right)t + O(\varepsilon^4 t).$$

Thus

$$u = e^{-\varepsilon t/2} \cos\left(1 - \frac{\varepsilon^2}{8}\right)t + O(\varepsilon),$$

is a uniformly valid statement for $t = O(1/\varepsilon^3)$.

Notice that if we are concerned only with uniformly valid leading order expansions then the second member of the bracket in 4.3.2 never contributes since it is uniformly of $O(\varepsilon)$ for all t.

Time scales

The basic idea behind the method of multiple scales is to avoid the introduction of nonuniformities associated with expansions of functions by preserving the combinations εt, $\varepsilon^2 t$, $\varepsilon^3 t$, etc. as variables on which these functions depend. These combinations are called scales and denoted as follows:

$$T_0 = t, \quad T_1 = \varepsilon t, \quad T_2 = \varepsilon^2 t, \quad \text{etc.}$$

To demonstrate the idea consider the following function:

$$f = \frac{1}{1-\varepsilon} \exp(t + \varepsilon t + \varepsilon^2 t + \varepsilon^3 t + \varepsilon^4 t).$$

Using one scale T_0 we have

$$\begin{aligned} f &\sim \left(1 + \varepsilon + \frac{\varepsilon^2}{2} + \cdots\right).\exp T_0.[1 + \varepsilon T_0 + \varepsilon^2 T_0 + \cdots \\ &\quad + \tfrac{1}{2}(\varepsilon T_0 + \varepsilon^2 T_0 + \cdots)^2 + \cdots] \\ &\sim \exp T_0 + \varepsilon(1 + T_0)\exp T_0 + \varepsilon^2(\tfrac{1}{2} + T_0 + \tfrac{3}{2}T_0^2)\exp T_0. \end{aligned}$$

The region of nonuniformity is $T_0 = O(1/\varepsilon)$ i.e. $t = O(1/\varepsilon)$. Using two scales,

$$f = \frac{1}{1-\varepsilon}\exp(T_0 + T_1 + \varepsilon^2 T_0 + \varepsilon^3 T_0 + \varepsilon^4 T_0)$$

$$\sim \left(1 + \varepsilon + \frac{\varepsilon^2}{2} + \cdots\right).\exp(T_0 + T_1).(1 + \varepsilon^2 T_0 + \cdots)$$

$$\sim \exp(T_0 + T_1) + \varepsilon \exp(T_0 + T_1) + \varepsilon^2(\tfrac{1}{2} + T_0).\exp(T_0 + T_1) + \cdots$$

If $T_0 = O(1/\varepsilon)$ the second and third members of the above expansion are of the same order, namely $O(\varepsilon)$. There is no point in preserving the second member of the above expansion since other terms are of this order, but we may state that

$$f = \exp(T_0 + T_1)\cdot(1 + O(\varepsilon, \varepsilon^2 T_0)),$$

where this expression is uniformly valid for $t = O(1/\varepsilon)$.

Using three scales we may either state

$$f = \exp(T_0 + T_1 + T_2)\cdot(1 + \varepsilon + O(\varepsilon^2, \varepsilon^3 T_0)),$$

which is uniformly valid for $t = O(1/\varepsilon)$, or we may state

$$f = \exp(T_0 + T_1 + T_2)\cdot(1 + O(\varepsilon, \varepsilon^2, \varepsilon^3 T_0)),$$

which is uniformly valid for $t = O(1/\varepsilon^2)$.

Introducing further scales allows either more terms to be included in the expansion for $t = O(1/\varepsilon)$ or fewer terms may be used with an extended region of uniformity.

We will restrict our application of the multiple scale technique to the introduction of two scales, T_0 and T_1, and obtain one-term uniformly valid expansions for $t = O(1/\varepsilon)$. Using the scales $T_0 = t$ and $T_1 = \varepsilon t$, the time derivatives become

$$\frac{\mathrm{d}}{\mathrm{d}t} = \frac{\mathrm{d}T_0}{\mathrm{d}t}\frac{\partial}{\partial T_0} + \frac{\mathrm{d}T_1}{\mathrm{d}t}\frac{\partial}{\partial T_1} = \frac{\partial}{\partial T_0} + \varepsilon\frac{\partial}{\partial T_1} \qquad 4.3.5a$$

$$\frac{\mathrm{d}^2}{\mathrm{d}t^2} = \frac{\partial^2}{\partial T_0} + 2\varepsilon\frac{\partial^2}{\partial T_0 \partial T_1} + \varepsilon^2\frac{\partial^2}{\partial T_1^2}. \qquad 4.3.5b$$

Returning to the example provided by equation 4.3.1 which becomes

$$\frac{\partial^2 u}{\partial T_0^2} + 2\varepsilon\frac{\partial^2 u}{\partial T_0 \partial T_1} + \varepsilon^2\frac{\partial^2 u}{\partial T_1^2} + u = -\varepsilon\frac{\partial u}{\partial T_0} - \varepsilon^2\frac{\partial u}{\partial T_1},$$

we expand u in the form

$$u \equiv u(T_0, T_1; \varepsilon) \sim u_0(T_0, T_1) + \varepsilon u_1(T_0, T_1) + \cdots$$

and obtain the following order equations

$$O(1){:}\quad \frac{\partial^2 u_0}{\partial T_0^2} + U_0 = 0 \qquad 4.3.6$$

$$O(\varepsilon): \quad \frac{\partial^2 u_1}{\partial T_0^2} + u_1 = -2\frac{\partial^2 u_0}{\partial T_0 \partial T_1} - \frac{\partial u_0}{\partial T_0}. \qquad 4.3.7$$

It is convenient to express the solution of the harmonic oscillator equation, 4.3.6, in complex form

$$u_0 = A(T_1)\exp(\mathrm{i}T_0) + A^*(T_1)\exp(-\mathrm{i}T_0), \qquad 4.3.8$$

where the * symbol indicates the complex conjugate and the function A depends on T_1 only. The complementary function associated with equation 4.3.7 has the same form as 4.3.8. The particular integral will contain secular terms if the right-hand side of 4.3.7 contains $\exp(\pm \mathrm{i}T_0)$ terms. The condition that these should be absent determines the function $A(T_1)$ in equation 4.3.8.

The right-hand side of 4.3.7 is

$$-2\mathrm{i}\left(\frac{\mathrm{d}A}{\mathrm{d}T_1}\exp(\mathrm{i}T_0) - \frac{\mathrm{d}A^*}{\mathrm{d}T_1}\exp(-\mathrm{i}T_0)\right) - \mathrm{i}[A\exp(\mathrm{i}T_0) - A^*\exp(-\mathrm{i}T_0)]$$

$$= -\mathrm{i}\left(2\frac{\mathrm{d}A}{\mathrm{d}T_1} + A\right)\exp(\mathrm{i}T_0) + \mathrm{i}\left(2\frac{\mathrm{d}A^*}{\mathrm{d}T_1} + A^*\right)\exp(-\mathrm{i}T_0).$$

Thus to avoid secular terms we require

$$2\frac{\mathrm{d}A}{\mathrm{d}T_1} + A = 0, \qquad 4.3.9a$$

and

$$2\frac{\mathrm{d}A^*}{\mathrm{d}T_1} + A^* = 0. \qquad 4.3.9b$$

Imposing 4.3.9a automatically ensures that 4.3.9b is satisfied.

It is convenient to express the complex function, $A(T_1)$, in polar form $R(T_1)\exp[\mathrm{i}\theta(T_1)]$, where $R(T_1)$ and $\theta(T_1)$ are real functions. Then

$$\frac{\mathrm{d}A}{\mathrm{d}T_1} = \left(\frac{\mathrm{d}R}{\mathrm{d}T_1} + \mathrm{i}R\frac{\mathrm{d}\theta}{\mathrm{d}T_1}\right).\exp(\mathrm{i}\theta),$$

so that 4.3.9a becomes

$$\left(2\frac{\mathrm{d}R}{\mathrm{d}T_1} + 2\mathrm{i}R\frac{\mathrm{d}\theta}{\mathrm{d}T_1} + R\right).\exp(\mathrm{i}\theta) = 0.$$

The bracketed term in the above equation must be zero. Equating the real part to zero yields

$$2\frac{\mathrm{d}R}{\mathrm{d}T_1} + R = 0, \qquad 4.3.10$$

while the imaginary part yields

$$R\frac{d\theta}{dT_1} = 0,$$

with nontrivial solution $\theta = \theta_0$ (a constant).

The solution of 4.3.10 is $R = R_0\exp(-T_1/2)$, where R_0 is a positive constant. Then

$$A(T_1) = R_0\exp(-T_1/2)\exp(i\theta_0),$$

and substituting into 4.3.8 leads to

$$\begin{aligned} u_0 &= R_0\exp(-T_1/2)[\exp i(T_0 + \theta_0) + \exp - i(T_0 + \theta_0)] \\ &= 2R_0\exp(-T_1/2)\cos(T_0 + \theta_0). \end{aligned} \qquad 4.3.11$$

At this stage the initial conditions are imposed. In this case we have $u(0) = 1$, $(du/dt)(0) = 0$ which must be expressed in terms of the scales T_0 and T_1. When $t = 0$ both T_0 and T_1 are zero, so the initial conditions become

$$u(T_0 = 0, T_1 = 0) = 1 \sim u_0(0, 0) + \varepsilon u_1(0, 0) + \cdots$$

and

$$\frac{\partial u}{\partial T_0}(0, 0) + \varepsilon\frac{\partial u}{\partial T_1}(0, 0) = 0 \sim \frac{\partial u_0}{\partial T_0}(0, 0) + \varepsilon\left(\frac{\partial u_1}{\partial T_0}(0, 0) + \frac{\partial u_0}{\partial T_1}(0, 0)\right) + \cdots.$$

Thus the conditions satisfied by u_0 are $u_0(0, 0) = 1$ and $(\partial u_0/\partial T_0)(0, 0) = 0$.

Imposing these on the expression 4.3.11 yields $2R_0\cos\theta_0 = 1$ and $-2R_0\sin\theta_0 = 0$ thus $\theta_0 = 0$ and $R_0 = 1/2$ so that

$$u_0 = \exp(-T_1/2)\cos T_0.$$

Just as in the strained coordinate method, we do not evaluate u_1 but merely ensure that secular terms are absent so that we may write

$$u = \exp(-T_1/2)\cos T_0 + O(\varepsilon),$$

where this expression is uniformly valid for $t = O(1/\varepsilon)$. At this stage the original variable is used and we have

$$u = e^{-\varepsilon t/2}\cos t + O(\varepsilon).$$

This agrees with the expansion 4.3.4 of the exact solution.

Van der Pol oscillator

For our next example we consider the van der Pol oscillator

$$\frac{d^2u}{dt^2} + u = \varepsilon(1 - u^2)\frac{du}{dt}, \quad u(0) = a, \quad \frac{du}{dt}(0) = 0. \qquad 4.3.12$$

Using the scales $T_0 = t$ and $T_1 = \varepsilon t$, this becomes

$$\frac{\partial^2 u}{\partial T_0^2} + 2\varepsilon \frac{\partial^2 u}{\partial T_0 \partial T_1} + \varepsilon^2 \frac{\partial^2 u}{\partial T_1^2} + u = \varepsilon(1 - u^2)\left(\frac{\partial u}{\partial T_0} + \varepsilon \frac{\partial u}{\partial T_1}\right).$$

Substituting $u \sim u_0(T_0, T_1) + \varepsilon u_1(T_0, T_1) + \cdots$ leads to the following order equations

$$\text{O}(1): \quad \frac{\partial^2 u_0}{\partial T_0^2} + u_0 = 0$$

$$\text{O}(\varepsilon): \quad \frac{\partial^2 u_1}{\partial T_0^2} + u_1 = -2\frac{\partial^2 u_0}{\partial T_0 \partial T_1} + (1 - u^2)\frac{\partial u_0}{\partial T_0}.$$

The O(1) equation has solution

$$u_0 = A(T_1)\exp(\mathrm{i}T_0) + A^*(T_1)\exp(-\mathrm{i}T_0),$$

so the right-hand side of the $\text{O}(\varepsilon)$ equation becomes

$$-2\mathrm{i}\left(\frac{\mathrm{d}A}{\mathrm{d}T_1}\exp(\mathrm{i}T_0) - \frac{\mathrm{d}A^*}{\mathrm{d}T_1}\exp(-\mathrm{i}T_0)\right) + \mathrm{i}(1 - A^2\exp(2\mathrm{i}T_0) - 2AA^*$$
$$+ A^{*2}\exp(-2\mathrm{i}T_0))\cdot(A\exp(\mathrm{i}T_0) - A^*\exp(-\mathrm{i}T_0)).$$

The coefficient of $\exp(\mathrm{i}T_0)$ is

$$-2\mathrm{i}\frac{\mathrm{d}A}{\mathrm{d}T_1} + \mathrm{i}A - \mathrm{i}A^2A^*,$$

and the coefficient of $\exp(-\mathrm{i}T_0)$ is the complex conjugate of this. To avoid secular terms appearing in u_1, the coefficients of $\exp(\pm\mathrm{i}T_0)$ must be zero. Using the polar form $A = R\mathrm{e}^{\mathrm{i}\theta}$ we require

$$-2\left(\frac{\mathrm{d}R}{\mathrm{d}T_1} + \mathrm{i}R\frac{\mathrm{d}\theta}{\mathrm{d}T_1}\right) + R - R^3 = 0.$$

The real and imaginary parts of this equation lead to

$$-2\frac{\mathrm{d}R}{\mathrm{d}T_1} + R - R^3 = 0 \qquad \text{4.3.13a}$$

$$\frac{\mathrm{d}\theta}{\mathrm{d}T_1} = 0. \qquad \text{4.3.13b}$$

The solution of 4.3.13b is $\theta = \theta_0$ (a constant). Equation 4.3.13a is of separable form,

$$\frac{1}{R(R^2 - 1)}\frac{\mathrm{d}R}{\mathrm{d}T_1} = -\frac{1}{2}.$$

Using partial fractions leads to

$$\int\left(-\frac{1}{R}+\frac{1}{2}\frac{1}{R-1}+\frac{1}{2}\frac{1}{R+1}\right)\mathrm{d}R=\int-\frac{1}{2}\,\mathrm{d}T_1,$$

and on integrating we obtain

$$-\ln|R|+\frac{1}{2}\ln|R-1|+\frac{1}{2}\ln|R+1|=-\frac{T_1}{2}+C$$

$$\ln\sqrt{\left|\frac{R^2-1}{R^2}\right|}=-\frac{T_1}{2}+C.$$

Thus

$$\frac{R^2-1}{R^2}=K\exp(-T_1),$$

where K is an arbitrary constant. Solving for R yields

$$R=\frac{1}{\sqrt{1-K\exp(-T_1)}}.$$

So we have

$$u_0=R\exp(\mathrm{i}\theta_0)\exp(\mathrm{i}T_0)+R\exp(-\mathrm{i}\theta_0)\exp(-\mathrm{i}T_0)=2R\cos(\theta_0+T_0).$$

The initial conditions require $u_0(0,0)=a$ and $(\partial u_0/\partial T_0)(0,0)=0$, thus

$$2R(0)\cos\theta_0=a \quad\text{and}\quad -2R(0)\sin\theta_0=0,$$

with solution $\theta_0=0$ and $R(0)=a/2$.
Then since $R=1/\sqrt{1-K\exp(-T_1)}$ we have $a/2=1/\sqrt{1-K}$, i.e. $K=1-4/a^2$ and finally

$$u_0=\frac{2}{\sqrt{1+\left(\dfrac{4}{a^2}-1\right)\exp(-T_1)}}\cos T_0,$$

with $u_1=\mathrm{O}(1)$ for all t.

Thus the uniformly valid one-term expansion of the solution of the van der Pol equation is

$$u=\frac{2}{\sqrt{1+\left(\dfrac{4}{a^2}-1\right)\mathrm{e}^{-\varepsilon t}}}\cos t+\mathrm{O}(\varepsilon). \qquad 4.3.14$$

This shows that as $t\to\infty$ the expansion tends to the limit cycle $u=2\cos t+\mathrm{O}(\varepsilon)$ for all initial values. Furthermore, it is only for the particular initial value $a=2$ that the solution is periodic.

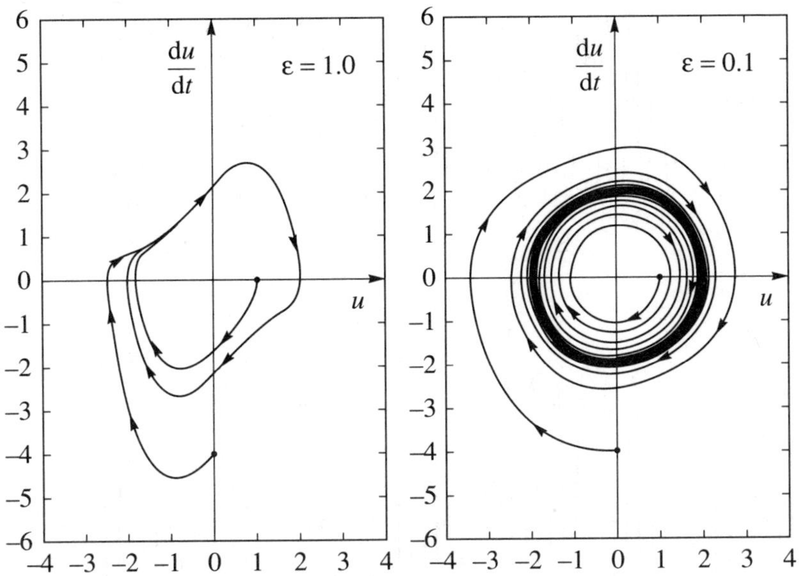

Fig. 4.4 Solutions of the van der Pol equation

Some solution trajectories for the van der Pol equation are shown in Fig. 4.4. These have been obtained using the improved Euler numerical scheme.

In Fig. 4.5 the variation with time of the numerical solution is shown for the case $\varepsilon = 0.1$ and initial conditions $u(0) = 1$, $(\mathrm{d}u/\mathrm{d}t)(0) = 0$. The straightforward Poincaré one-term expansion $u_p = \cos t$ and the multiple scale solution u_{ms} given by equation 4.3.14 are also shown in the figure. The straightforward expansion fails to follow the change in amplitude of the oscillation and is invalid after the first few oscillations. The multiple scale solution closely approximates the full solution over the time interval shown in the figure.

In Fig. 4.6 the corresponding expansions for Duffing's equation with $\varepsilon = 0.1$, $u(0) = 1$, $(\mathrm{d}u/\mathrm{d}t)(0) = 0$ are shown, namely

$$u_p = \cos t$$

$$u_{ms} = \cos\left(1 + \frac{3\varepsilon}{8}\right)t,$$

along with the numerically generated full solution.

The straightforward expansion deviates from the full solution (numerical) after the first few oscillations because it has the wrong frequency. The positions of every tenth maximum are marked and it can be seen how closely the multiple scale approximation follows the full solution.

The numerically generated solutions appear to render the asymptotic expansions redundant. In response to this the following points can be made.

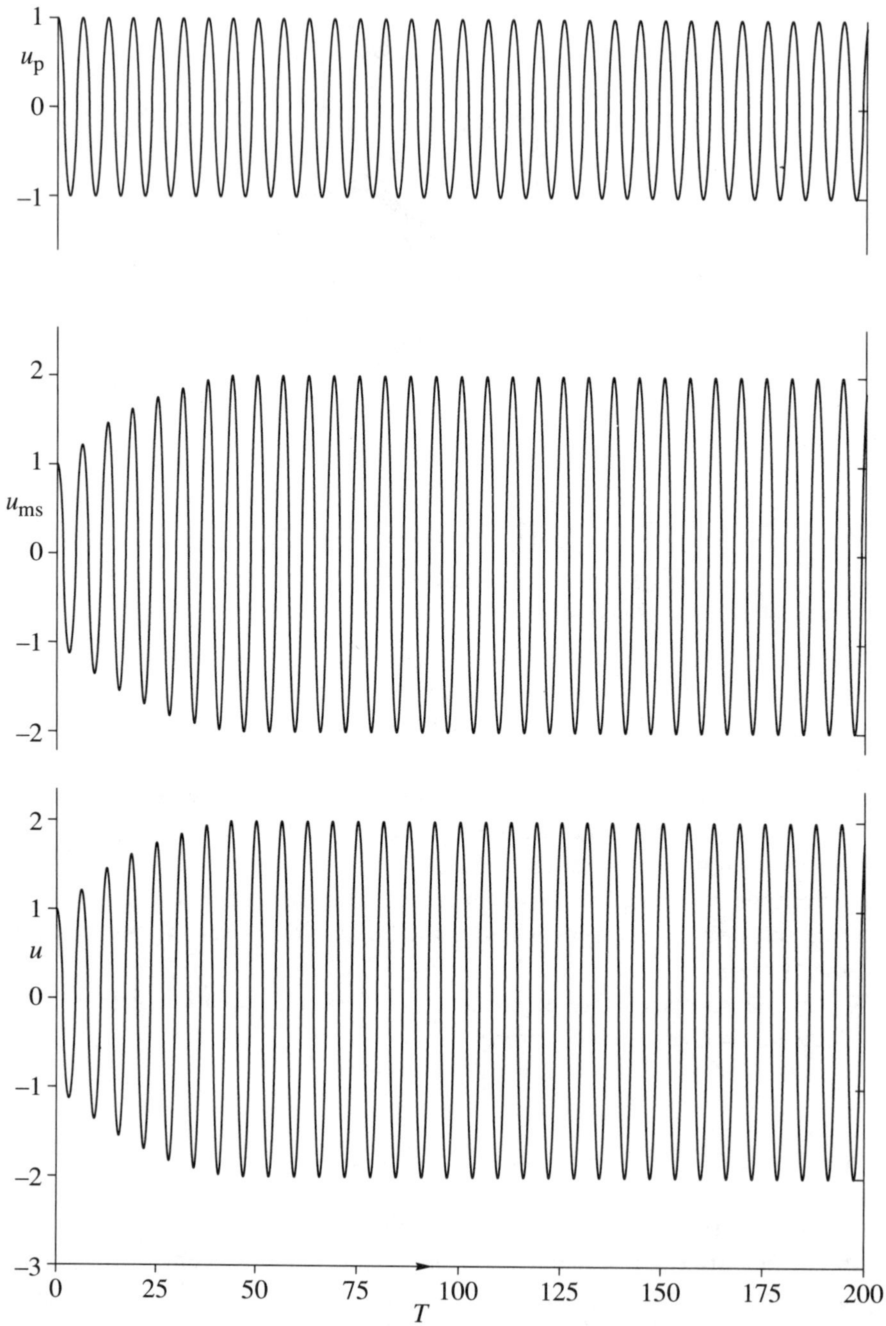

Fig. 4.5 Comparison of the numerical solution, u, the straightforward expansion, u_p, and the multiple scale expansion, u_{ms}, of the van der Pol equation with $\varepsilon = 0.1$.

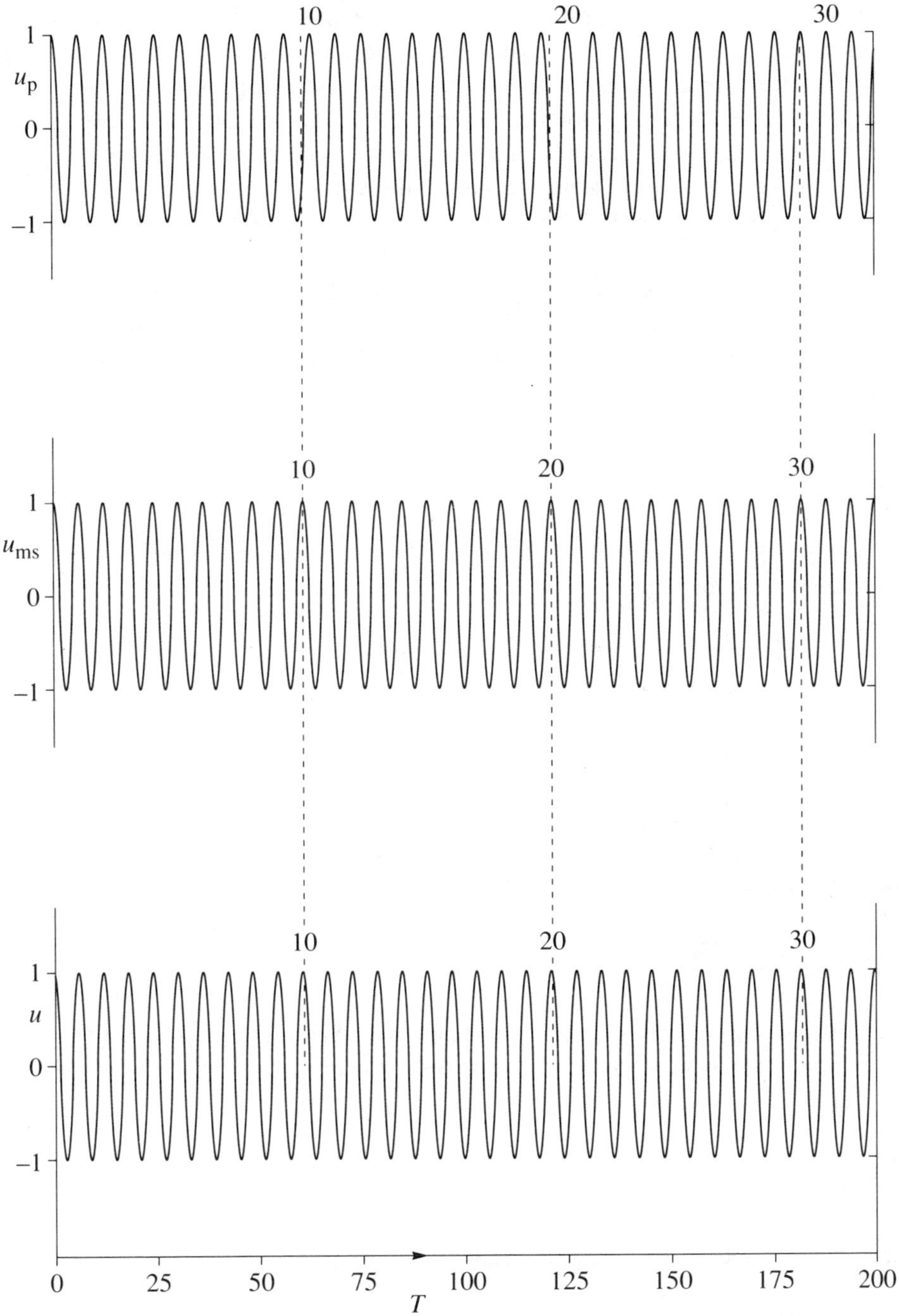

Fig. 4.6 Comparison of the numerical solution, u, the straightforward expansion, u_p, and the multiple scale expansion, u_{ms}, of Duffing's equation with $\varepsilon = 0.1$.

The asymptotic expansions provide an analytical expression for the dependence of the solution on the parameter and initial condition values. In contrast, separate numerical solutions must be generated for each value of the parameter and initial conditions. This limitation is particularly severe for multidimensional problems.

The numerical solution schemes have a limited accuracy. In the case of the improved Euler scheme used to generate the solutions shown in this chapter, the error at each time step is of order DT^3. Thus the numerical solution breaks down after N steps where $N = O(1/DT^3)$ as $DT \to 0$. If the perturbation parameter ε is very small its effect is not fully seen until a very large time interval has elapsed. This involves an accumulation of errors in the numerical scheme. The problem can be postponed by the use of a scheme with higher accuracy at each time step but there remains this ultimate limitation for large time intervals.

Worked example

Use the multiple scale technique with scales $T_0 = t$, $T_1 = \varepsilon t$ to obtain a uniformly valid one-term expansion of the solution of

$$\frac{d^2u}{dt^2} + 9u = \varepsilon\left((1-u^3)\frac{du}{dt} + u^3\right), \quad u(0) = 1, \quad \frac{du}{dt}(0) = 0.$$

Solution

The multiple scale procedure leads to the following order equations,

$$\frac{\partial^2 u_0}{\partial T_0^2} + 9u_0 = 0$$

$$\frac{\partial^2 u_1}{\partial T_0^2} + 9u_1 = -2\frac{\partial^2 u_0}{\partial T_1 \partial T_0} + (1-u_0^3)\frac{\partial u_0}{\partial T_0} + u_0^3.$$

The solution for u_0 is $A(T_1)\exp(3iT_0) + A^*(T_1)\exp(-3iT_0)$. Substituting this function into the right-hand side of the equation for u_1 yields, after some manipulation, the following coefficient of $\exp(3iT_0)$,

$$-6i\frac{dA}{dT_1} + 3iA + 3A^2A^*.$$

This expression must equal zero in order to avoid secular terms appearing in the function u_1. Letting $A = Re^{i\theta}$ we obtain the following equations,

$$6R\frac{d\theta}{dT_1} + 3R^3 = 0$$

$$-6\frac{dR}{dT_1} + 3R = 0.$$

The solution of the second of these equations is

$$R = K_1 \exp(T_1/2).$$

The first equation becomes

$$\frac{d\theta}{dT_1} = -\frac{K_1^2 \exp(T_1)}{2},$$

with solution

$$\theta = -\frac{K_1^2 \exp(T_1)}{2} + K_2.$$

Thus

$$u_0 = 2R\cos(3T_0 + \theta) = 2K_1 \exp(T_1/2)\cos(3T_0 - K_1^2 \exp(T_1)/2 + K_2).$$

The initial conditions are

$$u_0(0, 0) = 1 \qquad \text{and} \qquad \frac{\partial u_0}{\partial T_0}(0, 0) = 0,$$

i.e.

$$2K_1 \cos(-K_1^2/2 + K_2) = 1$$

$$-6K_1 \sin(-K_1^2/2 + K_2) = 0,$$

which yield $K_2 = K_1^2/2$ and $K_1 = 1/2$.

Thus the uniformly valid one-term expansion of the solution is

$$u = e^{\varepsilon t/2} \cos(3t - e^{\varepsilon t}/8 + 1/8) + O(\varepsilon).$$

Exercises

1 Use the multiple scale technique with scales $T_0 = t$, $T_1 = \varepsilon t$ to show that the uniformly valid one-term expansion for the solution of

$$\frac{d^2 u}{dt^2} + u = -\varepsilon\left[\left(\frac{du}{dt}\right)^3 + u\right], \quad u(0) = 1, \quad \frac{du}{dt}(0) = 0,$$

is

$$u = \frac{2}{\sqrt{4 + 3\varepsilon t}} \cos(t + \varepsilon t/2).$$

Write a computer program to solve the equation and compare the two solutions.

2 Use the multiple scale technique to obtain one-term uniformly valid expansions for the solutions of

(i) $$\frac{d^2 u}{dt^2} + u = -\varepsilon\left(\frac{du}{dt}\right)^3, \quad u(0) = a, \quad \frac{du}{dt}(0) = 0.$$

(ii) $$\frac{d^2u}{dt^2} + u = \varepsilon u\frac{du}{dt}\left(u + \frac{du}{dt}\right), \quad u(0) = 4, \quad \frac{du}{dt}(0) = 0.$$

(iii) $$\frac{d^2u}{dt^2} + u = \varepsilon\left[u^3 + 3\frac{du}{dt} - \left(\frac{du}{dt}\right)^3\right], \quad u(0) = 1, \quad \frac{du}{dt}(0) = 0.$$

(iv) $$\frac{d^2u}{dt^2} + u = \varepsilon\left(u^2\frac{du}{dt} + u\right), \quad u(0) = 0, \quad \frac{du}{dt}(0) = 1.$$

(v) $$\frac{d^2u}{dt^2} + u = \varepsilon\left(u^3 - u^2\frac{du}{dt} + 4\frac{du}{dt}\right), \quad u(0) = 2, \quad \frac{du}{dt}(0) = 0.$$

4.4 Surface roughness effects in lubricated bearings

In this section an application of the multiple scale technique to lubricated rough bearings is presented.

The surfaces of bearings are kept apart by a thin film of lubricating fluid. Commonly the lubricant is oil although sometimes water, air or other fluids are used. The film of lubricant is maintained by the motion of the bearing surfaces which, by means of viscous forces, drag the fluid into the bearing gap. This gap is a converging region such as that shown in Fig. 4.7 for roller and slider bearings. The process of dragging the lubricant into a converging gap generates a pressure in the lubricant which exceeds the surrounding pressure and thus enables the bearings to support loads without solid to solid contact occurring.

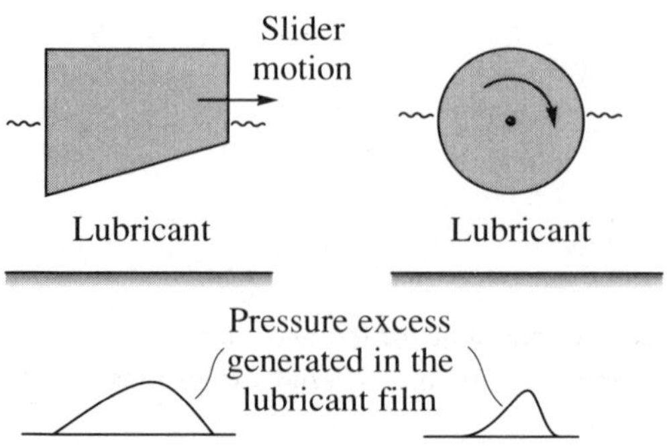

Fig. 4.7 Slider and roller bearings

Reynolds equation

For an incompressible lubricant the pressure, P, is related to the film thickness, H, by an equation derived by Osborne Reynolds,

$$\frac{\partial}{\partial X}\left(H^3\frac{\partial P}{\partial X}\right) = 6\mu(U_1 + U_2)\frac{\partial H}{\partial X} + 12\mu\frac{\partial H}{\partial T}. \tag{4.4.1}$$

Here μ is the lubricant viscosity, U_1 and U_2 are the bearing speeds in the X direction and T is time. A derivation of Reynolds' equation can be found in texts on viscous fluid flow. It is obtained from the basic fluid flow equations of Navier–Stokes for momentum and the continuity equation (see Section 5.6). The 'lubrication approximation' which is almost always valid is that the film thickness is thin compared with the bearing length. Thus slopes are small – those shown in the slider bearing figure are exaggerated and the bearing region for the roller is much smaller than indicated by the pressure curve so that in both cases the bearing regions are nearly parallel.

Surface roughness

Often the lubricant films are so thin that surface roughness effects must be considered. The average film thickness $\bar{H}$ along with the surface roughness heights H_1 and H_2 are shown in Fig. 4.8.

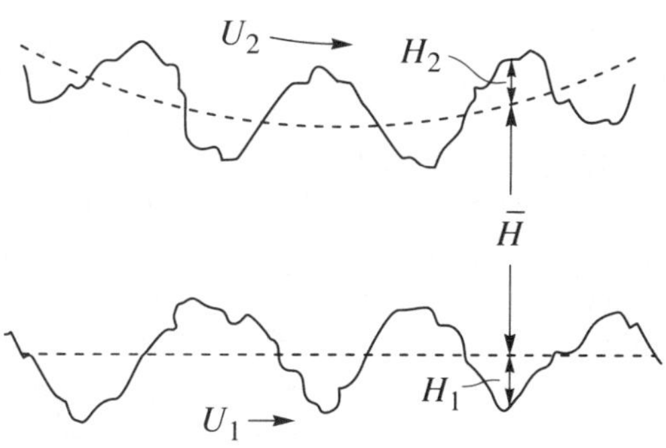

Fig. 4.8 Surface roughness in lubricated bearings

We will confine our study to bearings with a mean film thickness which is independent of time. This is the case if the coordinate reference frame is that of the slider or the axis of the roller. Then for two-dimensional 'infinitely wide' bearings where there is no film thickness variation or lubricant flow in the third direction the average film thickness is a function of X only.

The surface roughness profile is carried through the bearing region at the bearing speed and therefore varies with both X and T. However, there is a simple relation between the time and space derivatives since, if a roughness profile at $T = 0$ is given by $H_1 = F(X)$, then at any later time, T, the roughness is given by $H_1 = F(X - U_1 T)$ so that

$$\frac{\partial H_1}{\partial T} = -U_1 \frac{\partial H_1}{\partial X},$$

and similarly

$$\frac{\partial H_2}{\partial T} = -U_2 \frac{\partial H_2}{\partial X}.$$

Thus we need only consider spacial derivatives and at any instant of time Reynolds' equation (4.4.1) is

$$\frac{\mathrm{d}}{\mathrm{d}X}\left(H^3 \frac{\mathrm{d}P}{\mathrm{d}X}\right) = 6\mu(U_1 + U_2)\frac{\mathrm{d}\bar{H}}{\mathrm{d}X} + 6\mu(U_1 - U_2)\left(\frac{\mathrm{d}H_2}{\mathrm{d}X} - \frac{\mathrm{d}H_1}{\mathrm{d}X}\right), \quad 4.4.2$$

where $H = \bar{H} + H_1 + H_2$. The boundary conditions for the bearing excess pressure are $P = 0$ at $X = 0$ and $X = L$ where L is the bearing length.

On introducing the nondimensionalized variables

$$x = X/L, \quad \bar{h} = \bar{H}/H_0, \quad h_{1,2} = H_{1,2}/H_0, \quad p = PH_0^2/6\mu L(U_1 + U_2),$$

where H_0 is a characteristic value of the film thickness, Reynolds' equation becomes

$$\frac{\mathrm{d}}{\mathrm{d}x}\left(h^3 \frac{\mathrm{d}p}{\mathrm{d}x}\right) = \frac{\mathrm{d}\bar{h}}{\mathrm{d}x} + \frac{U_1 - U_2}{U_1 + U_2}\frac{\mathrm{d}}{\mathrm{d}x}(h_2 - h_1), \quad 4.4.3$$

with $h = \bar{h} + h_1 + h_2$. The boundary conditions are $p(0) = p(1) = 0$.

If the roughness is negligible then $h \simeq \bar{h}$ and the second member of the right-hand side of 4.4.3 is absent. In this case Reynolds' equation may be solved exactly for some simple film thickness functions, $\bar{h}(x)$, or if necessary a numerical discretization can be used to approximate the equation and a numerical solution scheme used to solve for the discretized pressures. The film shape $\bar{h}(x)$ varies smoothly in the bearing interval so that ten or twenty grid points are sufficient to adequately represent the pressure distribution.

For thin films when surface roughness cannot be neglected many more grid points are required. The length scale of the roughness fluctuations is typically one hundred times less than the bearing length scale itself, so that if ten discretization points are used for each fluctuation a suitable computational grid requires of the order of one thousand points. For a three-dimensional bearing the corresponding requirement is for a surface grid of one million points.

A solution of Reynolds' equation with fluctuating surface heights yields the corresponding fluctuating pressure distribution. In many applications the average pressure distribution through the bearing is all that is required. The nonlinear dependence of the pressure on the film thickness means, of course, that the average pressure is *not* determined by using the average film thickness in Reynolds' equation.

A simple demonstration of this essential point is provided by considering the function $f = 1/h^3$. Suppose we take three heights $h = 1, 2$ and 3. The average value of h, namely $\bar{h}$, is 2 and the average value of f is given by

$$\bar{f} = \tfrac{1}{3}(\tfrac{1}{1} + \tfrac{1}{8} + \tfrac{1}{27}).$$

Obviously the dominant contribution to $\bar{f}$ comes from the smallest value of h and clearly $\bar{f}$ is much greater than an erroneous estimate obtained from the value of $1/\bar{h}^3$ (which equals 1/8). Similarly it is invalid to attempt to obtain the average pressure from equation 4.4.3 by replacing the random height by its average value.

Average Reynolds equation

Our aim in this study is to derive an average Reynolds equation which governs the variation of the average pressure $\bar{p}(x)$. This can be solved on a relatively coarse grid of ten or twenty points rather than on the fine grid which is required if the random fluctuations in pressure are to be reproduced.

The following derivation of the average Reynolds equation is based on the work of Elrod[13] and his colleagues. They were the first to take full advantage of the multiple scale technique in dealing with the problem of surface roughness in lubrication.

Let λ be a characteristic roughness length and denote the small nondimensional ratio λ/L by ε. There are two length scales x_0 and x_1 where

$x_0 = x$ – the smooth or 'slow' length scale,

$x_1 = x/\varepsilon$ – the roughness or 'fast' length scale.

Average quantities will be independent of the roughness scale, thus

$$h = \bar{h}(x_0) + h_1(x_1) + h_2(x_1),$$

where the surface roughness h_1 and h_2 depend only on the roughness scale x_1. Thus, for example, a sinusoidal roughness with amplitude A and wavelength λ is

$$H_1 = A\sin(2\pi X/\lambda),$$

so that

$$h_1 = \frac{A}{H_0}\sin(2\pi x/\varepsilon) = \frac{A}{H_0}\sin(2\pi x_1),$$

which shows the x_1 dependence of $h_{1,2}$.

The random component of pressure, p_r, will be a function of the roughness variable x_1 and the smooth variable x_0. This is because the fluctuations in pressure are more severe when the average film thickness is at its minimum and the corresponding roughness influences are at their maximum. Thus the nature of the pressure fluctuations varies on the smooth length scale as well as the roughness length scale so that

$$p(x;\varepsilon) = \bar{p}(x_0;\varepsilon) + p_r(x_0, x_1;\varepsilon).$$

The spatial derivative expressed in terms of the scales x_0 and x_1 is

$$\frac{\mathrm{d}}{\mathrm{d}x} = \frac{\mathrm{d}x_0}{\mathrm{d}x}\frac{\partial}{\partial x_0} + \frac{\mathrm{d}x_1}{\mathrm{d}x}\frac{\partial}{\partial x_1} = \frac{\partial}{\partial x_0} + \frac{1}{\varepsilon}\frac{\partial}{\partial x_1},$$

and equation 4.4.3 becomes

$$\frac{1}{\varepsilon^2}\frac{\partial}{\partial x_1}\left(h^3\frac{\partial p}{\partial x_1}\right) + \frac{1}{\varepsilon}\left[\frac{\partial}{\partial x_1}\left(h^3\frac{\partial p}{\partial x_0}\right) + \frac{\partial}{\partial x_0}\left(h^3\frac{\partial p}{\partial x_1}\right)\right] + \frac{\partial}{\partial x_0}\left(h^3\frac{\partial p}{\partial x_0}\right)$$
$$= \frac{\partial \bar{h}}{\partial x_0} + \frac{1}{\varepsilon}\left(\frac{U_1 - U_2}{U_1 + U_2}\right)\cdot\frac{\partial}{\partial x_1}(h_2 - h_1). \qquad 4.4.4$$

We will assume that $\bar{p}$ and p_r possess asymptotic expansions involving the sequence $\{1, \varepsilon, \varepsilon^2, \ldots\}$ i.e.

$$\bar{p}(x_0;\varepsilon) \sim \bar{p}_0(x_0) + \varepsilon\bar{p}_1(x_0) + \varepsilon^2\bar{p}_2(x_0) + \cdots \qquad 4.4.5$$

$$p_r(x_0, x_1;\varepsilon) \sim p_{r_0}(x_0, x_1) + \varepsilon p_{r_1}(x_0, x_1) + \varepsilon^2 p_{r_2}(x_0, x_1) + \cdots \qquad 4.4.6$$

In the subsequent analysis average values of expressions will be denoted by angled brackets $\langle\ \rangle$ when the overbar is ambiguous. Thus for example the average value of $1/h^3$ is denoted by $\langle 1/h^3\rangle$.

We will use the fact that average values vary with the smooth scale x_0 only so that

$$\frac{\partial}{\partial x_1}\langle(\ )\rangle = 0.$$

The processes of differentiation and averaging may be interchanged since they are both linear operations so that

$$\left\langle\frac{\partial}{\partial x_1}(\ )\right\rangle = 0.$$

Substituting the expansions for p_r and $\bar{p}$ into 4.4.4 leads to the following order equations,

$$O(1/\varepsilon^2):\quad \frac{\partial}{\partial x_1}\left(h^3\frac{\partial p_{r0}}{\partial x_1}\right) = 0 \qquad 4.4.7$$

$$O(1/\varepsilon):\quad \frac{\partial}{\partial x_1}\left(h^3\frac{\partial p_{r_1}}{\partial x_1} + h^3\frac{\partial \bar{p}_0}{\partial x_0} + h^3\frac{\partial p_{r0}}{\partial x_0}\right) + \frac{\partial}{\partial x_0}\left(h^3\frac{\partial p_{r0}}{\partial x_1}\right)$$
$$= \frac{U_1 - U_2}{U_1 + U_2}\frac{\partial}{\partial x_1}(h_2 - h_1) \qquad 4.4.8$$

$$\text{O}(1): \quad \frac{\partial}{\partial x_1}\left(h^3\frac{\partial p_{r_2}}{\partial x_1} + h^3\frac{\partial \bar{p}_1}{\partial x_0} + h^3\frac{\partial p_{r_1}}{\partial x_0}\right) +$$

$$\frac{\partial}{\partial x_0}\left(h^3\frac{\partial p_{r_1}}{\partial x_1} + h^3\frac{\partial \bar{p}_0}{\partial x_0} + h^3\frac{\partial p_{r_0}}{\partial x_0}\right) = \frac{\partial \bar{h}}{\partial x_0}. \qquad 4.4.9$$

Our aim is to obtain an equation for the leading order average pressure distribution $\bar{p}_0$ which only involves functions of x_0. The above three equations are sufficient to achieve this result and we need not consider the lower order equations (ε, ε^2, etc).

From equation 4.4.7 we are able to prove that p_{r_0} is zero as follows. Integrating the equation yields

$$h^3\frac{\partial p_{r_0}}{\partial x_1} = f_0(x_0),$$

where f_0 is an arbitrary function. Thus

$$\frac{\partial p_{r_0}}{\partial x_1} = \frac{1}{h^3}f_0(x_0),$$

and on taking the average value of each side of this equation we have

$$\left\langle\frac{\partial p_{r_0}}{\partial x_1}\right\rangle = \left\langle\frac{1}{h^3}f_0(x_0)\right\rangle = \left\langle\frac{1}{h^3}\right\rangle f_0(x_0).$$

The second equality is obtained by noting that $f_0(x_0)$ is nonrandom since it varies on the smooth length scale only and may therefore be taken out of the expectation. The left-hand side of the above expression is the average value of an x_1 derivative which we know is zero. Thus since $\langle 1/h^3\rangle$ is nonzero we must have $f_0(x_0) = 0$. It follows that $\partial p_{r_0}/\partial x_1$ is zero so that p_{r_0} is a function of the smooth scale,

$$p_{r_0} = f_1(x_0).$$

Then on taking averages,

$$\langle p_{r_0}\rangle = \langle f_1(x_0)\rangle = f_1(x_0),$$

where the second equality results from $f_1(x_0)$ already being a nonrandom function. The random component of the pressure has zero mean and we conclude that $f_1(x_0)$ is zero and therefore p_{r_0} is zero.

This shows that the first term of the expansion 4.4.6 is absent, i.e. p_r is $\text{O}(\varepsilon)$. This does not imply that roughness is an $\text{O}(\varepsilon)$ effect, it is in fact an $\text{O}(1)$ effect. This results from the fact that although $p_r = \text{O}(\varepsilon)$, its derivative is of $\text{O}(1)$, i.e.

$$\frac{\mathrm{d}p_r}{\mathrm{d}x} = \frac{\partial p_r}{\partial x_0} + \frac{1}{\varepsilon}\frac{\partial p_r}{\partial x_1} = \frac{\partial}{\partial x_0}(\varepsilon p_{r_1} + \cdots) + \frac{1}{\varepsilon}\frac{\partial}{\partial x_1}(\varepsilon p_{r_1} + \cdots) = \frac{\partial p_{r_1}}{\partial x_1} + \text{O}(\varepsilon).$$

On removing the terms involving p_{r_0} from 4.4.8 we obtain the following

$$\frac{\partial}{\partial x_1}\left(h^3 \frac{\partial p_{r_1}}{\partial x_1} + h^3 \frac{\partial \bar{p}_0}{\partial x_0}\right) = \frac{U_1 - U_2}{U_1 + U_2} \frac{\partial}{\partial x_1}(h_2 - h_1).$$

Integrating with respect to x_1 yields

$$h^3\left(\frac{\partial p_{r_1}}{\partial x_1} + \frac{\partial \bar{p}_0}{\partial x_0}\right) = \frac{U_1 - U_2}{U_1 + U_2}(h_2 - h_1) + f_2(x_0), \qquad 4.4.10$$

where f_2 is an arbitrary function. To determine f_2 we divide 4.4.10 by h^3 and take average values to obtain

$$f_2(x_0) = \left(\frac{\partial \bar{p}_0}{\partial x_0} + \frac{U_1 - U_2}{U_1 + U_2}\left\langle \frac{h_1 - h_2}{h^3}\right\rangle\right) \Big/ \left\langle \frac{1}{h^3}\right\rangle. \qquad 4.4.11$$

Now consider equation 4.4.9 noting that p_{r_0} is zero. On taking averages the x_1 derivative is removed since $\left\langle \frac{\partial}{\partial x_1}(\ )\right\rangle = 0$ and we obtain

$$\frac{\partial}{\partial x_0}\left\{\left\langle h^3 \frac{\partial p_{r_1}}{\partial x_1}\right\rangle + \langle h^3 \rangle \frac{\partial \bar{p}_0}{\partial x_0}\right\} = \frac{\partial \bar{h}}{\partial x_0}. \qquad 4.4.12$$

The term in curly brackets in equation 4.4.12 is the average value of the left-hand side of equation 4.4.10. This is equal to $f_2(x_0)$ since the expectation of the first member of the right-hand side of 4.4.10 is zero ($\langle h_1 \rangle = \langle h_2 \rangle = 0$). Then on using the expression 4.4.11 for $f_2(x_0)$ we obtain the following

$$\frac{\partial}{\partial x_0}\left(\frac{1}{\left\langle \frac{1}{h^3}\right\rangle}\frac{\partial \bar{p}_0}{\partial x_0} + \frac{U_1 - U_2}{U_1 + U_2}\frac{\left\langle \frac{h_1 - h_2}{h^3}\right\rangle}{\left\langle \frac{1}{h^3}\right\rangle}\right) = \frac{\partial \bar{h}}{\partial x_0}.$$

All the terms in this equation are averages which vary on the smooth scale x_0 only. At this stage we omit the subscript on x to obtain the *average Reynolds equation* for the leading term in the expansion of the average pressure (which we simply denote by $\bar{p}$),

$$\frac{\mathrm{d}}{\mathrm{d}x}\left(\frac{1}{\left\langle \frac{1}{h^3}\right\rangle}\frac{\mathrm{d}\bar{p}}{\mathrm{d}x}\right) = \frac{\mathrm{d}\bar{h}}{\mathrm{d}x} + \frac{U_1 - U_2}{U_1 + U_2}\frac{\mathrm{d}}{\mathrm{d}x}\left(\frac{\left\langle \frac{h_2 - h_1}{h^3}\right\rangle}{\left\langle \frac{1}{h^3}\right\rangle}\right). \qquad 4.4.13$$

The achievement represented by this governing equation for the average pressure is that the surface roughness functions $\langle h_{1,2}/h^3 \rangle$ and $\langle 1/h^3 \rangle$ can be obtained by taking averages with respect to the roughness length scale yielding functions of the smooth variable. The average pressure can then be obtained from 4.4.13 using a coarse computational grid appropriate to the smooth length scale.

Effect of roughness on the average pressure distribution

As an example of evaluating the surface roughness functions consider the case of a slider bearing with average film thickness $\bar{h} = 1/(1 + x_0)$ and surface roughness $h_1 = a \sin 2\pi x_1$ on the lower surface (see Fig. 4.9).

The averages are taken over a period $0 < x_1 < 1$ on the fast scale. The variation of x_0 is negligible over such a distance so that x_0 may be taken as a constant, then

$$\left\langle \frac{1}{h^3} \right\rangle = \int_{x_1=0}^{1} \frac{1}{[1/(1 + x_0) + a \sin 2\pi x_1]^3} \, dx_1 .$$

Equation 4.4.13 allows the influence of various surface finishes and surface speeds on the average pressure to be determined. For example, suppose both surfaces have the same roughness distribution. Then we have

$$\left\langle \frac{h_1}{h^3} \right\rangle = \left\langle \frac{h_2}{h^3} \right\rangle ,$$

so that the last member of 4.4.13 is zero. The influence of the roughness in the Reynolds equation now only appears in the coefficient $1/\langle 1/h^3 \rangle$. The function $1/h^3$ has a dominant contribution from small values of $h(= \bar{h} + h_1 + h_2)$, i.e. negative values of $h_1 + h_2$ have a greater influence than positive values. Thus $\langle 1/h^3 \rangle$ is greater in value than $1/\bar{h}^3$. Therefore $1/\langle 1/h^3 \rangle$ is less than $\bar{h}^3$. Thus the coefficient of $d\bar{p}/dx$ is less for a rough bearing than for a smooth bearing with film thickness function $\bar{h}$.

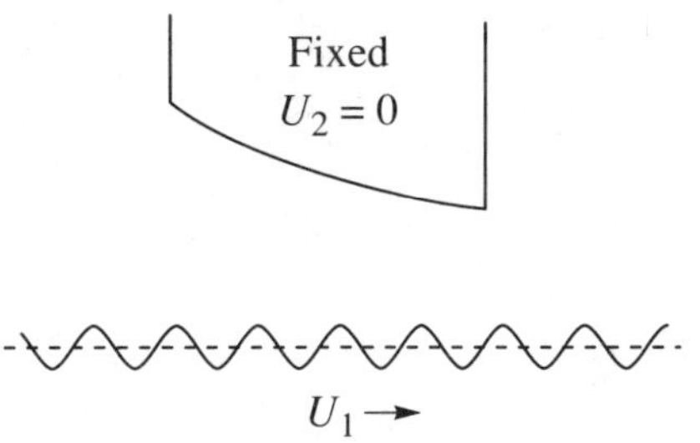

Fig. 4.9 A rough slider bearing

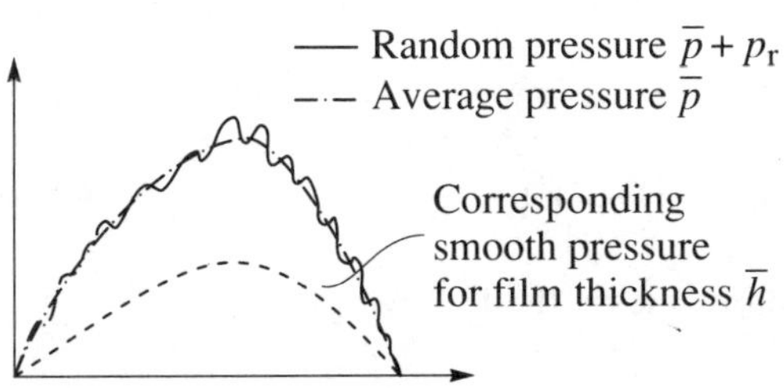

Fig. 4.10 Excess bearing pressures

Consequently for the same source term $d\bar{h}/dx$ on the right-hand side, the average pressure generated in a rough bearing is greater than that associated with a corresponding smooth bearing. This is represented in Fig. 4.10.

4.5 The method of averaging: the Krylov–Bogoliubov technique

The method of averaging provides an alternative to the method of multiple scales for dealing with certain classes of problems involving variations on different scales. We will consider the Krylov–Bogoliubov[14] method of averaging applied to weakly nonlinear oscillators,

$$\frac{d^2u}{dt^2} + \omega_0^2 u = \varepsilon F\left(u, \frac{du}{dt}\right). \tag{4.5.1}$$

The solution of the unperturbed oscillator equation is

$$u = a\cos(\omega_0 t + \theta), \tag{4.5.2}$$

where a and θ are constants whose values are determined by the initial conditions. The Krylov–Bogoliubov technique assumes that the solution of the perturbed equation maintains the form of 4.5.2 where a and θ are now functions of t. Since two functions of time have been introduced, any convenient constraint may be used in addition to the governing equation. The standard constraint is to impose the following,

$$\frac{du}{dt} = -a\,\omega_0 \sin(\omega_0 t + \theta).$$

Then since

$$\frac{du}{dt} = \frac{da}{dt}\cos(\omega_0 t + \theta) - a\sin(\omega_0 t + \theta)\cdot\left(\omega_0 + \frac{d\theta}{dt}\right),$$

the constraint becomes

$$\frac{da}{dt}\cos(\omega_0 t + \theta) - a\frac{d\theta}{dt}\sin(\omega_0 t + \theta) = 0. \tag{4.5.3}$$

The governing equation, 4.5.1, leads to

$$-\frac{da}{dt}\omega_0\sin(\omega_0 t + \theta) - a\,\omega_0\frac{d\theta}{dt}\cos(\omega_0 t + \theta) = \varepsilon F\left(u, \frac{du}{dt}\right). \tag{4.5.4}$$

Solving 4.5.3 and 4.5.4 for da/dt and $d\theta/dt$ yields the following:

$$\frac{da}{dt} = -\frac{\varepsilon}{\omega_0}\sin(\omega_0 t + \theta).F\left(u, \frac{du}{dt}\right) \tag{4.5.5}$$

$$\frac{d\theta}{dt} = -\frac{\varepsilon}{a\omega_0}\cos(\omega_0 t + \theta) . F\left(u, \frac{du}{dt}\right). \qquad 4.5.6$$

These equations show that both $a(t)$ and $\theta(t)$ are slowly varying functions

$$\left(\frac{da}{dt} \text{ and } \frac{d\theta}{dt} \text{ are } O(\varepsilon)\right).$$

Straightforward governing equations for a and θ can be obtained by replacing the right-hand sides of 4.5.5 and 4.5.6 by their average values over a time interval $2\pi/\omega_0$ which is the period of the unperturbed oscillator. The justification for this is that a and θ will not change very much in this interval since they are slowly varying functions. The error introduced is $O(\varepsilon)$, so the resulting solution is a leading order approximation.

The right-hand sides will be replaced by their averages as follows:

$$\text{RHS} \simeq \frac{\omega_0}{2\pi}\int_{t=0}^{2\pi/\omega_0} \text{RHS} . dt.$$

It is convenient to introduce the function ϕ where $\phi = \omega_0 t + \theta$. Then

$$\frac{d\phi}{dt} = \omega_0 + \frac{d\theta}{dt} = \omega_0 + O(\varepsilon),$$

so to leading order

$$\text{RHS} = \frac{1}{2\pi}\int_{\phi=0}^{2\pi} \text{RHS} . d\phi.$$

Equations 4.5.5 and 4.5.6 become

$$\frac{da}{dt} = -\frac{\varepsilon}{2\pi\omega_0}\int_0^{2\pi} \sin\phi . F(a\cos\phi, \ -a\,\omega_0 \sin\phi)\,d\phi \qquad 4.5.7$$

$$\frac{d\theta}{dt} = -\frac{\varepsilon}{2\pi a\omega_0}\int_0^{2\pi} \cos\phi . F(a\cos\phi, \ -a\,\omega_0 \sin\phi)\,d\phi. \qquad 4.5.8$$

The function a is treated as a constant while the integrals are evaluated. The right-hand sides of 4.5.7 and 4.5.8 are simply functions of a. Equation 4.5.7 is of separable form and can be solved for the function $a(t)$. Then equation 4.5.8 can be solved for the function $\theta(t)$.

Consider the van der Pol equation where

$$F = (1 - u^2)\frac{du}{dt} = (1 - a^2\cos^2\phi)(-a\omega_0 \sin\phi).$$

Then

$$\frac{da}{dt} = -\frac{\varepsilon}{2\pi\omega_0}\int_0^{2\pi} -a\omega_0 \sin^2\phi(1 - a^2\cos^2\phi)\,d\phi = \varepsilon a\left(\frac{1}{2} - \frac{a^2}{8}\right),$$

and

$$\frac{d\theta}{dt} = -\frac{\varepsilon}{2\pi a\omega_0}\int_0^{2\pi} -a\omega_0 \sin\phi\cos\phi(1 - a^2\cos^2\phi)\,d\phi = 0.$$

In this case $d\theta/dt = 0$ so that $\theta = \theta_0$ (a constant). The equation for a is

$$\frac{1}{a(a^2-4)}\frac{da}{dt} = -\frac{\varepsilon}{8}.$$

On introducing partial fractions we have

$$\int\left(-\frac{1}{4}\cdot\frac{1}{a} + \frac{1}{8}\cdot\frac{1}{a-2} + \frac{1}{8}\cdot\frac{1}{a+2}\right)da = \frac{(-\varepsilon t + c)}{8}.$$

Integrating leads to

$$\ln\left|\frac{a^2-4}{a^2}\right| = -\varepsilon t + c,$$

i.e.

$$a = \frac{2}{\sqrt{1 - Ke^{-\varepsilon t}}},$$

where $K = \pm e^c$ is an arbitrary constant.

Thus the solution of the van der Pol equation is

$$u = \frac{2}{\sqrt{1 - Ke^{-\varepsilon t}}}\cos(\omega_0 t + \theta_0) + O(\varepsilon).$$

If we impose the initial conditions $u(0) = a_0$ and $(du/dt)(0) = 0$ we obtain the following

$$\frac{2}{\sqrt{1-K}}\cos\theta_0 = a_0 \quad \text{(to leading order)}$$

$$-\frac{2\omega_0}{\sqrt{1-K}}\sin\theta_0 = 0 \quad \text{(to leading order).}$$

Therefore $\theta_0 = 0$ and $K = 1 - 4/a_0^2$ so that

$$u = \frac{2}{\sqrt{1 + \left(\frac{4}{a_0^2} - 1\right)e^{-\varepsilon t}}}\cos\omega_0 t + O(\varepsilon).$$

This is the same as the solution 4.3.14 obtained using the method of multiple scales.

Next consider Duffing's equation where

$$F = -u^3 = -a^3\cos^3\phi,$$

so that 4.5.7 and 4.5.8 become

$$\frac{da}{dt} = -\frac{\varepsilon}{2\pi\omega_0}\int_0^{2\pi} \sin\phi(-a^3\cos^3\phi)\,d\phi = 0$$

$$\frac{d\theta}{dt} = -\frac{\varepsilon}{2\pi a\omega_0}\int_0^{2\pi} \cos\phi(-a^3\cos^3\phi)\,d\phi = \frac{\varepsilon a^2}{\omega_0}\frac{3}{8}.$$

Therefore $a = a_0$ (a constant) and $\theta = (3a_0^2/8\omega_0)\varepsilon t + \theta_0$, where θ_0 is a constant.

Thus the solution of Duffing's equation is

$$u = a_0 \cos\left(\omega_0 t + \frac{3a_0^2}{8\omega_0}\varepsilon t + \theta_0\right) + O(\varepsilon).$$

The initial conditions $u_0(0) = 1$, $(du/dt)(0) = 0$ lead to $a_0 = 1$ and $\theta_0 = 0$ so that

$$u = \cos\left[\omega_0\left(1 + \frac{3\varepsilon}{8\omega_0^2}\right)t\right] + O(\varepsilon).$$

This is the same as the solution of Duffing's equation obtained by the method of strained coordinates.

Chapter 5
Boundary Layers

Boundary layers are regions in which a rapid change occurs in the value of a variable. Some physical examples of situations where boundary layers may occur are:

- the fluid velocity near a solid wall
- the velocity at the edge of a jet of fluid
- the temperature of a fluid near a solid wall
- solute concentration near an interface
- condensing vapor on a cool surface.

At the turn of the century Ludwig Prandtl pioneered the subject of boundary layer theory in his explanation of how a quantity as small as the viscosity of common fluids such as water and air could nevertheless play a crucial role in determining their flow. The situation at the end of the last century with regard to the mathematical theory of fluid flow was that many solutions were available for the ideal flow equations where viscosity is neglected. These solutions could often correctly describe the flow fields away from solid boundaries. However, they were not applicable to the flow near boundaries and in particular failed to explain fluid drag. The equations governing the behavior of real fluids include the viscous term and are far more difficult to solve.

The viscosity of many fluids is very small and yet taking account of this small quantity is vital. The essential point is that the viscous term involves higher order derivatives so that its omission necessitates the loss of a boundary condition. The ideal flow solutions allow slip to occur between a solid and fluid. In reality the tangential velocity of a fluid relative to a solid is zero. The fluid is brought to rest by the action of a tangential stress resulting from the viscous force.

For many flows, viscous effects are negligible except in the vicinity of solid boundaries where narrow regions of rapid change in the tangential component of fluid velocity occur. In Prandtl's work on the subject of boundary layers in fluids he took advantage of the fact that the viscous layer was thin. He reasoned that the well-known solutions of the ideal flow equations (in which the viscous term was absent) could be used to determine the fluid flow field away from solid boundaries. Furthermore Prandtl determined the behavior within the boundary region by the use of a system of viscous flow equations which are simplified due to the retention only of dominant derivative terms associated with the rapid variation of the tangential velocity component.

Mathematically the occurrence of boundary layers is associated with the presence of a small parameter multiplying the highest derivative in the governing equaton of a process. A straightforward perturbation expansion using an asymptotic sequence in the small parameter leads to differential equations of lower order than the original governing equation. In consequence not all of the boundary or initial conditions can be satisfied by the perturbation expansion. This is an example of what is commonly referred to as a singular perturbation problem. The technique for overcoming the difficulty is to combine the straightforward expansion, which is valid away from the boundary where a condition is not satisfied, with an expansion which is valid within a layer adjacent to this boundary. The straightforward expansion is referred to as the *outer expansion.* The *inner expansion* associated with the boundary layer region is expressed in terms of a stretched variable, rather than the original independent variable, which takes due account of the scale of certain derivative terms. The inner and outer expansions are matched over a region located at the edge of the boundary layer. The technique is called the method of *matched asymptotic expansions.*

We will first consider the method of matched asymptotic expansions applied to a model differential equation whose exact solution is available to guide us through the development of the technique. Subsequently we will see the power of the method in its application to problems involving nonconstant coefficients, nonlinear terms and the partial differential equations of fluid flow and heat transfer.

5.1 Model problem

The main features of the method of matched asymptotic expansions will be developed from a study of the model problem

$$\varepsilon\frac{d^2f}{dx^2}+\frac{df}{dx}=2x+1, \quad 0<x<1, \qquad 5.1.1$$

where ε is a small positive parameter and the value of the dependent variable is specified at the ends of the interval of interest. We choose the following boundary conditions

$$f(0)=1 \quad \text{and} \quad f(1)=4. \qquad 5.1.2$$

If we assume that f possesses a straightforward expansion in powers of ε,

$$f(x;\varepsilon)\sim f_0(x)+\varepsilon f_1(x)+\varepsilon^2 f_2(x)+\cdots \qquad 5.1.3$$

then the equations associated with powers of ε lead to

$$O(1): \quad \frac{df_0}{dx}=2x+1, \qquad 5.1.4$$

and

$$O(\varepsilon^n): \quad \frac{df_n}{dx} = -\frac{d^2 f_{n-1}}{dx^2} \quad \text{for } n = 1, 2, 3 \ldots . \tag{5.1.5}$$

The boundary conditions require

$$f_0(0) + \varepsilon f_1(0) + \cdots \sim 1 + \varepsilon 0 + \cdots ,$$

and

$$f_0(1) + \varepsilon f_1(1) + \cdots \sim 4 + \varepsilon 0 + \cdots ,$$

which lead to

$$f_0(0) = 1, \quad f_0(1) = 4,$$

and

$$f_n(0) = 0, \quad f_n(1) = 0 \quad n = 1, 2, 3 \ldots . \tag{5.1.6}$$

Equations 5.1.6 require that each $f_n(x)$ satisfy two boundary conditions. This is in general impossible since equations 5.1.4 and 5.1.5 governing each $f_n(x)$ are first order. We will discover that the boundary condition at $x = 0$ must be abandoned and consequently the expansion 5.1.3 is invalid near $x = 0$.

The general solution of 5.1.4 is $f_0 = x^2 + x + c$ where the constant, c, is determined from the boundary condition at $x = 1$, i.e. $f_0(1) = 4 = 2 + c$, so that $c = 2$ and

$$f_0 = x^2 + x + 2. \tag{5.1.7}$$

From 5.1.5 we obtain the governing equation for f_1,

$$\frac{df_1}{dx} = -2,$$

with solution satisfying $f_1(1) = 0$ given by

$$f_1 = -2(x - 1). \tag{5.1.8}$$

For this particular example all other terms f_2, f_3, etc. are zero. We will write

$$f^{\text{out}} = x^2 + x + 2 + \varepsilon 2(1 - x), \tag{5.1.9}$$

where the 'out' label is used to indicate that the solution is valid away from the region near $x = 0$. Clearly f^{out} fails to satisfy the boundary condition 5.1.2 at $x = 0$. The reason why the outer solution is of use is that it closely follows the exact solution of the problem except in a narrow region near $x = 0$ where the exact solution changes rapidly in order to satisfy the boundary condition. This rapid variation will be demonstrated by studying the exact solution of the boundary value problem.

Standard techniques for solving linear constant coefficient differential equations yield the following general solution of 5.1.1

$$f = A + B\exp(-x/\varepsilon) + x^2 + x(1 - 2\varepsilon). \tag{5.1.10}$$

The constants A and B are determined by the boundary conditions 5.1.2,

$$A + B = 1$$

$$A + B\exp(-1/\varepsilon) + 2 - 2\varepsilon = 4. \qquad 5.1.11$$

We have seen in our earlier study of the order of various functions that $\exp(-1/\varepsilon) = o(\varepsilon^N)$ as $\varepsilon \to 0$ for any N. This means that the exponential term tends to zero faster than any power of ε as ε tends to zero. It is called a transcendentally small term (T.S.T.) and can always be neglected since its contribution is asymptotically always less than any power of ε. Thus 5.1.11 has solution $A = 2(1 + \varepsilon)$, $B = -1 - 2\varepsilon$ giving the exact solution

$$f^{\text{exact}} = 2(1 + \varepsilon) - (1 + 2\varepsilon)\exp(-x/\varepsilon) + x^2 + x(1 - 2\varepsilon),$$

which becomes when arranged in asymptotic order,

$$f^{\text{exact}} = x^2 + x + 2 - \exp(-x/\varepsilon) + \varepsilon[2(1 - x) - 2\exp(-x/\varepsilon)]. \qquad 5.1.12$$

Comparing the exact solution with equation (5.1.9) shows that the terms involving $\exp(-x/\varepsilon)$ are absent in the outer expansion. The effect of these terms is negligible (T.S.T.) when $x = O(1)$. But when $x = O(\varepsilon)$ then $\exp(-x/\varepsilon) = O(1)$. This can be demonstrated by choosing particular values of ε and calculating the values of $\exp(-x/\varepsilon)$ for various values of x.

$\varepsilon = 0.1$

$x =$	0	0.1	0.2	0.3	0.4
$e^{-x/\varepsilon} =$	e^0	e^{-1}	e^{-2}	e^{-3}	e^{-4}
	$\Vert$	$\Vert$	$\Vert$	$\Vert$	$\Vert$
	1.00	0.368	0.135	0.0498	0.0183

$\varepsilon = 0.01$

$x =$	0	0.01	0.02	0.03	0.04
$e^{-x/\varepsilon} =$	e^0	e^{-1}	e^{-2}	e^{-3}	e^{-4}
	$\Vert$	$\Vert$	$\Vert$	$\Vert$	$\Vert$
	1.00	0.368	0.135	0.0498	0.0183.

If x is greater than 5ε then $\exp(-x/\varepsilon)$ becomes insignificant. However for values of x near to ε the exponential term is $O(1)$.

The behavior of the leading term in the exact solution, $f_0^{\text{ex}} = x^2 + x + 2 - \exp(-x/\varepsilon)$, is shown in Fig. 5.1 for various values of ε. The leading term in the outer expansion, $f_0^{\text{out}} = x^2 + x + 2$, is also shown in the figure. It is clear that as $\varepsilon \to 0$ the region in which the outer solution departs from the exact solution becomes arbitrarily close to $x = 0$ with a thickness $O(\varepsilon)$. This region is called a *boundary layer*.

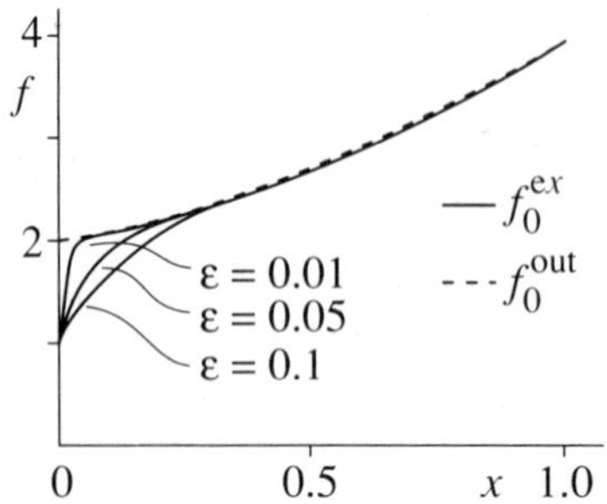

Fig. 5.1 The exact solution for various values of ε

The stretched variable and inner expansion

Figure 5.1 clearly shows the large gradient of the function in the boundary layer. By differentiating the leading term, f_0^{ex}, of the exact solution we have

$$f_0^{ex} = x^2 + x + 2 - \exp(-x/\varepsilon)$$

$$\frac{df_0^{ex}}{dx} = 2x + 1 + \frac{1}{\varepsilon}\exp(-x/\varepsilon)$$

$$\frac{d^2f_0^{ex}}{dx^2} = 2 - \frac{1}{\varepsilon^2}\exp(-x/\varepsilon).$$

Outside the boundary layer, i.e. for $x = O(1)$, we have $\exp(-x/\varepsilon) = o(\varepsilon^N)$ for all N so $(1/\varepsilon)\exp(-x/\varepsilon)$ and $(1/\varepsilon^2)\exp(-x/\varepsilon)$ are also transcendentally small. Within the boundary layer when $x = O(\varepsilon)$ we have $\exp(-x/\varepsilon) = O(1)$. The order of f_0^{ex} and its derivatives are summarized as follows:

	Outside Boundary Layer	Inside Boundary Layer
f_0^{ex}	$O(1)$	$O(1)$
$\frac{df_0^{ex}}{dx}$	$O(1)$	$O\left(\frac{1}{\varepsilon}\right)$
$\frac{d^2f_0^{ex}}{dx^2}$	$O(1)$	$O\left(\frac{1}{\varepsilon^2}\right)$

This indicates that x is the appropriate independent variable outside the boundary layer where f_0^{ex} and its derivatives are $O(1)$ quantities. However, within the boundary layer the appropriately scaled independent variable is $s = x/\varepsilon$ then $df/dx = (1/\varepsilon)\,df/ds$, $d^2f/dx^2 = (1/\varepsilon^2)\,d^2f/ds^2$, so that within the boundary layer $df/ds = O(1)$ and $d^2f/ds^2 = O(1)$. The variable $s = x/\varepsilon$ is called a stretched variable.

The differential equation becomes

$$\frac{\varepsilon}{\varepsilon^2}\frac{d^2f}{ds^2} + \frac{1}{\varepsilon}\frac{df}{ds} = 2\varepsilon s + 1,$$

and on multiplying through by ε we obtain the boundary layer equation

$$\frac{d^2f}{ds^2} + \frac{df}{ds} = \varepsilon + 2\varepsilon^2 s. \qquad 5.1.13$$

We assume a boundary layer expansion, called the *inner expansion*, of the form

$$f^{\text{in}}(s; \varepsilon) \sim f_0^{\text{in}}(s) + \varepsilon f_1^{\text{in}}(s) + \cdots \qquad 5.1.14$$

The inner expansion will satisfy the boundary condition at $x = s = 0$ namely $f_0^{\text{in}}(s = 0) = 1$ giving $f_0^{\text{in}}(0) = 1$ and $f_n^{\text{in}}(0) = 0$ for $n = 1, 2, 3 \ldots$. Substituting the expansion 5.1.14 into the differential equation 5.1.13 yields the equations

$$\begin{aligned}
O(1)&: \quad \frac{d^2f_0^{\text{in}}}{ds^2} + \frac{df_0^{\text{in}}}{ds} = 0, \quad f_0^{\text{in}}(0) = 1 \\
O(\varepsilon)&: \quad \frac{d^2f_1^{\text{in}}}{ds^2} + \frac{df_1^{\text{in}}}{ds} = 1, \quad f_1^{\text{in}}(0) = 0 \\
O(\varepsilon^2)&: \quad \frac{d^2f_2^{\text{in}}}{ds^2} + \frac{df_2^{\text{in}}}{ds} = 2s, \quad f_2^{\text{in}}(0) = 0 \\
O(\varepsilon^n)&: \quad \frac{d^2f_n^{\text{in}}}{ds^2} + \frac{df_n^{\text{in}}}{ds} = 0, \quad f_n^{\text{in}}(0) = 0, n = 3, 4, 5 \ldots
\end{aligned} \qquad 5.1.15$$

with solutions

$$\begin{aligned}
f_0^{\text{in}} &= A + (1 - A)e^{-s} \\
f_1^{\text{in}} &= B - Be^{-s} + s \\
f_2^{\text{in}} &= C - Ce^{-s} + s^2 - 2s \\
f_n^{\text{in}} &= D_n - D_n e^{-s}, n = 3, 4, 5 \ldots.
\end{aligned} \qquad 5.1.16$$

The boundary condition at $x = 1$ cannot be used to determine the constants appearing in these solutions because the differential equations 5.1.15 are only valid in the boundary layer. The constants in 5.1.16 are determined by matching the inner and outer expansions. We shall first restrict our attention to matching the leading order expansions f_0^{out} and f_0^{in}. The method which we shall apply is Prandtl's Matching Condition. In Section 5.4 we will see how this condition is refined to deal with higher order matching.

Prandtl's matching condition

The leading order terms in the inner and outer expansions are to be matched at the 'edge of the boundary layer'. Of course there is no precise edge of the boundary layer, we simply know that it has thickness of order ε. A plausible matching procedure would be to equate f_0^{in} and f_0^{out} at a value of x such that the region of rapid change has passed. We might choose to equate the terms at the point $x = 5\varepsilon$. The leading order expansions are

$$f_0^{\text{out}} = x^2 + x + 2$$

$$f_0^{\text{in}} = A + (1 - A)e^{-s} \quad (s = x/\varepsilon).$$

Equating at $x = 5\varepsilon$ gives the following,

$$2 + 5\varepsilon + 25\varepsilon^2 = A + (1 - A)e^{-5},$$

which when solved yields

$$A = \frac{2 + 5\varepsilon + 25\varepsilon^2 - e^{-5}}{1 - e^{-5}}.$$

If instead we choose to match at $x = 6\varepsilon$ then we obtain

$$A = \frac{2 + 6\varepsilon + 36\varepsilon^2 - e^{-6}}{1 - e^{-6}}.$$

These two expressions differ in the argument of the exponential and differ algebraically with 5ε replaced by 6ε. The exponential functions are approaching transcendentally small values so that their contribution can be neglected. The algebraic difference is of $O(\varepsilon)$. Thus the arbitrariness in the decision of the point at which we choose to equate the expansions leads to a difference of $O(\varepsilon)$. But we are only dealing with leading order expansions anyway. The difference between the exact solution and the leading order expansions will be $O(\varepsilon)$ so that an arbitrariness in f_0^{in} and f_0^{out} of $O(\varepsilon)$ is immaterial. Rather than choose between, for example, 5ε and 6ε as the value of x to evaluate f_0^{out} we may take the value at $x = 0$ since $f_0^{\text{out}}[x = O(\varepsilon)] = f_0^{\text{out}}(0) + O(\varepsilon)$ where the remainder is uniformly $O(\varepsilon)$ since the gradient of f_0^{out} is $O(1)$. For the inner expansion we are to ensure that the rapidly varying function has achieved its asymptotic value at the edge of the boundary layer. This means that the term $\exp(-x/\varepsilon)$ should be replaced by zero. This can be achieved by taking the limit $s \to \infty$. Thus rather than choosing a specific point to equate the inner and outer terms we are led to the condition

Prandtl's Matching Condition	$\lim_{x \to 0} f_0^{\text{out}}(x) = \lim_{s \to \infty} f_0^{\text{in}}(s).$

5.1.17

The limit $s \to \infty$ may appear rather dangerous since although it certainly removes the exponential term it could lead to an algebraically unbounded term. For example if $f_0^{\text{in}} = As + (1 - A)e^{-s}$ then the first member would be unbounded as $s \to \infty$. This possibility can be eliminated since the inner expansion must be of a form which varies rapidly for $x = O(\varepsilon)$ but not for $x = O(1)$, i.e. not for $s \to \infty$. Indeed if such an algebraic term occurred in f_0^{in} we would argue that the assumed location of the boundary layer or the stretching transformation was wrong. In practice if the boundary layer has been properly located and the correct inner variable used then Prandtl's matching condition is valid and elegantly avoids the need to choose an arbitrary 'edge' of the boundary layer. Applying this condition to the current example leads to

$$\lim_{x \to 0} (x^2 + x + 2) = \lim_{s \to \infty} [A + (1 - A)e^{-s}],$$

which becomes $2 = A$. Thus the leading order terms in the expansion of the solution are:

$$\text{Outer Region} \quad f_0^{\text{out}} = x^2 + x + 2 \quad \text{for} \quad x = O(1)$$

$$\text{Inner Region} \quad f_0^{\text{in}} = 2 - e^{-x/\varepsilon} \quad \text{for} \quad x = O(\varepsilon).$$

To prove that these are valid leading terms we consider f_0^{ex} for $x = O(1)$ and $x = O(\varepsilon)$.

If $x = O(1)$ then $f_0^{\text{ex}} = x^2 + x + 2 + \text{T.S.T.}$

If $x = O(\varepsilon)$ then $f_0^{\text{ex}} = 2 - e^{-x/\varepsilon} + O(\varepsilon)$.

We conclude that the matching condition has correctly predicted the leading order terms.

The composite expansion

A single composite expression for these leading order terms can be constructed using the combination

$$f_0^{\text{comp}} = f_0^{\text{in}} + f_0^{\text{out}} - f_0^{\text{match}}, \qquad 5.1.18$$

where f_0^{match} is given by 5.1.17. Then for $x = O(1)$, $f_0^{\text{in}} = f_0^{\text{match}} + \text{T.S.T.}$ so that $f_0^{\text{comp}} = f_0^{\text{out}} + \text{T.S.T.}$ Similarly for $x = O(\varepsilon)$, $f_0^{\text{out}} = f_0^{\text{match}} + O(\varepsilon)$, so that $f_0^{\text{comp}} = f_0^{\text{in}} + O(\varepsilon)$. Thus 5.1.18 provides a uniformly valid expression for the leading order term in the expansion of the solution over the entire range of interest.

For the current example $f_0^{\text{match}} = 2$ so the composite expression is

$$f_0^{\text{comp}} = x^2 + x + 2 - \exp(-x/\varepsilon). \qquad 5.1.19$$

Prandtl's matching condition can only be used for the leading order terms in the asymptotic expansions. We will see later how higher order terms are matched by considering the behavior of the inner and outer expansions in an intermediate region.

Worked example

Obtain the leading order composite expansion for the solution of

$$\varepsilon \frac{d^2 f}{dx} + (x+1)\frac{df}{dx} + f = 2x, \quad 0 < x < 1$$

$$0 < \varepsilon \ll 1, \quad f(0) = 1, \quad f(1) = 2. \qquad 5.1.20$$

Assume that the boundary layer occurs at $x = 0$ characterized by the stretching transformation $s = x/\varepsilon$ (we will see in the next section why this is true).

Solution

The outer expansion:

Let $f^{\text{out}} \sim f_0^{\text{out}} + \varepsilon f_1^{\text{out}} + \cdots$
where

$$(x+1)\frac{df_0^{\text{out}}}{dx} + f_0^{\text{out}} = 2x, \quad f_0^{\text{out}}(1) = 2.$$

Notice that the original second order equation has no convenient solution except as a power series in x. On the other hand the first order linear differential equation can easily be solved by the integrating factor method. In this example the left-hand side is already an exact dfferential so that

$$\frac{d}{dx}[(x+1)f_0^{\text{out}}] = 2x.$$

Integrating leads to

$$f_0^{\text{out}} = \frac{x^2 + c}{x+1}$$

and imposing the boundary condition at $x = 1$ requires $C = 3$.

The inner expansion:

Introduce the stretched variable $s = x/\varepsilon$ so that

$$\frac{df}{dx} = \frac{1}{\varepsilon}\frac{df}{ds} \quad \text{and} \quad \frac{d^2 f}{dx^2} = \frac{1}{\varepsilon^2}\frac{d^2 f}{ds^2}.$$

The differential equation becomes

$$\frac{1}{\varepsilon}\frac{d^2 f}{ds^2} + \frac{\varepsilon s + 1}{\varepsilon}\frac{df}{ds} + f = 2\varepsilon s. \qquad 5.1.21$$

Let $f^{\text{in}} \sim f_0^{\text{in}}(s) + \varepsilon f_1^{\text{in}}(s) + \cdots$ and substitute into 5.1.21 to obtain the leading order equation

$$\frac{d^2 f_0^{\text{in}}}{ds^2} + \frac{df_0^{\text{in}}}{ds} = 0.$$

The general solution of this equation is $f_0^{\text{in}} = A + Be^{-s}$ and the boundary condition at $s = 0$ is $f_0^{\text{in}}(0) = 1$ so that $f_0^{\text{in}} = A + (1 - A)e^{-s}$.

Applying Prandtl's matching condition,

$$\lim_{x \to 0} \left[\frac{x^2 + 3}{x + 1}\right] = \lim_{s \to \infty} [A + (1 - A)e^{-s}],$$

leads to $A = 3$, so that $f_0^{\text{in}} = 3 - 2e^{-s}$.

Then using $f_0^{\text{out}} = (x^2 + 3)/(x + 1)$ and $f_0^{\text{match}} = 3$ we obtain the leading order composite expansion

$$f_0^{\text{comp}} = \frac{x^2 + 3}{x + 1} - 2e^{-x/\varepsilon}.$$

5.2 Boundary layer location

In the two examples considered in the previous section the boundary layer occurred at the left-hand boundary $x = 0$. In general boundary layers may occur at either or both boundaries or at an interior point. To gain an understanding of the feature which determines the location of boundary layers we consider the following problem,

$$\varepsilon \frac{d^2 f}{dx^2} + (x - 2)\frac{df}{dx} + f = 0, \quad 0 < x < 1,$$

$$0 < \varepsilon \ll 1, \quad f(0) = 3, \quad f(1) = 2. \qquad 5.2.1$$

We follow the previous two examples and assume (erroneously) that the boundary layer occurs at $x = 0$. The leading order term in the outer expansion, f_0^{out}, will then satisfy the boundary condition at $x = 1$ and the first order equation

$$(x - 2)\frac{df_0^{\text{out}}}{dx} + f_0^{\text{out}} = 0. \qquad 5.2.2$$

The solution is easily found to be $f_0^{\text{out}} = 2/(2 - x)$.

The boundary layer is assumed to be of thickness $O(\varepsilon)$ so that the appropriate stretched variable is $s = x/\varepsilon$. The differential equation governing the inner expansion is then

$$\frac{\varepsilon}{\varepsilon^2}\frac{d^2 f^{\text{in}}}{ds^2} + (\varepsilon s - 2)\frac{1}{\varepsilon}\frac{df^{\text{in}}}{ds} + f^{\text{in}} = 0.$$

The leading order term, f_0^{in}, satisfies

$$\frac{d^2 f^{\text{in}}}{ds^2} - 2\frac{df_0^{\text{in}}}{ds} = 0, \quad f_0^{\text{in}}(0) = 3, \qquad 5.2.3$$

with solution

$$f_0^{\text{in}} = 3 - A + Ae^{2s}. \qquad 5.2.4$$

As s increases the exponential term in this function becomes arbitrarily large. This is not the correct form of boundary layer behavior and certainly cannot be matched to f_0^{out}. The difficulty has arisen because of the difference in sign between the coefficients of the first and second derivatives in equation 5.2.3. To obtain a decreasing exponential behavior both derivatives must have the same sign.

The above reasoning leads us to reject the assumption of a boundary layer occurring at $x = 0$. We next consider the possibility of a boundary layer at $x = 1$. The leading term in the outer expansion will again satisfy equation 5.2.2 but with boundary condition $f_0^{\text{out}}(0) = 3$. The following solution is easily found,

$$f_0^{\text{out}} = \frac{6}{2 - x}.$$

The inner variable, s, appropriate for the boundary layer at $x = 1$ is the difference $1 - x$ scaled by ε, i.e. $s = (1 - x)/\varepsilon$. If x differs from 1 by a quantity of order ε then s will be O(1). The derivatives become

$$\frac{df}{dx} = -\frac{1}{\varepsilon}\frac{df}{ds} \quad \text{and} \quad \frac{d^2f}{dx^2} = \frac{1}{\varepsilon^2}\frac{d^2f}{ds^2}.$$

We will see that the negative sign in the first derivative transformation is compensated by the negative coefficient of the first derivative in the differential equation 5.2.1. The inner expansion satisfies

$$\frac{\varepsilon}{\varepsilon^2}\frac{d^2f^{\text{in}}}{ds^2} + (1 - \varepsilon s - 2)\left(-\frac{1}{\varepsilon}\right)\frac{df^{\text{in}}}{ds} + f^{\text{in}} = 0.$$

The leading order term f_0^{in} satisfies

$$\frac{d^2f_0^{\text{in}}}{ds^2} + \frac{df_0^{\text{in}}}{ds} = 0, \qquad 5.2.5$$

and the boundary condition at $x = 1$ becomes $f_0^{\text{in}}(s = 0) = 2$. The solution

$$f_0^{\text{in}} = 2 - A + Ae^{-s}, \qquad 5.2.6$$

has the correct boundary layer behavior since as x decreases from unity, s increases so that the exponential term decays rapidly from the right-hand boundary. This crucial difference between the behavior of the functions in 5.2.4 and 5.2.6 depends on the sign difference in the governing equations 5.2.3 and 5.2.5.

Prandtl's matching condition applied to

$$f_0^{\text{out}} = \frac{6}{2 - x} \quad \text{and} \quad f_0^{\text{in}} = 2 - A + Ae^{-s},$$

is

$$\lim_{x \to 1}\left(\frac{6}{2 - x}\right) = \lim_{s \to \infty}(2 - A + Ae^{-s}),$$

where of course the outer solution is now evaluated at $x = 1$. This gives $A = -4$ and the composite leading order solution is

$$f_0^{\text{comp}} = \frac{6}{2-x} + \not{\phi} - 4\exp - \left(\frac{1-x}{\varepsilon}\right) - \not{\phi}.$$

The general linear equation

In the case of the linear equation

$$\varepsilon\frac{\mathrm{d}^2 f}{\mathrm{d}x^2} + a(x)\frac{\mathrm{d}f}{\mathrm{d}x} + b(x)f = c(x), \quad x_1 < x < x_2, \tag{5.2.7}$$

the following general statements can be made about the boundary layer location and the nature of the inner expansion.

I. If $a(x) > 0$ throughout $x_1 < x < x_2$ then the boundary layer will occur at $x = x_1$. The stretching transformation will be

$$s = (x - x_1)/\varepsilon,$$

and the one-term inner expansion will satisfy

$$\frac{\mathrm{d}^2 f_0^{\text{in}}}{\mathrm{d}s^2} + a(x_1)\frac{\mathrm{d}f_0^{\text{in}}}{\mathrm{d}s} = 0.$$

The solution of this equation is

$$f_0^{\text{in}} = A + B\exp\left(-a(x_1)\frac{(x - x_1)}{\varepsilon}\right), \tag{5.2.8}$$

where $A + B = f(x = x_1)$. The other condition to determine the constants A and B is obtained by matching with the value of the outer expansion at $x = x_1$.

II. If $a(x) < 0$ throughout $x_1 < x < x_2$ then the boundary layer will occur at $x = x_2$. The stretching transformation will be $s = (x_2 - x)/\varepsilon$ and the one-term inner expansion will involve the rapidly decaying function $\exp\left(a(x_2)\frac{(x_2 - x)}{\varepsilon}\right)$.

III. If $a(x)$ changes sign in the interval $x_1 < x < x_2$ then a boundary layer occurs at an interior point x_0 where $a(x_0) = 0$ and boundary layers may occur at both ends x_1 and x_2. In Section 5.3 we will consider an example of an interior boundary layer.

Far less is known about the behavior of nonlinear equations. However, it is reasonable to expect that these ideas on the location of the boundary layer will apply to the case where the coefficients a, b and c in equation 5.2.7 are functions of the dependent variable, f, itself. Care would be required if the coefficients were functions of df/dx because the essential features of boundary layers are associated with the rapidly varying behavior of derivatives in the boundary layer.

Exercises

Obtain a one-term composite expansion for the solutions of:

(i) $$\varepsilon \frac{d^2f}{dx^2} + \frac{df}{dx} + \frac{f}{x+1} = 2, \quad 0 < x < 1$$

$$0 < \varepsilon \ll 1, \quad f(0) = 0, \quad f(1) = 3.$$

(ii) $$\varepsilon \frac{d^2f}{dx^2} - \frac{df}{dx} + \frac{f}{x+1} = 2, \quad 0 < x < 1$$

$$0 < \varepsilon \ll 1, \quad f(0) = 0, \quad f(1) = 3.$$

(iii) $$\varepsilon \frac{d^2f}{dx^2} + x\frac{df}{dx} + f = 0, \quad 2 < x < 4$$

$$0 < \varepsilon \ll 1, \quad f(2) = 0, \quad f(4) = 1.$$

(iv) $$\varepsilon \frac{d^2f}{dx^2} - x\frac{df}{dx} + f = 0, \quad 2 < x < 4$$

$$0 < \varepsilon \ll 1, \quad f(2) = 0, \quad f(4) = 1.$$

(v) $$\varepsilon \frac{d^2f}{dx^2} + x\frac{df}{dx} + f = 0, \quad -4 < x < -2$$

$$0 < \varepsilon \ll 1, \quad f(-4) = 1, \quad f(-2) = 0.$$

Answers

(i) $$f_0^{\text{comp}} = \frac{x^2 + 2x + 3}{x+1} - 3e^{-x/\varepsilon}.$$

(ii) $$f_0^{\text{comp}} = -2(1+x)\ln(1+x) + (3 + 4\ln 2)\exp -\left(\frac{1-x}{\varepsilon}\right).$$

(iii) $$f_0^{\text{comp}} = \frac{4}{x} - 2\exp\left(-\frac{2(x-2)}{\varepsilon}\right).$$

(iv) $$f_0^{\text{comp}} = \exp\left(-4\frac{(4-x)}{\varepsilon}\right).$$

(v) $$f_0^{\text{comp}} = -\frac{4}{x} - 2\exp\left(\frac{2(2+x)}{\varepsilon}\right).$$

5.3 Boundary layer thickness and the principle of least degeneracy

The boundary layers which we have met so far have all had thickness $O(\varepsilon)$. By this we mean that a variation of $O(\varepsilon)$ in the independent variable will encompass the region of rapid change in the dependent variable. The associated stretched independent variable, s, appropriate for the boundary layer is related to x by a linear transformation involving division by ε as occurs for example in equation 5.2.8. There are practical examples which we will consider in Section 5.6 where the boundary layer thickness is $O(\varepsilon^{1/2})$ and $O(\varepsilon^{1/3})$. This means that if the boundary layer is located in the region $x = x_0$ the appropriate stretching transformation is $s = (x - x_0)/\varepsilon^p$ where p is 1, 1/2 or 1/3 depending on whether the thickness is $O(\varepsilon)$, $O(\varepsilon^{1/2})$ or $O(\varepsilon^{1/3})$ respectively. The choice of the power p and indeed more generally the choice of the function $\delta(\varepsilon)$ to use in the stretching transformation $s = (x - x_0)/\delta(\varepsilon)$ is determined by the need to represent the region of rapid change correctly. We must ensure that the boundary layer solution contains rapidly varying functions. The form of the governing equation in the boundary layer region must have sufficient structure to allow such solutions.

Consider the following example,

$$\varepsilon\frac{d^2f}{dx^2} + \frac{df}{dx} + f = x, \quad 0 < x < 1$$

$$0 < \varepsilon \ll 1, \quad f(0) = 1, \quad f(1) = 2. \qquad 5.3.1$$

Since the signs of the first and second derivatives are the same, we can assume, on the basis of our previous studies, that the boundary layer will occur at $x = 0$. We are not going to assume at the outset that the boundary layer thickness is $O(\varepsilon)$. Our intention is to deduce that the appropriate stretched variable is $s = x/\varepsilon$. The one-term outer expansion, f_0^{out} satisfies

$$\frac{df_0^{out}}{dx} + f_0^{out} = x, \quad f_0^{out}(1) = 2.$$

The solution is

$$f_0^{out} = 2e^{1-x} + x - 1. \qquad 5.3.2$$

To determine the inner expansion we first wrongly suppose that the boundary layer thickness is $O(\varepsilon^{1/2})$. The stretching transformation

$$s = x/\varepsilon^{1/2} \quad \text{leads to} \quad \frac{df}{dx} = \frac{1}{\varepsilon^{1/2}}\frac{df}{ds} \quad \text{and} \quad \frac{d^2f}{dx^2} = \frac{1}{\varepsilon}\frac{d^2f}{ds^2},$$

and the differential equation becomes

$$\frac{d^2f}{ds^2} + \frac{1}{\varepsilon^{1/2}}\frac{df}{ds} + f = \varepsilon^{1/2}s. \qquad 5.3.3$$

If the appropriate stretching transformation has been used for the boundary layer then df/ds and d^2f/ds^2 will be O(1) within it. The leading order expansion f_0^{in} will satisfy the dominant part of equation 5.3.3, i.e. the $O(\varepsilon^{-1/2})$ component.

$$\frac{df_0^{in}}{ds} = 0. \qquad 5.3.4$$

The solution satisfying the boundary condition at $x = s = 0$ is $f_0^{in} = 1$. This of course does not have the rapidly varying behavior which we anticipate in the boundary layer. Prandtl's matching condition cannot be satisfied since

$$\lim_{x\to 0}(2e^{1-x} + x - 1) = 2e - 1 \neq \lim_{s\to\infty}(f_0^{in}).$$

Thus we reject the assumption of a boundary layer of thickness $O(\varepsilon^{1/2})$.

Next we suppose that the boundary layer thickness is $O(\varepsilon^2)$ and again we will discover that this is incorrect because the corresponding inner expansion cannot be matched to the outer expansion. Proceeding with the analysis we introduce the stretching transformation $s = x/\varepsilon^2$ which leads to the equation

$$\frac{1}{\varepsilon^3}\frac{d^2f}{ds^2} + \frac{1}{\varepsilon^2}\frac{df}{ds} + f = \varepsilon^2 s.$$

Again we argue that if the appropriate stretching has been used then all derivatives are O(1) so that the governing equation for the leading term is $O(\varepsilon^{-3})$, namely

$$\frac{d^2f_0^{in}}{ds^2} = 0. \qquad 5.3.5$$

The solution satisfying $f_0^{in}(0) = 1$ is $f_0^{in} = 1 + As$ where A is a constant which we attempt to determine by matching. This solution is rapidly varying but the rapidity does not decay at the edge of the boundary layer (i.e. as $s \to \infty$). Indeed we cannot match f_0^{in} to the outer expansion because the term As becomes arbitrarily large as $s \to \infty$.

The correct choice of stretching transformation is $s = x/\varepsilon$ showing that the boundary layer thickness is $O(\varepsilon)$. The boundary layer equation becomes

$$\frac{1}{\varepsilon}\frac{d^2f}{ds^2} + \frac{1}{\varepsilon}\frac{df}{ds} + f = \varepsilon s.$$

The dominant equation satisfied by f_0^{in} is $O(1/\varepsilon)$, namely

$$\frac{d^2f_0^{in}}{ds^2} + \frac{df_0^{in}}{ds} = 0. \qquad 5.3.6$$

The solution satisfying $f_0^{in}(0) = 1$ is $f_0^{in} = 1 - A + Ae^{-s}$.

The last member provides the necessary rapid decay away from the point $x = s = 0$. Prandtl's matching condition requires

$$\lim_{x\to 0}(2e^{1-x} + x - 1) = \lim_{s\to\infty}(1 - A + Ae^{-s}),$$

which leads to $2e - 1 = 1 - A$ so that $A = 2 - 2e$ and $f_0^{\text{in}} = 2e - 1 + 2(1 - e)e^{-x/\varepsilon}$. The one-term composite expansion is

$$f_0^{\text{comp}} = 2e^{1-x} + x - 1 + 2e - 1 + 2(1 - e)e^{-x/\varepsilon} - (2e - 1). \qquad 5.3.7$$

The leading order boundary layer equation associated with the stretching transformation $s = x/\varepsilon$, equation 5.3.6, involves more terms than equation 5.3.4 associated with $s = x/\varepsilon^{1/2}$ and equation 5.3.5 associated with $s = x/\varepsilon^2$. The extra term in 5.3.6 allows sufficient structure in the solution to produce the required boundary layer behavior. An aid for choosing the boundary layer thickness is to seek a stretching transformation which retains the largest number of terms in the dominant equation governing f_0^{in}. This is referred to as the *principle of least degeneracy* by Van Dyke[10].

The composite expansion, 5.3.7 can be verified by comparison with the exact solution of 5.3.1 This is readily found since the linear equation has constant coefficients. The general solution is $f_{\text{gen}} = c_1 \exp(m_1 x) + c_2 \exp(m_2 x) + x - 1$ where m_1 and m_2 are the solutions of the equation $\varepsilon m^2 + m + 1 = 0$. Thus

$$m_1 = \frac{-1 + \sqrt{1 - 4\varepsilon}}{2\varepsilon} \quad \text{and} \quad m_2 = \frac{-1 - \sqrt{1 - 4\varepsilon}}{2\varepsilon}.$$

We expand $\sqrt{1 - 4\varepsilon}$ using the binomial series, $\sqrt{1 - 4\varepsilon} = 1 - 2\varepsilon + O(\varepsilon^2)$ then

$$m_1 = -1 + O(\varepsilon) \quad \text{and} \quad m_2 = -\frac{1}{\varepsilon} + 1 + O(\varepsilon) \quad \text{so that}$$

$$f_{\text{gen}} = c_1 e^{-x} + c_2 e^{-x/\varepsilon} e^{x} + x - 1 + O(\varepsilon). \qquad 5.3.8$$

The boundary conditions require

$$c_1 + c_2 - 1 = 1,$$

and

$$c_1 e^{-1} + c_2 e^{-1/\varepsilon} e^{1} = 2.$$

We can neglect the transcendentally small term $e^{-1/\varepsilon}$ to obtain $c_1 = 2e$ and $c_2 = 2(1 - e)$. Substituting into 5.3.8 leads to

$$f = 2e^{1-x} + 2(1 - e)e^{-x/\varepsilon} e^{x} + x - 1 + O(\varepsilon). \qquad 5.3.9$$

There is an apparent discrepancy between 5.3.9 and the composite expansion 5.3.7 in the coefficient of the $e^{-x/\varepsilon}$ term. There is an extra multiple of e^x in 5.3.9. However, the $e^{-x/\varepsilon}$ term only contributes in the boundary layer where $x = O(\varepsilon)$ so that the coefficient e^x may, to leading order, be replaced by unity. Thus the leading order composite expansion and the leading order term in the exact solution are in complete agreement.

A boundary layer of thickness $O(\sqrt{\varepsilon})$

A boundary layer of thickness $O(\sqrt{\varepsilon})$ occurs in the following example,

$$\varepsilon\frac{d^2f}{dx^2} + x^2\frac{df}{dx} - f = 0$$

$$0 < \varepsilon \ll 1, \quad f(0) = 1, \quad f(1) = 2. \qquad 5.3.10$$

We seek a one-term composite expansion. We will tentatively assume that a boundary layer occurs at $x = 0$ although the vanishing of the coefficient of the first derivative suggests the possibility of nonstandard behavior.

The one-term outer expansion satisfies

$$x^2\frac{df_0^{out}}{dx} - f_0^{out} = 0, \quad f_0^{out}(1) = 2.$$

This equation is of variable separable type so that

$$\int\frac{df_0^{out}}{f_0^{out}} = \int\frac{dx}{x^2}.$$

Integrating yields

$$\ln|f_0^{out}| = -\frac{1}{x} + K$$

so $f_0^{out} = Ce^{-1/x}$ where $C = \pm e^K$.

The boundary condition at $x = 1$ leads to $C = 2e$, so the one-term outer expansion is

$$f_0^{out} = 2e^{1-1/x}. \qquad 5.3.11$$

We suppose the boundary layer thickness is $O(\varepsilon^p)$ where p is to be determined from the principle of least degeneracy. The stretched variable is $s = x/\varepsilon^p$ and 5.3.10 becomes

$$\varepsilon^{1-2p}\frac{d^2f}{ds^2} + \varepsilon^p s^2\frac{df}{ds} - f = 0.$$

The second term is always dominated by the third, so the principle of least degeneracy requires the first term to be of the same order as the third term (i.e. O(1)). Thus $p = 1/2$ and the one-term inner expansion satisfies

$$\frac{d^2f_0^{in}}{ds^2} - f_0^{in} = 0, \quad f_0^{in}(0) = 1.$$

The solution is easily found to be

$$f_0^{\text{in}} = Ae^s + (1 - A)e^{-s}. \qquad 5.3.12$$

Prandtl's matching condition requires

$$\lim_{x \to 0} 2\exp(1 - 1/x) = \lim_{s \to \infty}[Ae^s + (1 - A)e^{-s}]$$

which yields $A = 0$. This example is rather special in that A will be zero for all boundary conditions.

The one-term composite expansion is

$$f_0^{\text{comp}} = 2e^{1-1/x} + e^{-x/\sqrt{\varepsilon}}.$$

We conclude this example with the observation that a choice for the value of the index p other than $p = 1/2$ leads to boundary layer equations with insufficient structure to generate the required rapidly decaying behavior.

Thus if $p > 1/2$ the dominant equation becomes $d^2f_0^{\text{in}}/ds^2 = 0$ with solution satisfying $f_0^{\text{in}}(0) = 1$ given by $f_0^{\text{in}} = 1 + As$. This obviously cannot be matched as $s \to \infty$ to $\lim_{x \to 0} f_0^{\text{out}}(x)$ which is zero. Whereas if $p < 1/2$ the dominant equation degenerates to $f_0^{\text{in}} = 0$ which does not satisfy the boundary condition at $s = 0$.

An interior boundary layer

In the following example a boundary layer occurs in the interior of a region.

$$\varepsilon\frac{d^2f}{dx^2} + x\frac{df}{dx} + xf = 0, \quad -1 < x < 1$$

$$0 < \varepsilon \ll 1, \quad f(-1) = e, \quad f(1) = 2e^{-1}. \qquad 5.3.13$$

The coefficient of the first derivative is positive in the range $0 < x < 1$ which indicates the occurrence of a boundary layer at the left-hand limit $x = 0$. While the corresponding coefficient is negative in the range $-1 < x < 0$ indicating a boundary layer located at the right-hand limit which again is $x = 0$. Thus we are led to expect two outer expansions for positive and negative x respectively and an inner expansion in the boundary layer located at $x = 0$. We denote the leading term in the outer expansion for positive x by f_0^+; it satisfies

$$\frac{df_0^+}{dx} + f_0^+ = 0, \quad f_0^+(1) = 2e^{-1},$$

with solution

$$f_0^+ = 2e^{-x}. \qquad 5.3.14$$

The outer expansion for negative x, f_0^-, satisfies

$$\frac{\mathrm{d}f_0^-}{\mathrm{d}x} + f_0^- = 0, \quad f_0^-(-1) = \mathrm{e},$$

with solution

$$f_0^- = \mathrm{e}^{-x}. \qquad 5.3.15$$

We suppose the boundary layer at $x = 0$ has thickness $\mathrm{O}(\varepsilon^p)$ and determine the index p using the principle of least degeneracy. Let $s = x/\varepsilon^p$ so that the differential equation becomes

$$\varepsilon^{1-2p}\frac{\mathrm{d}^2 f}{\mathrm{d}s^2} + s\frac{\mathrm{d}f}{\mathrm{d}s} + \varepsilon^p s f = 0.$$

The third term is dominated by the second term. The first term has the same order as the second term if $p = 1/2$. For this choice of p the leading term of the inner expansion, f_0^{in}, satisfies

$$\frac{\mathrm{d}^2 f_0^{\mathrm{in}}}{\mathrm{d}s^2} + s\frac{\mathrm{d}f_0^{\mathrm{in}}}{\mathrm{d}s} = 0. \qquad 5.3.16$$

To solve 5.3.16 we take advantage of the absence of f_0^{in} except in differentiated form and introduce the function $w = \mathrm{d}f_0^{\mathrm{in}}/\mathrm{d}s$. Then w satisfies the first order separable equation $\mathrm{d}w/\mathrm{d}s + sw = 0$ which has solution $w = A\exp(-s^2/2)$ where A is an arbitrary constant. Integrating leads to the expression

$$f_0^{\mathrm{in}} = A\int_0^s \exp(-t^2/2)\,\mathrm{d}t + f_0^{\mathrm{in}}(0),$$

which expressed in terms of the error function, $\operatorname{erf} z = (2/\sqrt{\pi})\int_0^z \exp(-u^2)\,\mathrm{d}u$, becomes

$$f_0^{\mathrm{in}} = B\operatorname{erf}(s/\sqrt{2}) + f_0^{\mathrm{in}}(0), \qquad 5.3.17$$

where B is a new constant. Prandtl's matching condition applied to the region $x > 0$ is

$$\lim_{s\to+\infty} f_0^{\mathrm{in}}(s) = \lim_{x\to 0^+} f_0^+(x)$$

and correspondingly for $x < 0$ we have

$$\lim_{s\to-\infty} f_0^{\mathrm{in}}(s) = \lim_{x\to 0^-} f_0^-(x).$$

Using the limiting values $\operatorname{erf}(\pm\infty) = \pm 1$ yields $B + f_0^{\mathrm{in}}(0) = 2$ and $-B + f_0^{\mathrm{in}}(0) = 1$ with solutions $f_0^{\mathrm{in}}(0) = 1.5$ and $B = 0.5$. The leading order terms over the whole region are

$$f_0^+ = 2e^{-x}, \qquad x > O(\sqrt{\varepsilon})$$
$$f_0^{in} = 0.5\,\mathrm{erf}(x/\sqrt{2\varepsilon}) + 1.5, \quad x = O(\sqrt{\varepsilon})$$
$$f_0^- = e^{-x}, \qquad x < -O(\sqrt{\varepsilon}).$$

A composite expansion cannot be formed in the standard way when there is more than one outer solution. However, the behavior of f_0^{in} for $|x| > O(\sqrt{\varepsilon})$ is as follows:

$$f_0^{in}[x > O(\sqrt{\varepsilon})] = 0.5 + 1.5 + \text{T.S.T.}$$

$$f_0^{in}[x < -O(\sqrt{\varepsilon})] = -0.5 + 1.5 + \text{T.S.T.}$$

Utilizing this enables a uniformly valid one-term composite expansion to be constructed which yields the correct coefficient of e^{-x} outside the boundary layer and the correct leading order behavior within the boundary layer. It is

$$f_0^{comp} = [0.5\,\mathrm{erf}(x/\sqrt{2\varepsilon}) + 1.5]e^{-x}. \qquad 5.3.18$$

In Fig. 5.2 the composite solution is shown for the case $\varepsilon = 0.005$. The rapid change within the interior boundary layer is clearly shown.

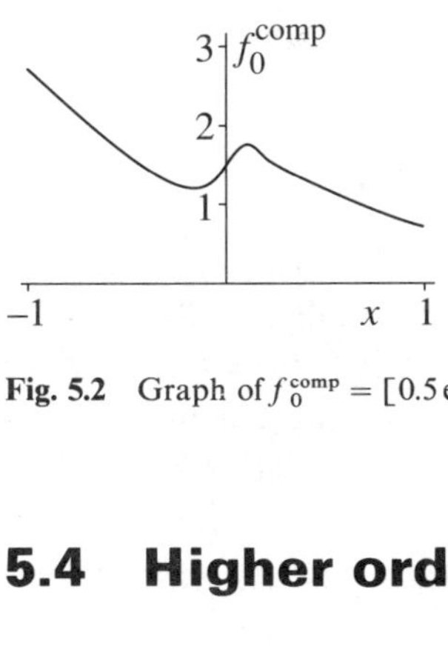

Fig. 5.2 Graph of $f_0^{comp} = [0.5\,\mathrm{erf}(x/\sqrt{2\varepsilon}) + 1.5]e^{-x}$ for $\varepsilon = 0.005$

5.4 Higher order matching

We return to the model problem studied in Section 5.1 to consider the nature of higher order terms in the inner and outer expansions and develop a matching condition. The problem specification is

$$\varepsilon\frac{d^2f}{dx^2} + \frac{df}{dx} = 2x + 1, \quad 0 < x < 1$$

$$0 < \varepsilon \ll 1, \quad f(0) = 1 \quad \text{and} \quad f(1) = 4.$$

The exact solution 5.1.12, is

$$f^{exact} = x^2 + x + 2 - \exp(-x/\varepsilon) + \varepsilon[2(1 - x) - 2\exp(-x/\varepsilon)].$$

The inner expansion, 5.1.16, obtained by solving the associated equations 5.1.15 for each term is

$$f^{\text{in}} \sim A + (1 - A)\mathrm{e}^{-s} + \varepsilon(B - B\mathrm{e}^{-s} + s) + \varepsilon^2(C - C\mathrm{e}^{-s} + s^2 - 2s)$$

$$+ \sum_{n=3}^{\infty} \varepsilon^n D_n(1 - \mathrm{e}^{-s}). \qquad 5.4.1$$

Clearly Prandtl's condition cannot be used for the higher order terms because of the occurrence of the algebraic terms εs and $\varepsilon^2(s^2 - 2s)$ which are unbounded as $s \to \infty$. These terms represent x and $x^2 - 2\varepsilon x$ respectively and can be identified as part of the exact solution. Indeed, if we replace x by εs in the exact solution we obtain

$$f^{\text{exact}} = \varepsilon^2 s^2 + \varepsilon s + 2 - \mathrm{e}^{-s} + \varepsilon[2(1 - \varepsilon s) - 2\mathrm{e}^{-s}],$$

which after arranging the terms in asymptotic order becomes

$$f^{\text{exact}} = 2 - \mathrm{e}^{-s} + \varepsilon(2 - 2\mathrm{e}^{-s} + s) + \varepsilon^2(s^2 - 2s).$$

Comparing this with the inner expansion yields the following values of the constants: $A = 2$, $B = 2$, $C = 0$, $D_n = 0$, $n = 3, 4 \ldots$.

Of course our concern is to determine these constants by matching f^{in} and f^{out} rather than by comparison with the exact solution. The outer expansion is given by equation 5.1.9,

$$f^{\text{out}} = x^2 + x + 2 + \varepsilon 2(1 - x). \qquad 5.4.2$$

The inner and outer expansions are not uniformly valid for all values of x. The outer expansion has been developed for $x = \mathrm{O}(1)$ and the transcendentally small term $\mathrm{e}^{-x/\varepsilon}$ did not occur in the derivation of 5.4.2. Thus f^{out} will not be valid for $x = \mathrm{O}(\varepsilon)$. On the other hand the inner expansion is not valid for $x = \mathrm{O}(1)$ because subsequent terms do not remain small corrections to previous terms if $x = \mathrm{O}(1)$. Thus, for example, if we write the one-term expansion of 5.4.1 as

$$f^{\text{in}} = A + (1 - A)\mathrm{e}^{-s} + \mathrm{O}(\varepsilon), \qquad 5.4.3$$

to indicate that the remainder tends to zero like ε, the statement is valid only for $s = \mathrm{O}(1)$, i.e. for $x = \mathrm{O}(\varepsilon)$. However for $x = \mathrm{O}(1)$ the term εs in the remainder of 5.4.3 is an $\mathrm{O}(1)$ term. The following far less demanding statement than 5.4.3,

$$f^{\text{in}} = A + (1 - A)\mathrm{e}^{-s} + \mathrm{o}(1) \qquad 5.4.4$$

is also invalid for $x = \mathrm{O}(1)$. However, 5.4.4 is valid in an intermediate region between $x = \mathrm{O}(1)$ and the boundary layer region $x = \mathrm{O}(\varepsilon)$. We could choose $x = \mathrm{O}(\varepsilon^{1/2})$ and 5.4.4 is valid although it can be simplified because $\exp[-\mathrm{O}(\varepsilon^{-1/2})]$ is T.S.T. Thus for $x = \mathrm{O}(\varepsilon^{1/2})$

$$f^{\text{in}} = A + \mathrm{o}(1). \qquad 5.4.5$$

If we now consider f^{out} for $x = \mathrm{O}(\varepsilon^{1/2})$ we have

$$f^{\text{out}} = 2 + \mathrm{o}(1). \tag{5.4.6}$$

Equating the two leading order expansions 5.4.5 and 5.4.6 in the intermediate region leads to the result $A = 2$. We have recovered Prandtl's condition by matching in an intermediate region. To match the two leading order terms the intermediate region could have been chosen to be $x = \mathrm{O}(\varepsilon^\alpha)$ for any α in the range $0 < \alpha < 1$ provided the inequalities are adhered to, i.e. $\alpha \neq 0$ and $\alpha \neq 1$. We will discover that stronger restrictions must be placed on the intermediate region $x = \mathrm{O}(\varepsilon^\alpha)$ to enable higher order expansions to be matched. The two-term outer expansion,

$$f^{\text{out}}_{\text{2term}} = x^2 + x + 2 + \varepsilon 2(1 + x),$$

happens to be the complete expansion in the outer region for this model problem but in general it will differ from the full outer expansion by a term of $\mathrm{O}(\varepsilon^2)$. We introduce an intermediate variable $t = x/\varepsilon^\alpha$ and expand $f^{\text{out}}_{\text{2term}}$ in the intermediate region by fixing t and arranging terms in powers of ε.

$$f^{\text{out}}_{\text{2term}} = 2 + \varepsilon^\alpha t + \varepsilon^{2\alpha} t^2 + 2\varepsilon - 2\varepsilon^{1+\alpha} t. \tag{5.4.7}$$

The next step is to express the two-term inner expansion in terms of the intermediate variable t, using $s = t/\varepsilon^{1-\alpha}$, and expand this for fixed t. The exponential term, e^{-s}, becomes $\exp(-t/\varepsilon^{1-\alpha})$ which is transcendentally small as $\varepsilon \to 0$ for $\alpha < 1$. The two-term inner expansion becomes

$$\begin{aligned} f^{\text{in}}_{\text{2term}} &= A + \varepsilon(B + t/\varepsilon^{1-\alpha}) + \text{T.S.T.} \\ &= A + \varepsilon^\alpha t + \varepsilon B + \text{T.S.T.} \end{aligned} \tag{5.4.8}$$

Matching 5.4.7 and 5.4.8 to $\mathrm{O}(\varepsilon)$ is performed by equating them to this order. This is only possible if $\varepsilon^{2\alpha} = \mathrm{o}(\varepsilon)$ i.e. $\alpha > 1/2$. Then

$$f^{\text{out}}_{\text{2term}} = 2 + \varepsilon^\alpha t + 2\varepsilon + \mathrm{o}(\varepsilon) \tag{5.4.9}$$

and

$$f^{\text{in}}_{\text{2term}} = A + \varepsilon^\alpha t + \varepsilon B + \mathrm{o}(\varepsilon), \tag{5.4.10}$$

which when equated yield $A = 2$, $B = 2$.

The uniformly valid two-term composite expansion is formed as follows:

$$f^{\text{comp}}_{\text{2term}} = f^{\text{out}}_{\text{2term}} + f^{\text{in}}_{\text{2term}} - f^{\text{match}}_{\text{2term}}$$

where $f^{\text{match}}_{\text{2term}}$ is given by the terms up to $\mathrm{O}(\varepsilon)$ appearing in 5.4.9 or equivalently 5.4.10 with $\varepsilon^\alpha t$ replaced by x. The resulting expression is

$$\begin{aligned} f^{\text{comp}}_{\text{2term}} &= x^2 + x + 2 + \varepsilon 2(1 - x) + 2 - \mathrm{e}^{-x/\varepsilon} + \varepsilon\left(2 - 2\mathrm{e}^{-x/\varepsilon} + \frac{x}{\varepsilon}\right) - 2 - x - 2\varepsilon \\ &= x^2 + x + 2 - \mathrm{e}^{-x/\varepsilon} + \varepsilon(2 - 2x - 2\mathrm{e}^{-x/\varepsilon}). \end{aligned}$$

The two-term composite expansion will in general differ from the exact solution by a quantity which is uniformly $o(\varepsilon)$ throughout the region of interest ($0 < x < 1$). The constant C appearing in the three-term inner expansion, 5.4.1, can be evaluated by matching the three-term inner and outer expansion in an intermediate region. In this example the third and all subsequent terms in the outer expansion happen to be zero. Matching leads to $C = 0$ and higher order matching (4 terms and above) leads to the constants $D_n = 0$ for $n = 3, 4, 5 \ldots$.

Worked example

Obtain a two-term composite expansion for the solution of

$$\varepsilon \frac{d^2 f}{dx^2} + (x+1)\frac{df}{dx} + f = 2x, \quad 0 < x < 1$$

$$0 < \varepsilon \ll 1, \quad f(0) = 1, \quad f(1) = 2.$$

Solution

The leading order composite expansion for this problem was obtained in Section 5.1. The boundary layer thickness is $O(\varepsilon)$ located at $x = 0$. The two-term outer expansion

$$f_{2\text{term}}^{\text{out}} = f_0^{\text{out}} + \varepsilon f_1^{\text{out}}$$

is determined from the following system of equations

$$O(1): (x+1)\frac{df_0^{\text{out}}}{dx} + f_0^{\text{out}} = 2x, \quad f_0^{\text{out}}(1) = 2$$

$$O(\varepsilon): (x+1)\frac{df_1^{\text{out}}}{dx} + f_1^{\text{out}} = -\frac{d^2 f_0^{\text{out}}}{dx^2}, \quad f_1^{\text{out}}(1) = 0.$$

The $O(1)$ equation can be expressed in the form

$$\frac{d}{dx}[(x+1)f_0^{\text{out}}] = 2x$$

with solution satisfying $f_0^{\text{out}}(1) = 2$ given by

$$f_0^{\text{out}} = \frac{x^2+3}{x+1}.$$

The $O(\varepsilon)$ equation may be written as

$$\frac{d}{dx}[(x+1)f_1^{\text{out}}] = -\frac{d}{dx}\left[\frac{2x}{x+1} - \frac{(x^2+3)}{(x+1)^2}\right]$$

with solution satisfying $f_1^{\text{out}}(1) = 0$ given by

$$f_1^{\text{out}} = -\frac{2x}{(x+1)^2} + \frac{(x^2+3)}{(x+1)^3}.$$

The stretching transformation $s = x/\varepsilon$ leads to the differential equation

$$\frac{d^2f}{ds^2} + (1 + \varepsilon s)\frac{df}{ds} + \varepsilon f = 2\varepsilon^2 s.$$

The two-term inner expansion

$$f^{\text{in}}_{\text{2term}} = f^{\text{in}}_0(s) + \varepsilon f^{\text{in}}_1(s)$$

is determined by the equations

$$\mathrm{O}(1): \frac{d^2 f^{\text{in}}_0}{ds^2} + \frac{df^{\text{in}}_0}{ds} = 0, \quad f^{\text{in}}_0(0) = 1$$

$$\mathrm{O}(\varepsilon): \frac{d^2 f^{\text{in}}_1}{ds^2} + \frac{df^{\text{in}}_1}{ds} = -s\frac{df^{\text{in}}_0}{ds} - f^{\text{in}}_0, \quad f^{\text{in}}_1(0) = 0.$$

The solution of the O(1) equation is $f^{\text{in}}_0 = A + (1 - A)e^{-s}$.

The $\mathrm{O}(\varepsilon)$ equation becomes

$$\frac{d^2 f^{\text{in}}_1}{ds^2} + \frac{df^{\text{in}}_1}{ds} = s(1 - A)e^{-s} - A - (1 - A)e^{-s}.$$

The complementary function is $C + De^{-s}$ and the particular integral will contain the sum of terms $\alpha s e^{-s}$, $\beta s^2 e^{-s}$ and γs. Substituting into the differential equation yields the following values for the constants, $\alpha = 0$, $\beta = \frac{1}{2}(A - 1)$ and $\gamma = -A$. The general solution is

$$f^{\text{in}}_1 = C + De^{-s} - As + \tfrac{1}{2}(A - 1)s^2 e^{-s}$$

and the boundary condition $f^{\text{in}}_1(0) = 0$ requires $D = -C$.

The two-term inner expansion is

$$f^{\text{in}}_{\text{2term}} = A + (1 - A)e^{-s} + \varepsilon[C(1 - e^{-s}) - As + \tfrac{1}{2}(A - 1)s^2 e^{-s}].$$

The two-term outer expansion is

$$f^{\text{out}}_{\text{2term}} = \frac{x^2 + 3}{x + 1} + \varepsilon\left(-\frac{2x}{(x + 1)^2} + \frac{(x^2 + 3)}{(x + 1)^3}\right).$$

We choose an intermediate region $x = \mathrm{O}(\varepsilon^\alpha)$, $0 < \alpha < 1$, in which both expansions are assumed to be valid in that they differ from the exact solution by a quantity which is $\mathrm{o}(\varepsilon)$. Let $t = x/\varepsilon^\alpha$ so that $s = \varepsilon^{\alpha - 1}t$. Now the exponential term e^{-s} becomes $\exp(-\varepsilon^{\alpha - 1}t)$ and for $\alpha < 1$ this is T.S.T. as $\varepsilon \to 0$ for t fixed. The inner expansion for fixed t is therefore

$$f^{\text{in}}_{\text{2term}} = A + \varepsilon(C - A\varepsilon^{\alpha - 1}t) + \text{T.S.T.}$$

The outer expansion, expressed in terms of the intermediate variable, becomes

$$\begin{aligned} f^{\text{out}}_{\text{2term}} &= \frac{3 + \varepsilon^{2\alpha}t^2}{1 + \varepsilon^\alpha t} + \varepsilon\left(-\frac{2\varepsilon^\alpha t}{(1 + \varepsilon^\alpha t)^2} + \frac{(3 + \varepsilon^{2\alpha}t^2)}{(1 + \varepsilon^\alpha t)^3}\right) \\ &= (3 + \varepsilon^{2\alpha}t^2)(1 - \varepsilon^\alpha t + \varepsilon^{2\alpha}t^2 - \cdots) + \varepsilon[3 + \mathrm{O}(\varepsilon^\alpha)] \\ &= 3 - 3\varepsilon^\alpha t + 3\varepsilon + \mathrm{O}(\varepsilon^{2\alpha}). \end{aligned}$$

This can only be matched to the inner expansion up to and including terms of $O(\varepsilon)$ if $\alpha > 1/2$. Then $A - A\varepsilon^\alpha t + \varepsilon C = 3 - 3\varepsilon^\alpha t + 3\varepsilon$ with solution $A = 3$ and $C = 3$. The two-term uniformly valid composite expansion is given by

$$f^{\text{comp}}_{\text{2term}} = \frac{x^2+3}{x+1} + \varepsilon\left(-\frac{2x}{(x+1)^2} + \frac{x^2+3}{(x+1)^3}\right) + 3 - 2\varepsilon^{-x/\varepsilon}$$

$$+ \varepsilon\left(3(1 - e^{-x/\varepsilon}) - \frac{3x}{\varepsilon} + \frac{x^2}{\varepsilon^2}e^{-x/\varepsilon}\right) - 3 + 3x - 3\varepsilon.$$

Exercises

Obtain a uniformly valid two-term composite expansion for the solutions of:

(i) $\varepsilon\dfrac{d^2f}{dx^2} + (2+x)\dfrac{df}{dx} + f = 1, \qquad 0 < x < 1$

$0 < \varepsilon \ll 1, \quad f(0) = 2, \quad f(1) = 0.$

(ii) $\varepsilon\dfrac{d^2f}{dx^2} + \dfrac{df}{dx} + \dfrac{f}{x+1} = 3(1+x), \qquad 0 < x < 1$

$0 < \varepsilon \ll 1, \quad f(0) = 1, \quad f(1) = 2.$

(iii) $\varepsilon\dfrac{d^2f}{dx^2} + (x-2)\dfrac{df}{dx} + f = 2 - x, \qquad 0 < x < 1$

$0 < \varepsilon \ll 1, \quad f(0) = 2, \quad f(1) = 1.$

(iv) $\varepsilon\dfrac{d^2f}{dx^2} - \dfrac{df}{dx} + \dfrac{f}{2-x} = 2, \qquad -1 < x < 0$

$0 < \varepsilon \ll 1, \quad f(-1) = 2, \quad f(0) = 1.$

Answers

(i) $$f^{\text{comp}}_{\text{2term}} = \frac{x-1}{2+x} + \varepsilon\left(\frac{1}{3(2+x)} - \frac{3}{(2+x)^3}\right) + \frac{5}{2}e^{-2x/\varepsilon}$$

$$+ \varepsilon\left(\frac{5}{24} - \frac{5}{4}\frac{x^2}{\varepsilon^2}\right)e^{-2x/\varepsilon}.$$

(ii) $$f^{\text{comp}}_{\text{2term}} = (x+1)^2 - \frac{4}{x+1} + \varepsilon\left(\frac{8}{x+1} - \frac{8}{(x+1)^2} - (x+1)\right) + 4e^{-x/\varepsilon}$$

$$+ \varepsilon\left(1 + \frac{4x}{\varepsilon}\right)e^{-x/\varepsilon}.$$

(iii) $$f^{\text{comp}}_{\text{2term}} = \frac{2-x}{2} + \frac{2}{2-x} + \varepsilon\left(\frac{2}{(2-x)^3} - \frac{1}{2(2-x)}\right) - \frac{3}{2}\exp -\left(\frac{1-x}{\varepsilon}\right)$$
$$+\,\varepsilon\left[-\frac{3}{2} + \frac{3}{4}\left(\frac{1-x}{\varepsilon}\right)^2\right]\exp -\left(\frac{1-x}{\varepsilon}\right).$$

(iv) $$f^{\text{comp}}_{\text{2term}} = 2 - x - \frac{3}{2-x} + \varepsilon\left(\frac{2}{2-x} - \frac{6}{(2-x)^2}\right) + \frac{1}{2}e^{x/\varepsilon} + \varepsilon\left(\frac{1}{2} - \frac{x}{4\varepsilon}\right)e^{x/\varepsilon}.$$

5.5 Nonlinear examples

Most nonlinear second order differential equations cannot be solved by exact techniques. However, if a small parameter multiplies the highest derivative then the method of matched asymptotic expansions can be used and the resulting equations may be exactly solvable because they are of lower order.

Consider the following example:

$$\varepsilon\frac{d^2f}{dx^2} + \frac{df}{dx} + f^2 = 0, \quad 0 < x < 1 \qquad 5.5.1$$

$$0 < \varepsilon \ll 1, \quad f(0) = 2, \quad f(1) = \tfrac{1}{2}.$$

The coefficients of the first and second order derivatives have the same sign, so the boundary layer will occur at the lower boundary $x = 0$. The one-term outer expansion satisfies

$$\frac{df^{\text{out}}_0}{dx} + (f^{\text{out}}_0)^2 = 0, \quad f^{\text{out}}_0(1) = \tfrac{1}{2}.$$

Although this is nonlinear it is of first order separable form and can easily be solved by exact techniques. We have

$$\int \frac{df^{\text{out}}_0}{(f^{\text{out}}_0)^2} = -\int dx$$

so that on integrating we obtain

$$\frac{1}{f^{\text{out}}_0} = x + c$$

and the boundary condition yields $c = 1$, thus

$$f^{\text{out}}_0 = \frac{1}{x+1}. \qquad 5.5.2$$

The stretching transformation for the inner region will be $s = x/\varepsilon$ and therefore the inner expansion satisfies

$$\frac{d^2 f^{in}}{ds^2} + \frac{df^{in}}{ds} + \varepsilon(f^{in})^2 = 0.$$

The one-term inner expansion f_0^{in} satisfies the dominant part of this equation

$$\frac{d^2 f_0^{in}}{ds^2} + \frac{df_0^{in}}{ds} = 0.$$

The solution is $f_0^{in} = A + Be^{-s}$ and the boundary condition $f_0^{in}(0) = 2$ yields $A + B = 2$. Thus

$$f_0^{in} = A + (2 - A)e^{-s}. \qquad 5.5.3$$

Prandtl's matching condition applied to 5.5.2 and 5.5.3 is

$$\lim_{x \to 0} f_0^{out} = \lim_{s \to \infty} f_0^{in}$$

i.e.

$$\frac{1}{0 + 1} = A + (2 - A)e^{-\infty}.$$

Thus $A = 1$ and the composite one-term uniformly valid expansion is

$$f^{comp}_{1\,term} = \frac{1}{x + 1} + \cancel{1} + e^{-x/\varepsilon} - \cancel{1}.$$

Next consider the nonlinear example

$$\varepsilon \frac{d^2 f}{dx^2} + 2f\frac{df}{dx} - 4f = 0, \quad 0 < x < 1, \quad 0 < \varepsilon \ll 1, \quad f(0) = 0, \quad f(1) = 4. \qquad 5.5.4$$

The nonlinearity is associated with the first derivative. The location of the boundary layer depends on the relative sign of the first and second derivative coefficients. If we assume that the dependent variable is nonnegative throughout the interval $0 < x < 1$ then the boundary layer will occur at $x = 0$. The one-term outer expansion satisfies

$$2f_0^{out} \cdot \frac{df_0^{out}}{dx} - 4f_0^{out} = 0$$

and for nonzero f_0^{out} this reduces to

$$\frac{df_0^{out}}{dx} = 2$$

with solution satisfying the $x = 1$ boundary condition given by

$$f_0^{out} = 2x + 2. \qquad 5.5.5$$

Assuming that the boundary layer stretching transformation is $s = x/\varepsilon$ we obtain the following dominant order equation for the one-term inner expansion

$$\frac{d^2 f_0^{in}}{ds^2} + 2f_0^{in}\frac{df_0^{in}}{ds} = 0.$$

This may be written in the form

$$\frac{d}{ds}\left[\frac{df_0^{in}}{ds} + (f_0^{in})^2\right] = 0$$

so that

$$\frac{df_0^{in}}{ds} + (f_0^{in})^2 = c. \qquad 5.5.6$$

The boundary condition at $x = 0$ becomes $f_0^{in}(s = 0) = 0$. The matching value for f_0^{in} at the edge of the boundary layer is $\lim_{x\to 0} f_0^{out} = 2$. It is reasonable to assume (and later verify) that f_0^{in} increases monotonically through the boundary layer so that df_0^{in}/ds is positive. Then the constant c in equation 5.5.6 will be positive and we denote it by a^2. The equation is separable and we have

$$\int \frac{df_0^{in}}{a^2 - (f_0^{in})^2} = \int ds.$$

Using partial fractions leads to

$$-\frac{1}{2a}\int \frac{df_0^{in}}{f_0^{in} - a} + \frac{1}{2a}\int \frac{df_0^{in}}{f_0^{in} + a} = s + b,$$

therefore

$$-\ln|f_0^{in} - a| + \ln|f_0^{in} + a| = 2as + 2ab.$$

Thus

$$\frac{f_0^{in} + a}{f_0^{in} - a} = Ae^{2as},$$

where the constant A may be positive or negative ($A = \pm e^{2ab}$). The boundary condition requires that $A = -1$ and solving for f_0^{in} leads to

$$f_0^{in} = a\frac{e^{2as} - 1}{e^{2as} + 1} = a \tanh as.$$

Prandtl's matching condition requires

$$\lim_{s\to\infty} f_0^{in} = \lim_{x\to 0} f_0^{out}$$

i.e. $\qquad a = 2.$

Thus $f_0^{\text{in}} = 2\tanh 2s$ and the composite one-term uniformly valid expansion is

$$f_{1\text{term}}^{\text{comp}} = 2x + \not{2} + 2\tanh(2x/\varepsilon) - \not{2}.$$

At this stage we can confirm that the solution is nonnegative and monotonically increasing in the interval $0 < x < 1$ as was assumed during the solution process.

Exercises

Obtain one-term uniformly valid composite expansions for the solutions of

(i) $\varepsilon\dfrac{d^2f}{dx^2} - \dfrac{df}{dx} + \dfrac{1}{f} = 0, \quad 0 < x < 1$

$0 < \varepsilon \ll 1, \quad f(0) = 2, \quad f(1) = 1.$

(ii) $\varepsilon\dfrac{d^2f}{dx^2} + \dfrac{df}{dx} + e^f = 0, \quad 0 < x < 1$

$0 < \varepsilon \ll 1, \quad f(0) = 1, \quad f(1) = -\ln 2.$

(iii) $\varepsilon\dfrac{d^2f}{dx^2} + 2\dfrac{df}{dx} + xf^2 = 0, \quad 0 < x < 1$

$0 < \varepsilon \ll 1, \quad f(0) = 0, \quad f(1) = 2.$

(iv) $\varepsilon\dfrac{d^2f}{dx^2} + \dfrac{df}{dx} - 2xe^{-f} = 0, \quad 0 < x < 1$

$0 < \varepsilon \ll 1, \quad f(0) = 0, \quad f(1) = 1.$

Answers

(i) $f_{1\text{term}}^{\text{comp}} = \sqrt{2x+4} + (1 - \sqrt{6})e^{(x-1)/\varepsilon}.$

(ii) $f_{1\text{term}}^{\text{comp}} = -\ln(1+x) + e^{-x/\varepsilon}.$

(iii) $f_{1\text{term}}^{\text{comp}} = \dfrac{4}{1+x^2} - 4e^{-2x/\varepsilon}.$

(iv) $f_{1\text{term}}^{\text{comp}} = \ln(x^2 + e - 1) - \ln(e-1).e^{-x/\varepsilon}.$

5.6 Practical applications

We now consider examples of thermal and velocity boundary layers which occur in fluid flow. Heat and momentum transfer both involve the combined effect of convection and diffusion. The derivation of the governing equations of heat transfer and fluid flow can be found in standard textbooks on fluid dynamics. In vector form the heat equation is

$$\rho c\left(\frac{\partial T}{\partial \tau} + \mathbf{V}.\nabla T\right) - k\nabla^2 T = Q, \qquad 5.6.1$$

Convection — Conduction (diffusion of heat) — Source

and the viscous fluid flow equations (the Navier–Stokes equations) are

$$\rho\left(\frac{\partial \mathbf{V}}{\partial \tau} + \mathbf{V}.\nabla\mathbf{V}\right) = -\nabla P + \mu\nabla^2\mathbf{V} + \mathbf{B}. \qquad 5.6.2$$

Convection — Pressure force — Viscous force (diffusion of momentum) — Body force

To these must be added the equation of mass continuity

$$\frac{\partial \rho}{\partial \tau} + \nabla.(\rho\mathbf{V}) = 0 \qquad 5.6.3$$

which for incompressible flow becomes

$$\nabla.\mathbf{V} = 0. \qquad 5.6.4$$

The quantities which appear in these equations are:
T = Temperature
$\mathbf{V}$ = Velocity with Cartesian components U, V, W.
P = pressure
τ = time
$\nabla = \hat{\mathbf{X}}\frac{\partial}{\partial X} + \hat{\mathbf{Y}}\frac{\partial}{\partial Y} + \hat{\mathbf{Z}}\frac{\partial}{\partial Z}$ in Cartesian components where X, Y, Z are the space coordinates and $\hat{}$ denotes a unit vector.
ρ = density
c = specific heat
k = thermal conductivity
μ = viscosity
} These fluid properties will be taken to be constant.
$\mathbf{B}$ = body force
Q = heat source
} There will be no heat source or body force in subsequent applications.

The two-dimensional steady state version of these equations are

$$\text{Heat: } U\frac{\partial T}{\partial X} + V\frac{\partial T}{\partial Y} = \alpha\left(\frac{\partial^2 T}{\partial X^2} + \frac{\partial^2 T}{\partial Y^2}\right) \qquad 5.6.5$$

$$X \text{ momentum: } U\frac{\partial U}{\partial X} + V\frac{\partial U}{\partial Y} = -\frac{1}{\rho}\frac{\partial P}{\partial X} + \nu\left(\frac{\partial^2 U}{\partial X^2} + \frac{\partial^2 U}{\partial Y^2}\right) \qquad 5.6.6$$

Y momentum: $U\dfrac{\partial V}{\partial X} + V\dfrac{\partial V}{\partial Y} = -\dfrac{1}{\rho}\dfrac{\partial P}{\partial Y} + \nu\left(\dfrac{\partial^2 V}{\partial X^2} + \dfrac{\partial^2 V}{\partial Y^2}\right)$ 5.6.7

Mass continuity: $\dfrac{\partial U}{\partial X} + \dfrac{\partial V}{\partial Y} = 0$ 5.6.8

where α is the thermal diffusivity ($\alpha = k/\rho c$) and ν is the kinematic viscosity ($\nu = \mu/\rho$). Upper case letters have been used to denote variables in dimensional form. In the remainder of this chapter we will reserve lower case letters for nondimensionalized variables.

Péclet and Reynolds numbers

If a characteristic velocity is U_0 and a characteristic length is L then the appropriate nondimensionalized variables are: $x = X/L$, $y = Y/L$, $u = U/U_0$, $v = V/U_0$ and it is convenient to nondimensionalize pressure by dividing by ρU_0^2, $p = P/\rho U_0^2$. Then 5.6.5–5.6.8 become

$$u\frac{\partial T}{\partial x} + v\frac{\partial T}{\partial y} - \frac{1}{Pe}\left(\frac{\partial^2 T}{\partial x^2} + \frac{\partial^2 T}{\partial y^2}\right) \qquad 5.6.5'$$

$$u\frac{\partial u}{\partial x} + v\frac{\partial u}{\partial y} = -\frac{\partial p}{\partial x} + \frac{1}{Re}\left(\frac{\partial^2 u}{\partial x^2} + \frac{\partial^2 u}{\partial y^2}\right) \qquad 5.6.6'$$

$$u\frac{\partial v}{\partial x} + v\frac{\partial v}{\partial y} = -\frac{\partial p}{\partial y} + \frac{1}{Re}\left(\frac{\partial^2 v}{\partial x^2} + \frac{\partial^2 v}{\partial y^2}\right) \qquad 5.6.7'$$

$$\frac{\partial u}{\partial x} + \frac{\partial y}{\partial y} = 0. \qquad 5.6.8'$$

The nondimensional quantities Pe and Re are respectively the *Péclet number* and *Reynolds number* for the flow, where $Pe = U_0L/\alpha$ and $Re = U_0L/\nu$.

For water at room temperature $\alpha = 1.43 \times 10^{-7}$ m^2/s and $\nu = 1.0 \times 10^{-6}$ m^2/s. These diffusivities are extremely small in comparison with the product of the characteristic length and velocity of many flows. Indeed for the flow of most fluids α and ν are sufficiently small in comparison with the product U_0L to make Pe and Re large. The reciprocals of these numbers are associated with the diffusion terms in the heat and momentum equations. Thus for many flows convection is dominant and the diffusion term can be neglected in the bulk of flow. However, the small parameters $1/Pe$ and $1/Re$ multiply the highest order derivatives so that boundary layer effects are to be anticipated in convection dominated flows.

The incompressible fluid flow equations 5.6.6′, 7′ and 8′ are independent of the temperature and may be separately solved for the velocity and pressure field. The temperature field can then be obtained from equation 5.6.5′ using the calculated velocity field. This stage of the solution process is far easier because 5.6.5′ is a linear differential equation whereas the momentum equations are nonlinear. For our first example of a physical boundary layer we consider the case where the velocity field is given and determine the temperature field by the method of matched asymptotic expansions.

Flow between parallel planes with heat transfer

Consider the flow of fluid between parallel planes of length L separated by a distance H and held at constant temperatures $T = T_w$. Suppose a fluid flows between the planes with a uniform velocity profile $U = U_0$ and $V = 0$ as shown in Fig. 5.3. This is an idealization because the fluid will be brought to rest at the solid boundaries. Nevertheless there are applications where the fluid boundary layer is much thinner than the thermal boundary layer which allows the narrow region of rapid decay in the fluid velocity to be neglected. This point will be returned to later where we shall see that it depends on ν being much smaller than α (which is the case for liquid metals). At this stage we simply assume that U takes the constant value U_0 everywhere and V is zero.

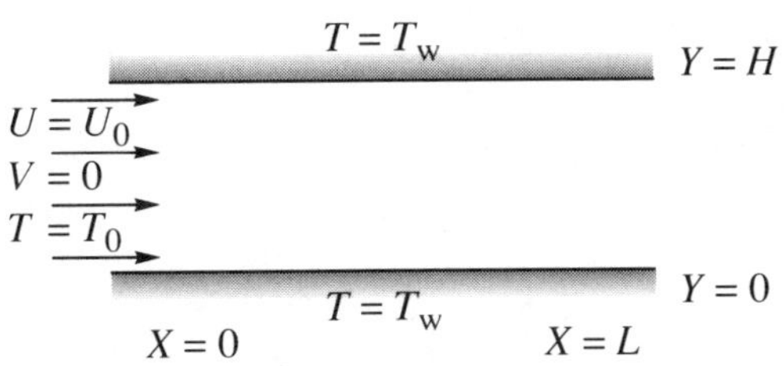

Fig. 5.3 Heat transfer by conduction and convection

Equation 5.6.5′ becomes

$$\frac{\partial T}{\partial x} = \frac{1}{Pe}\left(\frac{\partial^2 T}{\partial x^2} + \frac{\partial^2 T}{\partial y^2}\right) \quad 0 < x < 1, \quad 0 < y < h \qquad 5.6.9$$

where $h = H/L$. The boundary conditions are

$$T(0, y) = T_0 \quad 0 < y < h$$

$$T(x, 0) = T_w \quad 0 < x < 1$$

$$T(x, h) = T_w \quad 0 < x < 1.$$

Here T_0 is the inlet temperature of the fluid which is taken to be a constant value and for definiteness we suppose that $T_0 > T_w$ so that heat is extracted from the fluid by the cooler walls.

A further boundary condition is required to specify the problem uniquely. The exit temperature profile is usually unknown. We will choose to set its value at T_w,

$$T(1, y) = T_w \quad 0 < y < h.$$

This might be arranged by placing a fine mesh at the outlet with its temperature maintained at this value. In fact the outlet boundary condition has little effect on the temperature between the plates (its contribution is transcendentally small) so our choice of outlet condition is not important.

The small parameter, $1/Pe$ which we denote by ε, indicates the presence of boundary layers. They occur on both walls ($y = 0, h$) and at the outlet ($x = 1$). To study the outlet boundary layer we first consider the one-dimensional problem

$$\frac{dT}{dx} = \varepsilon \frac{d^2 T}{dx^2}, \quad 0 < x < 1$$

$$T(0) = T_0 \quad \text{and} \quad T(1) = T_w.$$

The solution could be obtained by matched asymptotic expansions but this is not necessary since the exact solution of the equation is easily found. The general solution is $T = A + Be^{x/\varepsilon}$ and the boundary conditions require

$$A + B = T_0, \quad A + Be^{1/\varepsilon} = T_w$$

which have solution

$$A = \frac{T_0 - T_w e^{-1/\varepsilon}}{1 - e^{-1/\varepsilon}}, \quad B = \frac{(T_w - T_0)e^{-1/\varepsilon}}{1 - e^{-1/\varepsilon}}$$

so that

$$T = \frac{T_0 - T_w e^{-1/\varepsilon} + (T_w - T_0)e^{-(1-x)/\varepsilon}}{1 - e^{-1/\varepsilon}}$$

which simplifies to become

$$T = T_0 + (T_w - T_0)e^{-(1-x)/\varepsilon} + \text{T.S.T.}$$

This shows that a boundary layer of thickness $O(\varepsilon)$ occurs at the outlet $x = 1$ where the temperature changes rapidly from its previous constant value of T_0, the inlet value, to satisfy the outlet boundary condition, $T = T_w$. The choice of the outlet boundary condition only influences the solution in the outlet boundary layer and is unimportant for the bulk of the region to the left of the outlet.

We now return to the two-dimensional problem. The walls $y = 0$ and $y = h$ will have boundary layers which we subsequently show are parabolic in profile with thickness $O(\sqrt{\varepsilon})$. This result has been anticipated in Fig. 5.4 which shows the boundary layer locations.

The 'outer region' is the region which is geometrically interior but it is outer in the sense that it is outside the boundary layers. The temperature in the outer region is the

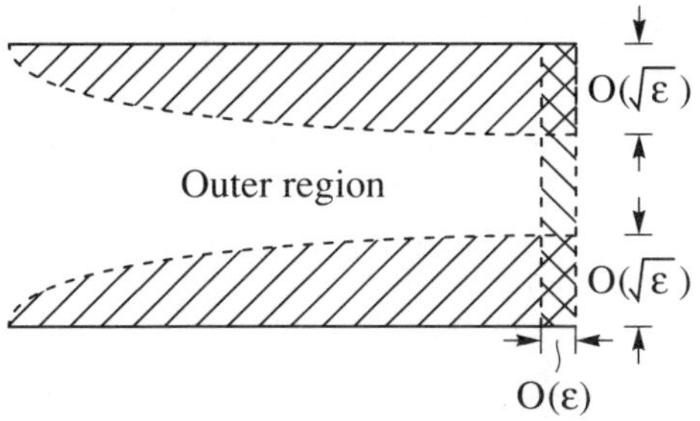

Fig. 5.4 Boundary layer thickness and locations

same as in the one-dimensional case namely T_0. Formally the one-term outer expansion of the solution of 5.6.9, $T_0^{\text{out}}(x, y)$, satisfies

$$\frac{\partial T_0^{\text{out}}}{\partial x} = 0$$

with solution $T_0^{\text{out}} = f(y)$ where f is an arbitrary function which can accommodate any inlet temperature profile at $x = 0$. In this example

$$T_0^{\text{out}} = T_0.$$

It is only necessary to consider the boundary layer at $y = 0$, the $y = h$ boundary layer will be a reflection of it. We assume the stretching transformation is of the form $s = y/\varepsilon^p$ so that equation 5.6.9 becomes

$$\frac{\partial T}{\partial x} = \varepsilon \frac{\partial^2 T}{\partial x^2} + \varepsilon^{1-2p} \frac{\partial^2 T}{\partial s^2}.$$

The left-hand side is O(1) and the first member of the right-hand side is O(ε). The principle of least degeneracy requires that p be chosen so that the second term on the right-hand side contributes to the dominant order, i.e. has O(1). Thus $p = 1/2$ and the one-term inner expansion $T_0^{\text{in}}(x, s)$ satisfies

$$\frac{\partial T_0^{\text{in}}}{\partial x} = \frac{\partial^2 T_0^{\text{in}}}{\partial s^2} \qquad 5.6.10$$

with the boundary condition $T_0^{\text{in}}(s = 0) = T_{\text{w}}$.

In general it is not possible to obtain the exact solution of partial differential equations. In this case, however, an appropriate solution is known, it is

$$T_0^{\text{in}} = T_{\text{w}} + A\,\text{erf}(s/2\sqrt{x}), \qquad 5.6.11$$

where A is an arbitrary constant and the error function is defined by

$$\text{erf}(s/2\sqrt{x}) = \frac{2}{\sqrt{\pi}} \int_0^{s/2\sqrt{x}} \exp(-t^2)\,\mathrm{d}t.$$

z	erf z
0.0	0.0
0.5	0.5205
1.0	0.8427
1.5	0.9661
2.0	0.9953
∞	1.0

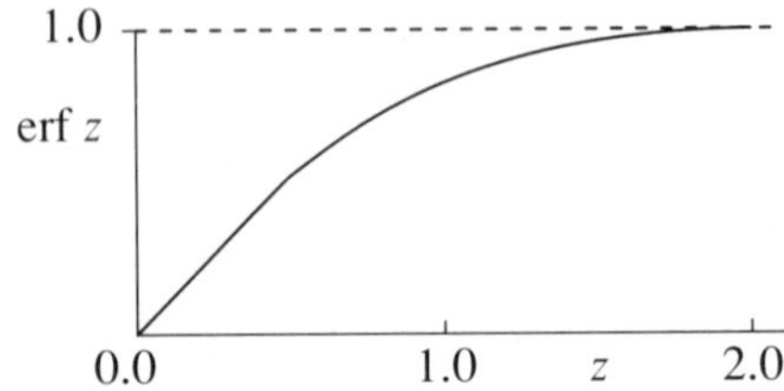

Fig. 5.5 The error function

Some values of the error function are given in the table in Fig. 5.5. The graph of the function shows its approach to the asymptotic value $\text{erf}(\infty) = 1$. For practical purposes the asymptotic value is reached when the argument exceeds 2.

To verify that 5.6.11 is a solution of 5.6.10 the technique of differentiating an integral with respect to a limit of integration must be applied. This yields

$$\frac{\partial T_0^{\text{in}}}{\partial x} = A\frac{2}{\sqrt{\pi}}\frac{\partial}{\partial x}\left(\frac{s}{2\sqrt{x}}\right)\exp(-t^2)|_{t=s/2\sqrt{x}}$$

$$= \frac{2A}{\sqrt{\pi}}\left(-\frac{s}{4x^{3/2}}\right)\exp\left(-\frac{s^2}{4x}\right)$$

and

$$\frac{\partial T_0^{\text{in}}}{\partial s} = \frac{2A}{\sqrt{\pi}}\left(\frac{1}{2\sqrt{x}}\right)\exp\left(-\frac{s^2}{4x}\right)$$

so that

$$\frac{\partial^2 T_0^{\text{in}}}{\partial s^2} = \frac{2A}{\sqrt{\pi}}\left(\frac{1}{2\sqrt{x}}\right)\left(-\frac{2s}{4x}\right)\exp\left(-\frac{s^2}{4x}\right),$$

which provides the required verification.

The boundary conditions satisfied by 5.6.11 are

$$T_0^{\text{in}}(x, s = 0) = T_w \qquad \text{for } x > 0 \qquad 5.6.12\text{a}$$

$$T_0^{\text{in}}(x, s \to \infty) = T_w + A \qquad \text{for } x > 0 \qquad 5.6.12\text{b}$$

$$T_0^{\text{in}}(x \to 0, s) = T_w + A \qquad \text{for } s > 0. \qquad 5.6.12\text{c}$$

Prandtl's matching condition is

$$\lim_{s\to\infty} T_0^{\text{in}}(x, s) = \lim_{y\to 0} T_0^{\text{out}}(x, y).$$

The left-hand side of this condition is provided by 5.6.12b and the right-hand side is T_0 so that $A = T_0 - T_w$. The inner solution expressed as a function of x and y is therefore

$$T_0^{\text{in}} = T_w + (T_0 - T_w)\text{erf}(y/2\sqrt{\varepsilon x}). \qquad 5.6.13$$

The isotherms are those curves for which $y/2\sqrt{\varepsilon x}$ is constant. Examples of these parabolic isotherms are shown in Fig. 5.6 for the cases $y/2\sqrt{\varepsilon x} = 0.5$, 1.0, 2.0.

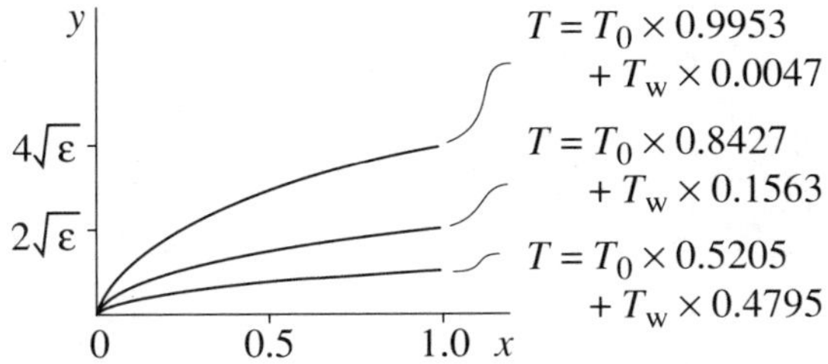

Fig. 5.6 Parabolic isotherms

This solution satisfies the inlet boundary condition since the right-hand side of 5.6.12c is T_0. The outlet boundary condition at $x = 1$ is not satisfied. A special treatment of the overlap region between the wall boundary layer and the outlet boundary layer is required where both x and y are stretched. However, this small region has no effect on the rest of the temperature region and we will neglect it.

An unphysical aspect of this solution is that the point (0, 0) has no unique temperature. The step change in the temperature boundary condition at this point cannot occur in practice. Instead there will be a local region where the temperature varies continuously between T_0 and T_w. This local imperfection at (0, 0) and (0, h) does not impair the derived solution away from these points and we will accept this imperfection in preference to undertaking the complexities of a rigorous treatment of these points.

The composite one-term approximation is formed as follows:

$$\begin{aligned} T_0^{\text{comp}} &= T_0^{\text{out}} + T_0^{\text{in}}\ (x = 1 \text{ boundary layer}) - T_0^{\text{match}} \\ &\quad + T_0^{\text{in}}\ (y = 0 \text{ boundary layer}) - T_0^{\text{match}} \\ &\quad + T_0^{\text{in}}\ (y = h \text{ boundary layer}) - T_0^{\text{match}} \end{aligned}$$

For all the boundary layers T_0^{match} is T_0 so that

$$\begin{aligned} T_0^{\text{comp}} &= T_0 + (T_w - T_0)e^{-(1-x)/\varepsilon} \\ &\quad + T_w + (T_0 - T_w)\operatorname{erf}(y/2\sqrt{\varepsilon x}) - T_0 \\ &\quad + T_w + (T_0 - T_w)\operatorname{erf}[(h - y)/2\sqrt{\varepsilon x}] - T_0. \end{aligned} \qquad 5.6.14$$

Heat transfer rate

A quantity of great practical importance is the rate of transfer of heat at the surface of the walls. Consider the wall $y = 0$ and neglect the outlet boundary layer at $x = 1$. The expression for the temperature in the vicinity of the wall is

$$T = T_w + (T_0 - T_w)\,\mathrm{erf}\left(\frac{Y}{2}\sqrt{\frac{Pe}{LX}}\right),$$

where dimensional lengths have been used. The heat transfer rate per unit area to the wall is given by $k\left.\frac{\partial T}{\partial Y}\right|_{Y=0}$ which becomes

$$k\left.\frac{\partial T}{\partial Y}\right|_{Y=0} = k(T_0 - T_w)\frac{2}{\sqrt{\pi}}\cdot\frac{1}{2}\sqrt{\frac{Pe}{LX}}.$$

The rate of flow of heat into a strip of length dX and unit depth (in the Z direction) is

$$k\left.\frac{\partial T}{\partial Y}\right|_{Y=0}.\mathrm{d}X.1,$$

so the rate of flow of heat into the plate of length L and unit depth is

$$\int_0^L k\left.\frac{\partial T}{\partial Y}\right|_{Y=0}\mathrm{d}X = 2k(T_0 - T_w)\sqrt{\frac{Pe}{\pi}}.$$

The finite difference solution

A numerical solution for the temperature field between the plates will be obtained using a finite difference approximation on a computational mesh consisting of the points of intersection of the lines $x = I\delta x$, $y = J\delta y$ for $I = 0, 1, 2 \ldots N$ and $J = 0, 1, 2 \ldots M$. The value of $h\,(=H/L)$ must be specified (it has been set to 2 in the following FORTRAN program.) It is important that the grid spacings $\delta x = 1.0/N$ and $\delta y = 2.0/M$ are sufficiently small to allow the regions of rapid change to be represented. The results shown are for $\varepsilon = 1.0$, 0.1 and 0.04 with $N = 200$ and $M = 40$. This grid spacing allows 8 intervals in the x direction for every ε increment and 4 intervals in the y direction for every $\sqrt{\varepsilon}$ increment when $\varepsilon = 0.04$.

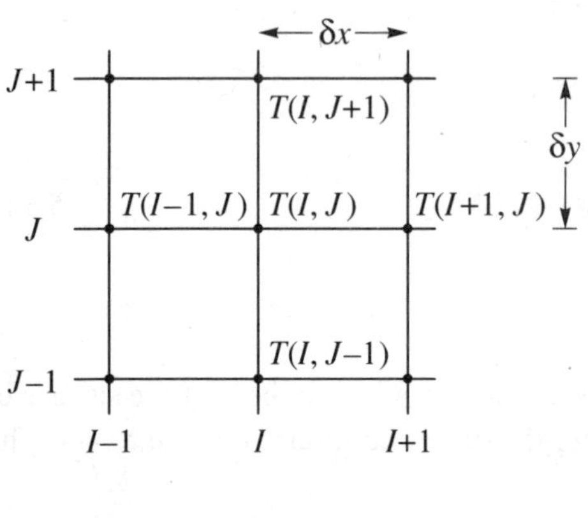

At the point $I\delta x$, $J\delta y$, the temperature $T(I\delta x, J\delta y)$ is denoted by $T(I, J)$. Standard texts on numerical techniques derive the following finite difference approximations for derivatives at the (I, J) point,

$$\frac{\partial T}{\partial x} = \frac{T(I+1,J) - T(I-1,J)}{2\delta x} + O(\delta x^2)$$

$$\frac{\partial^2 T}{\partial x^2} = \frac{T(I+1,J) - 2T(I,J) + T(I-1,J)}{\delta x^2} + O(\delta x^2)$$

$$\frac{\partial^2 T}{\partial y^2} = \frac{T(I,J+1) - 2T(I,J) + T(I,J-1)}{\delta y^2} + O(\delta y^2).$$

Substituting these expressions into the governing equation 5.6.9 yields the algebraic equation

$$\text{CO} \,.\, T(I, J) = \text{CN} \,.\, T(I, J+1) + \text{CS} \,.\, T(I, J-1) + \text{CE} \,.\, T(I+1, J) + \text{CW} \,.\, T(I-1, J), \qquad 5.6.15$$

with an error of order δx^2 and δy^2. The coefficients are given by

$$\text{CO} = 2\varepsilon\left(\frac{1}{\delta x^2} + \frac{1}{\delta y^2}\right)$$

$$\text{CN} = \text{CS} = \frac{\varepsilon}{\delta y^2}$$

$$\text{CE} = \frac{\varepsilon}{\delta x^2} - \frac{0.5}{\delta x}$$

$$\text{CW} = \frac{\varepsilon}{\delta x^2} + \frac{0.5}{\delta x}.$$

Equation 5.6.15 is linear in the temperatures and could be solved directly by elimination. Instead an iterative solution procedure – the Gauss–Seidel technique – has been used in the computer program listed on p. 182. This choice is due to the simplicity of the algorithm in comparison with direct solution techniques. The boundary conditions $T_0 = 1.0$ and $T_w = 0.0$ have been chosen without loss of generality since any other values merely involve a rescaling of the temperature field.

The results are printed as an array of 11 × 21 temperature values with a corresponding space interval of 0.1. The three sets of results for $\varepsilon = 1.0$, 0.1 and 0.04 (pages 183–5) show the boundary layer develop as ε becomes small. The shaded portions indicate the penetration of the $T = 0.8$ isotherm.

The computed solution provides the full solution up to the accuracy of the finite difference scheme. The analytic solution given by equation 5.6.14 is the leading term in an asymptotic expansion. In general when the boundary layer thickness is $O(\sqrt{\varepsilon})$ the asymptotic expansion for the solution involves half powers of ε so that the one-term

```
CCC   PROGRAM TO SOLVE DT/DX=EPSILON*(D2T/DX2+D2T/DY2)
      PARAMETER(N=200,M=40)
      DIMENSION T(0:N,0:M)
CCC   READING EPSILON AND CREATING DX AND DY ------------------------------
      PRINT*,'EPS=?'
      READ*,EPS
      DX=1.0/N
      DY=2.0/M
CCC   ---------------------------------------------------------------------
CCC   THE ITERATION STOPS IF THE DIFFERENCE BETWEEN SUCCESIVE ITERATES IS
CCC   LESS THAN TOL.
CCC   ITMAX IS THE MAXIMUM NUMBER OF ITERATIONS ALLOWED --------------------
      TOL=0.0001
      ITMAX=5000
CCC   ---------------------------------------------------------------------
CCC   CREATING THE COEFFICIENTS FOR THE ITERATION --------------------------
      CO=2.0*EPS*(1.0/DX/DX+1.0/DY/DY)
      CN=EPS/DY/DY
      CS=CN
      CE=EPS/DX/DX-0.5/DX
      CW=EPS/DX/DX+0.5/DX
CCC   ---------------------------------------------------------------------
CCC   SETTING THE INTERIOR VALUES TO ZERO INITIALLY AND
CCC   SETTING THE WALL AND OUTLET VALUES TO ZERO ---------------------------
      DO 10 I=0,N
      DO 10 J=0,M
      T(I,J)=0.0
 10   CONTINUE
CCC   ---------------------------------------------------------------------
CCC   SETTING THE INLET TEMPERATURE ----------------------------------------
      DO 20 J=1,M-1
      T(0,J)=1.0
 20   CONTINUE
CCC   ---------------------------------------------------------------------
      ITCOUNT=0
 30   CONTINUE
      ITCOUNT=ITCOUNT+1
      ICONV=0
CCC   CREATING THE NEW ITERATIVE VALUES AND TESTING FOR CONVERGENCE---------
      DO 100 I=1,N-1
      DO 100 J=1,M-1
      TNEW=(CN*T(I,J+1)+CS*T(I,J-1)+CE*T(I+1,J)+CW*T(I-1,J))/CO
      ABSDIF=ABS(TNEW-T(I,J))
      IF(ABSDIF.GT.TOL)ICONV=ICONV+1
      T(I,J)=TNEW
100   CONTINUE
      IF(ITCOUNT.LT.ITMAX.AND.ICONV.GT.0)GOTO 30
CCC   ---------------------------------------------------------------------
CCC   WRITING THE RESULTS --------------------------------------------------
      WRITE(1,1000) ITCOUNT,EPS
      PRINT*,'  T E M P E R A T U R E S '
      DO 200 J=M,0,-2
      WRITE(1,1001)(T(I,J),I=0,N,20)
200   CONTINUE
1000  FORMAT(1X,' NUMBER OF ITERATIONS =',I6,'     EPSILON =',F6.2)
1001  FORMAT(1X,21F5.2)
CCC   ---------------------------------------------------------------------
      END
```

FORTRAN program to solve $\dfrac{\partial T}{\partial x} = \varepsilon\left(\dfrac{\partial^2 T}{\partial x^2} + \dfrac{\partial^2 T}{\partial y^2}\right)$.

EPS = ?
1.0
NUMBER OF ITERATIONS = 2806 EPSILON = 1.00
TEMPERATURES

0.00	0.00	0.00	0.00	0.00	0.00	0.00	0.00	0.00	0.00	0.00
1.00	0.50	0.28	0.17	0.11	0.07	0.04	0.02	0.01	0.01	0.00
1.00	0.69	0.46	0.30	0.19	0.12	0.07	0.04	0.02	0.01	0.00
1.00	0.77	0.55	0.38	0.26	0.16	0.10	0.06	0.03	0.01	0.00
1.00	0.80	0.60	0.43	0.30	0.19	0.12	0.07	0.04	0.02	0.00
1.00	0.81	0.63	0.46	0.32	0.21	0.13	0.08	0.04	0.02	0.00
1.00	0.82	0.64	0.47	0.33	0.22	0.14	0.08	0.04	0.02	0.00
1.00	0.82	0.65	0.48	0.34	0.23	0.14	0.09	0.05	0.02	0.00
1.00	0.83	0.65	0.49	0.34	0.23	0.15	0.09	0.05	0.02	0.00
1.00	0.83	0.65	0.49	0.35	0.23	0.15	0.09	0.05	0.02	0.00
1.00	0.83	0.65	0.49	0.35	0.23	0.15	0.09	0.05	0.02	0.00
1.00	0.83	0.65	0.49	0.35	0.23	0.15	0.09	0.05	0.02	0.00
1.00	0.83	0.65	0.49	0.34	0.23	0.15	0.09	0.05	0.02	0.00
1.00	0.82	0.65	0.48	0.34	0.23	0.14	0.09	0.05	0.02	0.00
1.00	0.82	0.64	0.47	0.33	0.22	0.14	0.08	0.04	0.02	0.00
1.00	0.81	0.63	0.46	0.32	0.21	0.13	0.08	0.04	0.02	0.00
1.00	0.80	0.60	0.43	0.30	0.19	0.12	0.07	0.04	0.02	0.00
1.00	0.77	0.55	0.38	0.26	0.16	0.10	0.06	0.03	0.01	0.00
1.00	0.69	0.46	0.30	0.19	0.12	0.07	0.04	0.02	0.01	0.00
1.00	0.50	0.28	0.17	0.11	0.07	0.04	0.02	0.01	0.01	0.00
0.00	0.00	0.00	0.00	0.00	0.00	0.00	0.00	0.00	0.00	0.00

Temperature field for $\varepsilon = 1.0$

EPS = ?
0.1
NUMBER OF ITERATIONS = 4208 EPSILON = 0.10
TEMPERATURES

0.00	0.00	0.00	0.00	0.00	0.00	0.00	0.00	0.00	0.00	0.00
1.00	0.65	0.47	0.37	0.31	0.27	0.24	0.21	0.17	0.12	0.00
1.00	0.88	0.75	0.64	0.56	0.50	0.44	0.39	0.33	0.22	0.00
1.00	0.95	0.88	0.81	0.73	0.67	0.60	0.54	0.45	0.31	0.00
1.00	0.98	0.94	0.89	0.84	0.78	0.72	0.65	0.55	0.38	0.00
1.00	0.99	0.97	0.94	0.90	0.85	0.79	0.72	0.62	0.43	0.00
1.00	0.99	0.98	0.96	0.93	0.89	0.84	0.77	0.67	0.47	0.00
1.00	1.00	0.99	0.97	0.95	0.91	0.87	0.80	0.69	0.49	0.00
1.00	1.00	0.99	0.98	0.96	0.92	0.88	0.82	0.71	0.50	0.00
1.00	1.00	0.99	0.98	0.96	0.93	0.89	0.82	0.72	0.51	0.00
1.00	1.00	0.99	0.98	0.96	0.93	0.89	0.83	0.72	0.51	0.00
1.00	1.00	0.99	0.98	0.96	0.93	0.89	0.82	0.72	0.51	0.00
1.00	1.00	0.99	0.98	0.96	0.92	0.88	0.82	0.71	0.50	0.00
1.00	1.00	0.99	0.97	0.95	0.91	0.87	0.80	0.69	0.49	0.00
1.00	0.99	0.98	0.96	0.93	0.89	0.34	0.77	0.67	0.47	0.00
1.00	0.99	0.97	0.94	0.90	0.85	0.79	0.72	0.62	0.43	0.00
1.00	0.98	0.94	0.89	0.84	0.78	0.72	0.65	0.55	0.38	0.00
1.00	0.95	0.88	0.81	0.73	0.67	0.60	0.54	0.45	0.31	0.00
1.00	0.88	0.75	0.64	0.56	0.50	0.44	0.39	0.33	0.22	0.00
1.00	0.65	0.47	0.37	0.31	0.27	0.24	0.20	0.17	0.12	0.00
0.00	0.00	0.00	0.00	0.00	0.00	0.00	0.00	0.00	0.00	0.00

Temperature field for $\varepsilon = 0.1$

EPS = ?
0.04
NUMBER OF ITERATIONS = 2240   EPSILON = 0.04
TEMPERATURES

0.00	0.00	0.00	0.00	0.00	0.00	0.00	0.00	0.00	0.00	0.00
1.00	0.78	0.61	0.51	0.45	0.40	0.37	0.34	0.31	0.27	0.00
1.00	0.96	0.89	0.82	0.75	0.70	0.65	0.61	0.57	0.50	0.00
1.00	0.99	0.97	0.94	0.91	0.87	0.83	0.80	0.76	0.67	0.00
1.00	1.00	0.99	0.98	0.97	0.95	0.93	0.90	0.87	0.78	0.00
1.00	1.00	1.00	0.99	0.99	0.98	0.97	0.95	0.93	0.84	0.00
1.00	1.00	1.00	1.00	1.00	0.99	0.99	0.98	0.96	0.87	0.00
1.00	1.00	1.00	1.00	1.00	1.00	0.99	0.99	0.97	0.88	0.00
1.00	1.00	1.00	1.00	1.00	1.00	0.99	0.99	0.97	0.89	0.00
1.00	1.00	1.00	1.00	1.00	1.00	1.00	0.99	0.97	0.89	0.00
1.00	1.00	1.00	1.00	1.00	1.00	1.00	0.99	0.97	0.89	0.00
1.00	1.00	1.00	1.00	1.00	1.00	1.00	0.99	0.97	0.89	0.00
1.00	1.00	1.00	1.00	1.00	1.00	0.99	0.99	0.97	0.89	0.00
1.00	1.00	1.00	1.00	1.00	1.00	0.99	0.99	0.97	0.88	0.00
1.00	1.00	1.00	1.00	1.00	0.99	0.99	0.98	0.96	0.87	0.00
1.00	1.00	1.00	0.99	0.99	0.98	0.97	0.95	0.93	0.84	0.00
1.00	1.00	0.99	0.98	0.97	0.95	0.93	0.90	0.87	0.78	0.00
1.00	0.99	0.97	0.94	0.91	0.87	0.83	0.80	0.76	0.67	0.00
1.00	0.96	0.89	0.82	0.75	0.70	0.65	0.61	0.57	0.50	0.00
1.00	0.78	0.61	0.51	0.45	0.40	0.37	0.34	0.31	0.27	0.00
0.00	0.00	0.00	0.00	0.00	0.00	0.00	0.00	0.00	0.00	0.00

Temperature field for $\varepsilon = 0.04$

expansion has an error of order $\sqrt{\varepsilon}$. In this case it can be shown that the $\sqrt{\varepsilon}$ term is absent and the error is $O(\varepsilon)$. A comparison between the one-term expansion and the finite difference solution for the case $\varepsilon = 0.04$ shows that they differ by less than ε. For example at $x = 0.5$ and $y = 0.2$ in the case of $\varepsilon = 0.04$ the finite difference solution is $T = 0.70$ while 5.6.14 gives $T_0^{\text{comp}} = 0.68$. Having tested the computer program against the analytic solution the program can be used to determine temperature fields for different boundary conditions or velocity profiles where analytic solutions are not available.

Exercise

Modify the FORTRAN program to solve the convection diffusion problem for the velocity fields

(i) $U = U_0 y \qquad 0 < y < 2$

(ii) $U = 1.5U_0 y(2 - y) \quad 0 < y < 2.$

Note that in both cases the total material flow is the same as the uniform velocity profile $U = U_0$ $(0 < y < 2.)$ Case (i) corresponds to fully developed Couette flow with the upper plate moving at speed $2U_0$ while case (ii) corresponds to fully developed Poiseuille flow.

Estimation of boundary layer thickness

The thermal boundary layer considered in the previous pages was associated with a uniform velocity field. Usually both a momentum and thermal boundary layer must be considered. It is possible to predict the thickness of the momentum and thermal boundary layers by an algebraic method based on orders of magnitude estimates. We consider the flow over a plate of length L. The undisturbed fluid velocity is U_0. The fluid velocity on the surface of the plate is zero. The velocity field is governed by the equations 5.6.6–8. The continuity equation can be used in the momentum boundary layer to determine the order of magnitude of the vertical velocity V by replacing the derivatives by their order of magnitudes, i.e.

$$\frac{\partial U}{\partial X} = O\left(\frac{U_0}{L}\right), \quad \frac{\partial V}{\partial Y} = \frac{O(V)}{\delta_m},$$

where δ_m is the momentum boundary layer thickness. The continuity equation requires

$$O\left(\frac{U_0}{L}\right) = \frac{O(V)}{\delta_m},$$

so that

$$V = \mathrm{O}\left(\frac{\delta_m U_0}{L}\right). \qquad 5.6.16$$

The order of magnitude of the pressure gradients is not known at the outset of the analysis but will either be of the same order as the dominant velocity terms or of lower order – they could only be dominant if they balanced a body force such as gravity in hydrostatics. We neglect gravity or assume that it has been compensated by a hydrostatic pressure variation which has already been subtracted from the pressure field which appears in the equations. The terms making up equation 5.6.6 have the following orders of magnitude,

$$\underbrace{U\frac{\partial U}{\partial X}}_{\mathrm{O}\left(\frac{U_0^2}{L}\right)} + \underbrace{V\frac{\partial U}{\partial Y}}_{\mathrm{O}\left(\frac{\delta_m U_0}{L}\frac{U_0}{\delta_m}\right)} = -\frac{1}{\rho}\frac{\partial P}{\partial X} + \nu\left(\underbrace{\frac{\partial^2 U}{\partial X^2}}_{\mathrm{O}\left(\frac{\nu U_0}{L^2}\right)} + \underbrace{\frac{\partial^2 U}{\partial Y^2}}_{\mathrm{O}\left(\frac{\nu U_0}{\delta_m^2}\right)}\right). \qquad 5.6.17$$

Clearly the term $\partial^2 U/\partial X^2$ is dominated by $\partial^2 U/\partial Y^2$ and may be omitted from the dominant equation. The principle of least degeneracy requires that

$$\mathrm{O}\left(\frac{U_0^2}{L}\right) = \mathrm{O}\left(\frac{\nu U_0}{\delta_m^2}\right),$$

so that in terms of the Reynolds number, $Re = U_0 L/\nu$, we have

$$\delta_m = \mathrm{O}(L/\sqrt{Re}). \qquad 5.6.18$$

This shows that the momentum boundary layer thickness has a square root dependence on the reciprocal of the Reynolds number in the same way as the thermal boundary layer of the previous example varies with the Péclet number. Indeed we will see in our detailed study of flow over a plate that the momentum boundary layer has the same parabolic nature as the thermal example.

The Prandtl number

The ratio of the Péclet number and Reynolds number for fluids is called the *Prandtl number*, *Pr*.

$$Pr = \frac{Pe}{Re} = \frac{U_0 L/\alpha}{U_0 L/\nu} = \frac{\nu}{\alpha}.$$

The Prandtl number for air is 0.71 while for water at 20°C it is almost ten times this value. Fluids which are good conductors of heat will have a relatively large value of α;

this occurs for liquid metals whose Prandtl numbers are correspondingly small. Mercury, for example, has a Prandtl number of 0.023. Fluids which are particularly viscous will have a relatively large value of ν and correspondingly a large Prandtl number. Lubricating oils have Prandtl numbers which can exceed 10^4.

For a fluid with a Prandtl number close to unity the thermal and momentum boundary layers have a similar thickness. For small Prandtl numbers the thermal boundary layer is much thicker than the momentum boundary layer, whereas for large Prandtl numbers the converse applies (see Fig. 5.7).

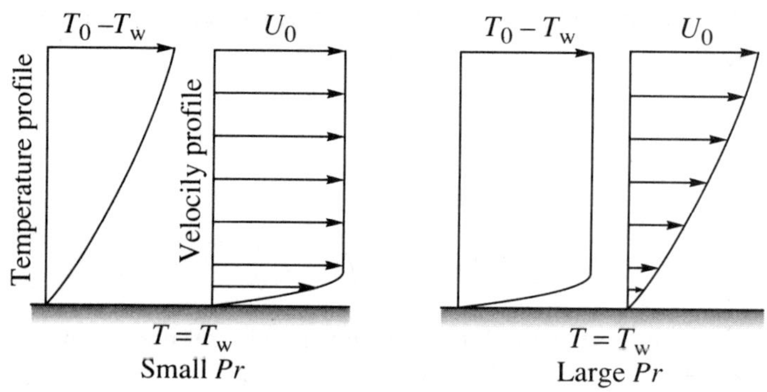

Fig. 5.7 Thermal and momentum boundary layers

In the case of $Pr \simeq 1$ or Pr small the characteristic velocity throughout the thermal boundary layer is U_0 and an order of magnitude analysis of equation 5.6.5 yields the thermal boundary layer thickness δ_T as follows. Consider the order of magnitude of the terms

$$U\frac{\partial T}{\partial X} + V\frac{\partial T}{\partial Y} = \alpha\left(\frac{\partial^2 T}{\partial X^2} + \frac{\partial^2 T}{\partial Y^2}\right)$$

$$\mathrm{O}\left(\frac{U_0(T_0 - T_w)}{L}\right) \quad \mathrm{O}\left(\frac{\alpha(T_0 - T_w)}{L^2}\right) \quad \mathrm{O}\left(\frac{\alpha(T_0 - T_w)}{\delta_T^2}\right).$$

The second member of the right-hand side dominates the first member and the principle of least degeneracy yields

$$\delta_T^2 = \mathrm{O}\left(\frac{\alpha L^2}{U_0 L}\right),$$

so that $\delta_T = \mathrm{O}(L/\sqrt{Pe})$.

This argument does not apply to the case of large Prandtl number because the thermal boundary layer occurs in a region which is narrow on the momentum boundary layer scale. Throughout the thermal boundary layer the velocity is small and as we shall see in our subsequent study, the velocity variation normal to the wall is linear. The velocity profile can be reasonably represented by $U = U_0 Y/\delta_m$ where δ_m is a measure of the momentum boundary layer thickness and is of order $L/\sqrt{Re}$. The convection terms in the heat equation are of order $(U/L)(T_0 - T_w)$ which is to be equated to the order of the term representing diffusion in the Y direction. The appropriate scale for Y in the thermal boundary layer is δ_T giving

$$O\left(U_0 \frac{\delta_T}{\delta_m} \frac{(T_0 - T_w)}{L}\right) = O\left(\alpha \frac{(T_0 - T_w)}{\delta_T^2}\right),$$

so that

$$\delta_T^3 = O\left(\frac{\alpha}{U_0 L} \cdot \frac{\delta_m}{L} \cdot L^3\right).$$

This involves the Péclet and Reynolds numbers in the following way:

$$\delta_T = O\left(\frac{L}{Pe^{1/3} Re^{1/6}}\right).$$

In terms of the Prandtl number this becomes

$$\delta_T = O\left(\frac{L Pr^{1/6}}{Pe^{1/2}}\right).$$

Exercise

Fully developed Poiseuille flow develops over a long inlet region between parallel walls held at the inlet temperature of $T = T_0$. Then at $X = 0$ there is a step increase in the wall temperature to $T = T_w$ (see Fig. 5.8). Show that the thermal boundary layer thickness, δ_T, has order $L(\alpha H/U_0 L^2)^{1/3}$ [continued overleaf].

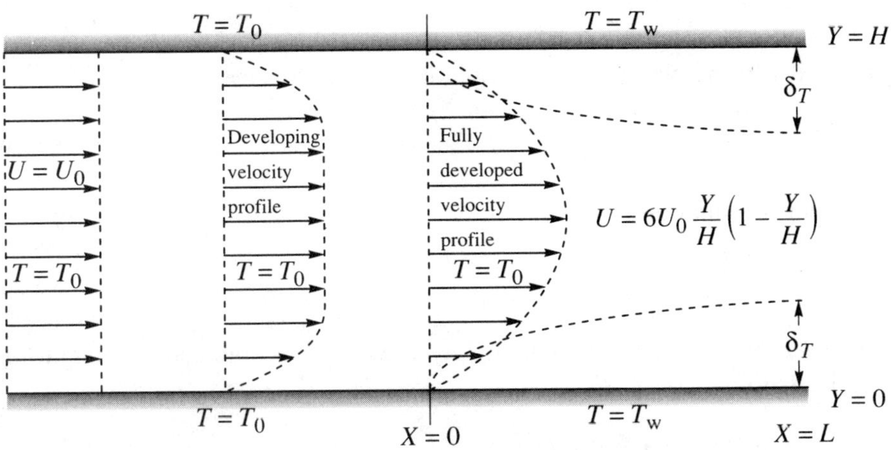

Fig. 5.8 Specification of the thermal boundary layer exercise

(*Hint*: Consider the region $0 < X < L$ and use $U = 6U_0 Y/H$ for the velocity near $Y = 0$.)
Compare this prediction of the boundary layer thickness with the finite difference solution for the temperature distribution with Poiseuille flow.

A remark on turbulence

The values of the Prandtl numbers quoted have been based on the so-called molecular thermal diffusivity and kinematic viscosity. Both processes involve the transfer of a quantity (heat or momentum) by interactions over a molecular length scale. Mathematical models of turbulence replace the random velocity, pressure and temperature fields by their values averaged over the length scale of a typical turbulent eddy. The effect of the random turbulent eddies is represented by an enhanced diffusion. Typically α and ν are increased by a factor of 10^3. The mechanism of heat and momentum diffusion in turbulent flow modeled in this way is similar. The Prandtl number for all fluids in turbulent flow reflects this and is reasonably close to unity in value.

Prandtl's momentum boundary layer equations

We continue our consideration of flow over a plate where U_0 is the characteristic velocity parallel to the plate and L is a characteristic length in this direction. Equation 5.6.16 provides the order of magnitude of the normal velocity in the boundary layer, namely $V = \mathrm{O}(\delta_{\mathrm{m}} U_0/L)$ where the momentum boundary layer thickness is given by equation 5.6.18, namely

$$\delta_{\mathrm{m}} = \mathrm{O}(L/\sqrt{Re}).$$

The X momentum equation provides the order of magnitude of the pressure gradient parallel to the plate, and from equation 5.6.17 we have

$$\frac{1}{\rho}\frac{\partial P}{\partial X} = \mathrm{O}\left(\frac{U_0^2}{L}\right). \qquad 5.6.19$$

The Y momentum equation leads to

$$\frac{1}{\rho}\frac{\partial P}{\partial Y} = \mathrm{O}\left(\frac{\delta_{\mathrm{m}} U_0^2}{L^2}\right). \qquad 5.6.20$$

The pressure variation across the boundary layer, $\Delta P_{\mathrm{normal}}$, is obtained by integrating 5.6.20 and leads to

$$\Delta P_{\mathrm{normal}} = \mathrm{O}\left(\frac{\rho U_0^2}{L^2}\,\delta_{\mathrm{m}}^2\right) = \mathrm{O}\left(\frac{\rho U_0^2}{Re}\right).$$

Integrating 5.6.19 along the boundary layer leads to the pressure variation along the boundary layer,

$$\Delta P_{\text{tangential}} = O(\rho U_0^2).$$

Thus the variation in pressure across the boundary layer is order Re^{-1} which to leading order in an expansion for large Re may be neglected. Therefore the tangential pressure gradient in the boundary layer may be replaced by its value outside the boundary layer. This is related to the velocity field outside the boundary layer by the momentum equations with the viscous term omitted. The inviscid velocity and pressure fields satisfy *Euler's equations of motion* and the continuity equation. For steady two-dimensional flow with no body force the equations governing inviscid flow are

$$U_{\text{Inv}} \frac{\partial U_{\text{Inv}}}{\partial X} + V_{\text{Inv}} \frac{\partial U_{\text{Inv}}}{\partial Y} = -\frac{1}{\rho} \frac{\partial P_{\text{Inv}}}{\partial X} \tag{5.6.21}$$

$$U_{\text{Inv}} \frac{\partial V_{\text{Inv}}}{\partial X} + V_{\text{Inv}} \frac{\partial V_{\text{Inv}}}{\partial Y} = -\frac{1}{\rho} \frac{\partial P_{\text{Inv}}}{\partial Y} \tag{5.6.22}$$

$$\frac{\partial U_{\text{Inv}}}{\partial X} + \frac{\partial V_{\text{Inv}}}{\partial Y} = 0. \tag{5.6.23}$$

These are the counterparts of 5.6.6, 5.6.7 and 5.6.8 respectively for viscous fluids. The inviscid momentum equations are of lower order and require that a boundary condition be relaxed. Viscosity provides the mechanism for transmitting the tangential component of stress and its absence necessitates that fluid be allowed to slip at the surface of a solid. The condition of no penetration, i.e. zero normal velocity at a solid surface is, of course, required of the inviscid solution.

The inviscid equations provide the outer velocity and pressure field to which the boundary layer fields are matched. Strictly speaking, equations 5.6.21 and 5.6.22 govern the leading order terms in an outer expansion because the viscous term which is of order Re^{-1} has been omitted. We will restrict our analysis to the leading order terms.

If the viscous region is confined to a narrow boundary layer region close to solid surfaces then, to leading order, the inviscid field can be determined by neglecting the boundary layer. The solid boundaries are surfaces on which the normal component of the inviscid velocity is set to zero. For some flows there are complications due to boundary layer separation where backflow occurs causing a sudden thickening of the boundary layer. We will not consider the case of boundary layer separation. We will assume the boundary layers are narrow regions close to solid walls.

The right-hand member of equation 5.6.21 evaluated at the solid surface provides the viscous pressure gradient. In the case of flow over a plate orientated along the X

axis, V_{Inv} is zero on the plate surface so that

$$-\frac{1}{\rho}\frac{\partial P}{\partial X}\bigg|_{\text{In boundary layer}} = -\frac{1}{\rho}\frac{\partial P_{\text{Inv}}}{\partial X}\bigg|_{Y=0} = U_{\text{Inv}}\frac{\partial U_{\text{Inv}}}{\partial X}\bigg|_{Y=0}. \qquad 5.6.24$$

We will denote the last term by $U_{\text{I}}(\mathrm{d}U_{\text{I}}/\mathrm{d}X)$ with the understanding, that it refers to the inviscid slip velocity at the plate.

The steady flow boundary layer equations become

$$U\frac{\partial U}{\partial X} + V\frac{\partial U}{\partial Y} = U_{\text{I}}\frac{\mathrm{d}U_{\text{I}}}{\mathrm{d}X} + \nu\frac{\partial^2 U}{\partial Y^2} \qquad 5.6.25$$

$$\frac{\partial U}{\partial X} + \frac{\partial V}{\partial Y} = 0, \qquad 5.6.26$$

where to leading order the viscous term involves only the normal derivative.

While these equations have been derived for flow over a flat plate they remain valid for flow over a curved surface provided the radius of curvature of the surface is large in comparison with the boundary layer thickness. In this case X and U are the tangential length and velocity variables while Y and V are the normal variables.

Equations 5.6.25 and 5.6.26 are Prandtl's boundary layer equations. They are a significant simplification of the full Navier–Stokes equations. We will assume that the inviscid flow field U_{I} has already been obtained from the solution of the Euler equations of inviscid flow. The pressure in the boundary layer is then known from equation 5.6.24. Our task is to determine the leading order velocity fields, U and V, in the boundary layer.

Introducing nondimensional variables $x = X/L$, $y = Y/L$, $u = U/U_0$, $v = V/U_0$, $u_{\text{I}} = U_{\text{I}}/U_0$ leads to the following form of the boundary layer equations

$$u\frac{\partial u}{\partial x} + v\frac{\partial u}{\partial y} = u_{\text{I}}\frac{\mathrm{d}u_{\text{I}}}{\mathrm{d}x} + \frac{1}{Re}\frac{\partial^2 u}{\partial y^2} \qquad 5.6.27$$

$$\frac{\partial u}{\partial x} + \frac{\partial v}{\partial y} = 0. \qquad 5.6.28$$

The stretched normal variable, s, is determined by the principle of least degeneracy to be $s = y\sqrt{Re}$. Throughout the boundary layer v is order $1/\sqrt{Re}$. Although this is small it is important to retain it because the term $\partial u/\partial y$ is large ($O(\sqrt{Re})$) so that the product $v(\partial u/\partial y)$ appearing in equation 5.6.27 is of order 1. We use the appropriately scaled normal velocity, $v^* = v\sqrt{Re}$, and the boundary layer equations become

$$u\frac{\partial u}{\partial x} + v^*\frac{\partial u}{\partial s} = u_{\text{I}}\frac{\mathrm{d}u_{\text{I}}}{\mathrm{d}x} + \frac{\partial^2 u}{\partial s^2} \qquad 5.6.29$$

$$\frac{\partial u}{\partial x} + \frac{\partial v^*}{\partial s} = 0. \qquad 5.6.30$$

It should be remembered that these are the governing equations for the leading order terms in an asymptotic expansion for large *Re*. The form of the stretching transformation suggests that the expansions should involve gauge functions which form the sequence $\{1, 1/\sqrt{Re}, 1/Re, \ldots\}$ i.e.

$$u^{\text{outer}}(x, y, Re) \sim u_0^{\text{outer}}(x, y) + \frac{1}{\sqrt{Re}}\, u_1^{\text{outer}}(x, y) + \cdots$$

$$u^{\text{inner}}(x, s, Re) \sim u_0^{\text{inner}}(x, s) + \frac{1}{\sqrt{Re}}\, u_1^{\text{inner}}(x, s) + \cdots.$$

The leading term of the outer expansion is governed by the Euler equations and provides the field u_{I} which appears in the governing system of equations 5.6.29 and 5.6.30 for the leading terms u and v^* of the inner expansions.

The boundary conditions on the solid surface $y = s = 0$ are $u(x, 0) = v^*(x, 0) = 0$. Prandtl's matching condition provides the second boundary condition for u, namely $u(x, s \to \infty) = u_{\text{I}}(x)$. It is also necessary to impose a velocity profile at some value of x. We will choose $x = 0$ and specify $u(0, s)$ and $v^*(0, s)$ for $s > 0$ (see Fig. 5.9).

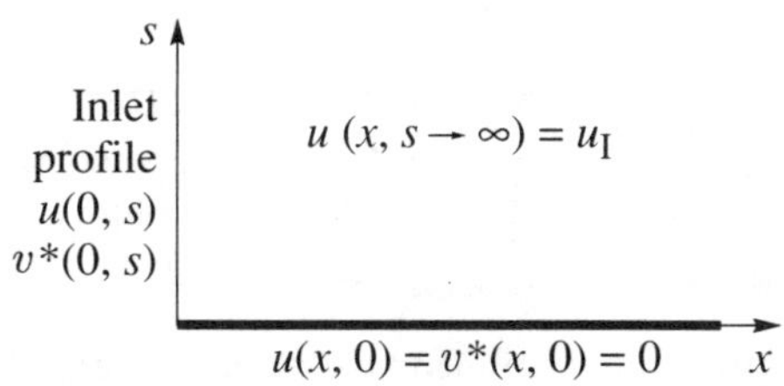

Fig. 5.9 Boundary conditions for flow over a plate

Although the boundary layer equations are significantly simpler than the full Navier–Stokes equations, there remains the essential nonlinearity in the convection terms. For this reason only very few exact solutions exist. There are three commonly used approximate solution techniques available, they are

the boundary integral equation method
the partial differential equation method
the similarity solution method.

The boundary integral equation method

In this technique the equations are integrated across the boundary layer using polynomial or other approximations to represent the dependence of the fields on the normal length variable. This results in ordinary differential equations with the tangential length as the independent variable. These can sometimes be integrated exactly but usually they are integrated numerically. Details of the method and its applications can be found in fluid dynamics texts.

The partial differential equation method

This method involves the direct solution of the boundary layer equations by a numerical technique such as finite differences or finite elements. It may appear that if a partial differential equation solution algorithm is to be devised, then it might just as well be applied to the full Navier–Stokes equations. In fact there are considerable advantages to be gained by solving the equations in the boundary layer form, namely:

— The pressure is known so there is one less equation to solve.
— Appropriate scaling of the variables is intrinsic in the boundary layer equations. A full Navier–Stokes solution procedure must allow for regions of rapid change near boundaries of the flow.
— Boundary layers on curved surfaces can be treated using the Cartesian form of the equations provided the radius of curvature of the surface is large in comparison with the boundary layer thickness.
— The dominance of the normal derivative in the diffusion term leads to a parabolic rather than an elliptic partial differential equation. Straightforward numerical procedures can be devised for parabolic equations based on 'marching' forward in the tangential direction from known inlet values.

Some examples of the finite difference approximate solutions will be presented later in this section.

The similarity solution method

For certain flows with special geometries it is possible to introduce a new independent variable, called the similarity variable, such that the momentum equation becomes an ordinary differential equation. Although this equation is usually nonlinear it is easily integrated numerically. There follows an example of a similarity solution applied to the case of uniform flow over an infinite plate. (For an explanation of the symmetry group concepts associated with this general approach to the solution of differential equations see Bluman and Cole[15].)

Flow over an infinite plate – the Blasius solution

Blasius considered the flow of a uniform stream $U = U_0$, $V = 0$ over an infinite plate occupying the region $X > 0$, $Y = 0$ (Fig. 5.10). In the absence of viscosity a plate of

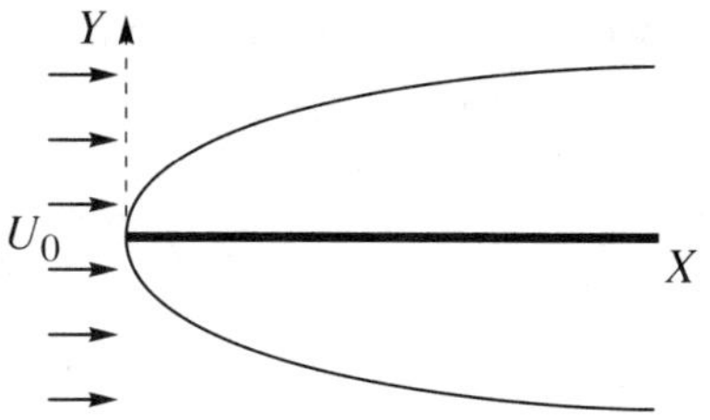

Fig. 5.10 Blasius flow

vanishing thickness will not influence the flow. The inviscid solution is $U_I = U_0$. The pressure is constant since dU_I/dX is zero.

The governing equations in dimensional form are

$$U\frac{\partial U}{\partial X} + V\frac{\partial U}{\partial Y} = \nu\frac{\partial^2 U}{\partial Y^2} \qquad 5.6.31$$

$$\frac{\partial U}{\partial X} + \frac{\partial V}{\partial Y} = 0. \qquad 5.6.32$$

Consider the region $X > 0$, $Y > 0$, with the boundary conditions

$$U(X, 0) = V(X, 0) = 0, X > 0$$

$$U(0, Y) = U_0, V(0, Y) = 0, Y > 0$$

$$U(X, Y \to \infty) = U_0, X > 0.$$

There is no length scale in the case of an infinite plate. It turns out to be fruitful if the Reynolds number is defined at any given location based on the tangential length coordinate, X, itself. Thus Re is the varying quantity U_0X/ν. We tentatively assume that the boundary layer thickness will be of order $X/\sqrt{Re}$, i.e. $O(X/\sqrt{U_0X/\nu})$. The similarity assumption is that the variation of the tangential velocity, U, between the value zero on the plate and the asymptotic value, U_0, is dependent on a single appropriately scaled normal variable. This *similarity variable* is the ratio of the normal length variable Y and the boundary layer thickness. Thus we assume that

$$U(X, Y) = U_0 \,.\, \text{Function of}\left(\frac{Y}{X/\sqrt{U_0X/\nu}}\right).$$

We choose to define the similarity variable, η, by the equation $\eta = Y\sqrt{U_0/2X\nu}$ where the factor of 2 is introduced merely for convenience.

Next we define a function $f(\eta)$ such that

$$U = U_0\frac{df}{d\eta}. \qquad 5.6.33$$

This derivative form is used so that integrals do not appear when the continuity equation is integrated to determine V. The definition of $f(\eta)$ involves an arbitrary constant which allows the condition $f(\eta = 0) = 0$ to be imposed.

The similarity variable takes the place of the normal length variable. The tangential variable is unaltered in the transformation of independent variables and is merely relabeled as ξ so that

$$\xi = X, \eta = Y\sqrt{\frac{U_0}{2X\nu}}$$

$$\frac{\partial}{\partial X} \equiv \frac{\partial \xi}{\partial X}\frac{\partial}{\partial \xi} + \frac{\partial \eta}{\partial X}\frac{\partial}{\partial \eta} \equiv \frac{\partial}{\partial \xi} - \frac{\eta}{2\xi}\frac{\partial}{\partial \eta}$$

$$\frac{\partial}{\partial Y} \equiv \frac{\partial \xi}{\partial Y}\frac{\partial}{\partial \xi} + \frac{\partial \eta}{\partial Y}\frac{\partial}{\partial \eta} \equiv \sqrt{\frac{U_0}{2\nu\xi}}\frac{\partial}{\partial \eta}$$

$$\frac{\partial^2}{\partial Y^2} \equiv \frac{U_0}{2\nu\xi}\frac{\partial^2}{\partial \eta^2}.$$

The continuity equation (5.6.32) becomes

$$\frac{\partial U}{\partial \xi} - \frac{\eta}{2\xi}\frac{\partial U}{\partial \eta} + \sqrt{\frac{U_0}{2\nu\xi}}\frac{\partial V}{\partial \eta} = 0.$$

The first member is zero on the assumption that U is only dependent on η and from 5.6.33 we obtain

$$\frac{\partial V}{\partial \eta} = \sqrt{\frac{U_0\nu}{2\xi}}\,\eta\frac{\mathrm{d}^2 f}{\mathrm{d}\eta^2}.$$

Integrating yields

$$V = \sqrt{\frac{U_0\nu}{2\xi}}\left(\eta\frac{\mathrm{d}f}{\mathrm{d}\eta} - f\right) + g(\xi), \qquad 5.6.34$$

where g is an arbitrary function of ξ. The boundary condition $V(X, Y = 0) = 0$ becomes $V(\xi, \eta = 0) = 0$ which, along with our choice of $f(0) = 0$, leads to $g(\xi) = 0$. Substituting the expressions 5.6.33 and 5.6.34 for U and V into the momentum equation (5.6.31) leads, after some manipulation, to the equation

$$\frac{\mathrm{d}^3 f}{\mathrm{d}\eta^3} + f\frac{\mathrm{d}^2 f}{\mathrm{d}\eta^2} = 0.$$

The result that this governing equation for $f(\eta)$ does not involve the variable ξ is a partial justification of the original assumption concerning the form of the similarity variable. The justification is completed by confirming that the boundary conditions are consistent with the form of the η variable. We require U to take the value U_0 when $Y \to \infty$ for $X > 0$ and also when $X = 0$ for $Y > 0$. Both these conditions correspond to $\eta \to \infty$. The boundary condition at the wall is $U = 0$ when $Y = 0$ for $X > 0$ which corresponds to $U = 0$ when $\eta = 0$.

Thus the assumed form of the similarity solution is consistent both with the governing equations and boundary conditions and leads to the ordinary differential equation

$$\frac{d^3f}{d\eta^3} + f\frac{d^2f}{d\eta^2} = 0, \qquad 5.6.35$$

with boundary conditions

$$f(0) = \frac{df}{d\eta}(0) = 0, \quad \frac{df}{d\eta}(\eta \to \infty) = 1. \qquad 5.6.36$$

This nonlinear equation cannot be integrated by exact techniques. A numerical solution is easily obtained by any of the standard integration schemes for ordinary differential equations. The problem is slightly complicated by the fact that the conditions 5.6.36 do not include the value of $(d^2f/d\eta^2)(0)$ and instead the gradient of f is specified at infinity. The procedure is to choose trial values of $(d^2f/d\eta^2)(0)$ and to solve for increasing values of η until the value of $df/d\eta$ is constant. In practice it is sufficient to solve up to $\eta = 10$ to obtain constant values of $df/d\eta$. The required value of $df/d\eta$ for large η is unity and the value of $(d^2f/d\eta^2)(0)$ must be adjusted to achieve this.

The FORTRAN program listing below is the implementation of the Euler integration scheme. A step size of 0.002 has been used and the results are printed after

```
CCC   TO SOLVE D3F/DS3 +F*D2F/DS2=0
CCC   SETTING THE STEP SIZE, NUMBER OF STEPS AND PRINTING FREQUENCY ---------
        DS=0.002
        N=5000
        IPFREQ=100
CCC   ---------------------------------------------------------------------
CCC   READING THE TRIAL VALUE OF D2F/DS2 -----------------------------------
        PRINT*,'TRIAL D2F/DS2 ='
        READ*,H
CCC   ---------------------------------------------------------------------
CCC   SETTING INITIAL VALUES OF F AND DF/DS(=G) ----------------------------
        F=0.0
        G=0.0
CCC   ---------------------------------------------------------------------
        PRINT*,'    S           F           DF/DS       D2F/DS2 '
        IP=0
CCC   SOLVING FOR THE VARIABLES AND WRITING THE RESULTS --------------------
        DO 10 I=0,N
        S=I*DS
        IF(IP.EQ.0)THEN
        WRITE(1,100)S,F,G,H
        ENDIF
        IP=IP+1
        IF(IP.EQ.IPFREQ)THEN
        IP=0
        ENDIF
        F=F+DS*G
        G=G+DS*H
        H=H-DS*F*H
  10    CONTINUE
 100    FORMAT(1X,4F12.6)
CCC   ---------------------------------------------------------------------
        END
```

FORTRAN program to solve $\frac{d^3f}{d\eta^3} + f\frac{d^2f}{d\eta^2} = 0$

```
TRIAL D2F/DS2 =
0.47
        S            F            DF/DS        D2F/DS2
    0.000000     0.000000     0.000000     0.470000
    0.200000     0.009305     0.093986     0.469706
    0.400000     0.037394     0.187767     0.467650
    0.600000     0.084178     0.280821     0.462121
    0.800000     0.149433     0.372295     0.451557
    1.000000     0.232735     0.461052     0.434715
    1.200000     0.333410     0.545752     0.410856
    1.400000     0.450507     0.624975     0.379922
    1.600000     0.582792     0.697365     0.342644
    1.800000     0.728784     0.761788     0.300521
    2.000000     0.886804     0.817477     0.255662
    2.200000     1.055068     0.864122     0.210496
    2.400000     1.231777     0.901903     0.167414
    2.600000     1.415212     0.931444     0.128423
    2.800000     1.603817     0.953712     0.094903
    3.000000     1.796251     0.969876     0.067500
    3.200000     1.991415     0.981167     0.046177
    3.400000     2.188452     0.988752     0.030370
    3.600000     2.386725     0.993651     0.019196
    3.800000     2.585782     0.996692     0.011658
    4.000000     2.785316     0.998505     0.006802
    4.200000     2.985130     0.999545     0.003813
    4.400000     3.185101     1.000117     0.002053
    4.600000     3.385158     1.000419     0.001062
    4.800000     3.585258     1.000573     0.000527
    5.000000     3.785381     1.000648     0.000252
    5.200000     3.985515     1.000683     0.000115
    5.400000     4.185653     1.000699     0.000051
    5.600000     4.385794     1.000706     0.000021
    5.800000     4.585935     1.000709     0.000009
    6.000000     4.786077     1.000710     0.000003
    6.200000     4.986219     1.000710     0.000001
    6.400000     5.186362     1.000711     0.000000
    6.600000     5.386504     1.000711     0.000000
    6.800000     5.586646     1.000711     0.000000
    7.000000     5.786788     1.000711     0.000000
    7.200000     5.986930     1.000711     0.000000
    7.400000     6.187072     1.000711     0.000000
    7.600000     6.387214     1.000711     0.000000
    7.800000     6.587357     1.000711     0.000000
    8.000000     6.787499     1.000711     0.000000
    8.200000     6.987641     1.000711     0.000000
    8.400000     7.187783     1.000711     0.000000
    8.600000     7.387925     1.000711     0.000000
    8.800000     7.588067     1.000711     0.000000
    9.000000     7.788209     1.000711     0.000000
    9.200000     7.988352     1.000711     0.000000
    9.400000     8.188494     1.000711     0.000000
    9.600000     8.388636     1.000711     0.000000
    9.800000     8.588778     1.000711     0.000000
   10.000000     8.788920     1.000711     0.000000
```

Solution for $\dfrac{d^2f}{d\eta^2}(0) = 0.47$

Table 5.1

$\dfrac{d^2f}{d\eta^2}(0)$	$\dfrac{df}{d\eta}(\eta \to \infty)$
1.0	1.6555
0.5	1.0429
0.45	0.9721
0.47	1.0007

every hundred steps (see opposite). The sequence of trial $(d^2f/d\eta^2)(0)$ and resulting $(df/d\eta)(\eta \to \infty)$ values shown in Table 5.1 lead to the result $(d^2f/d\eta^2)(0) = 0.47$. The notation used in the program is:

$$\eta = S$$

$$f = F$$

$$\frac{df}{d\eta} = \frac{dF}{dS} = G$$

$$\frac{d^2f}{d\eta^2} = \frac{dG}{dS} = H$$

and equation 5.6.35 becomes $dH/dS + FH = 0$.

The system of first order equations

$$\frac{dF}{dS} = G$$

$$\frac{dG}{dS} = H$$

$$\frac{dH}{dS} = -FH$$

is solved using a truncated Taylor series approximation

$$F(S + DS) \simeq F(S) + DS . \frac{dF}{dS}(S) = F + DS . G$$

$$G(S + DS) \simeq G(S) + DS . \frac{dG}{dS}(S) = G + DS . H$$

$$H(S + DS) \simeq H(S) + DS . \frac{dH}{dS}(S) = H - DS . F . H.$$

A small value of DS must be chosen to achieve accurate values as the integration proceeds.

The tangential velocity U/U_0 is given by the column of $df/d\eta$ values ($= dF/dS$). This is shown in graphical form in Fig. 5.11. The value of U/U_0 is fixed for a given value of η. Thus curves of constant η correspond to curves on which U is constant. The curves of constant η are parabolas in the X, Y plane. Some representative constant velocity curves are shown in Fig. 5.11.

An important quantity which can be obtained from this solution is the drag force exerted on the plate. It was the ability to predict the drag and the accuracy of this prediction which provided one of the early successes and validations of Prandtl's

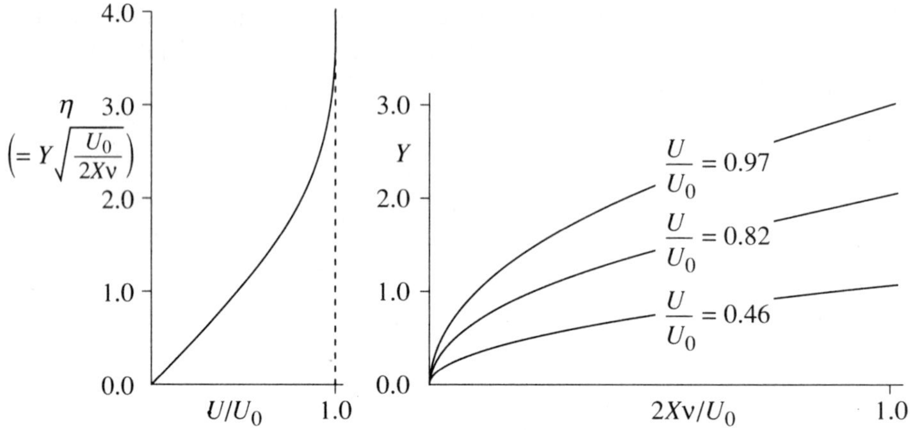

Fig. 5.11 The variation of U/U_0 with η and some parabolas of constant U/U_0

boundary layer ideas. The shear stress on the plate, τ_0, is given by the equation

$$\tau_0 = \mu \left.\frac{\partial U}{\partial Y}\right|_{Y=0}$$

where μ is the dynamic viscosity ($\mu = \rho\nu$). In terms of the similarity variables this becomes

$$\tau_0 = \mu\sqrt{\frac{U_0}{2\nu\xi}}\, U_0 \frac{d^2 f}{d\eta^2}(0)$$

$$= 0.47\mu U_0\sqrt{\frac{U_0}{2\nu X}},$$

The drag force suffered by an infinitesimal strip of width dX and unit depth in the Z direction is $\tau_0 \,.\, dX \,.\, 1$. An area of the plate between $X = 0$ and $X = L$ and unit depth suffers a drag

$$D = 2\int_0^L 0.47\mu U_0 \sqrt{\frac{U_0}{2\nu X}}\, dX,$$

where the factor 2 takes into account the drag on the upper and lower surfaces of the plate. Integrating leads to the following expression for the drag per unit depth,

$$D = 1.88\mu U_0\sqrt{\frac{U_0 L}{2\nu}}.$$

In practice this formula can be used for a plate of finite size because the edge effects are confined to narrow regions and have only a transcendentally small influence on the velocity profile over the remaining area of the plate.

A feature of the Blasius solution which is interesting to note is the asymptotic behavior of the normal velocity field. From equation 5.6.34 we have

$$V(X, \eta \to \infty) = \sqrt{\frac{U_0 \nu}{2X}} \left(\eta \frac{df}{d\eta} - f \right) \bigg|_{\eta \to \infty} .$$

The asymptotic value of the bracketed expression can be obtained from the computed values of f and $df/d\eta$ for large η and is found to be 1.218 so that V tends to the value $V_\infty(X) = 1.218\sqrt{U_0 \nu / 2X}$. This differs from the inviscid solution which is $V_{\mathrm{Inv}}(X, Y) = 0$ for all X and Y. The difference is of order $Re^{-1/2}$ so that in the outer (inviscid region) there is agreement to leading order between the inviscid solution and the asymptotic limit of the boundary layer solution. Within the boundary layer the normal velocity component is important because although it is of order $Re^{-1/2}$ the term $V(\partial U/\partial Y)$ as a whole is of order one. The small but nonzero value of V_∞ can be anticipated because as the boundary layer thickens, fluid is displaced away from the plate.

A finite difference solution procedure for the momentum boundary layer equations

The technique used in the Blasius solution can be extended to certain other flows such as those for which the inviscid slip velocity is proportional to a power of the tangential length coordinate. However, for general flows there does not exist a similarity variable which will reduce the boundary layer equations to an ordinary differential equation. Thus we are faced with the need to solve coupled partial differential equations to determine the velocity fields in the boundary layer. A finite difference solution technique will now be described.

The nondimensional form of Prandtl's boundary layer equations, 5.6.29 and 5.6.30, are repeated here for convenience,

$$u \frac{\partial u}{\partial x} + v^* \frac{\partial u}{\partial s} = u_{\mathrm{I}} \frac{du_{\mathrm{I}}}{dx} + \frac{\partial^2 u}{\partial s^2} \qquad 5.6.29'$$

$$\frac{\partial u}{\partial x} + \frac{\partial v^*}{\partial s} = 0. \qquad 5.6.30'$$

We will restrict our consideration to flow over a flat plate of unit nondimensional length occupying the region $0 < x < 1$, $s = 0$. The boundary conditions are

$$u(x, 0) = v^*(x, 0) = 0 \quad \text{for } 0 < x < 1,$$

$$u(x, s \to \infty) = u_{\mathrm{I}}(x) \quad \text{for } 0 < x < 1,$$

along with a specified inlet velocity field $u(0, s)$ and $v^*(0, s)$ for $s > 0$. Here $u_{\mathrm{I}}(x)$ is the nondimensional inviscid 'slip' velocity.

The Blasius solution will be used to check the accuracy of the finite difference solution scheme. Then various forms of incoming inviscid fields can be considered for which no similarity solution is available but to which the finite difference scheme is equally applicable.

The nondimensional normal variable, s, is related to the dimensional variable, Y, by the expression $s = Y\sqrt{Re}/L$. The upper limit for the value of s which must be chosen so that the velocity fields become sufficiently close to their asymptotic values can be determined by trial. It is found that taking $s = 8$ as the upper boundary of the finite difference computational grid is satisfactory.

In the computer program the region $0 < x < 1$, $0 < s < 8$ is divided into a rectangular mesh of N intervals in the x direction and M intervals in the s direction. The corresponding space intervals are $\delta x = 1.0/N$ and $\delta s = 8.0/M$. The velocity fields at the point $x = I\delta x$ and $s = J\delta s$ are denoted by $u(I, J)$ and $v(I, J)$ where for convenience the δx and δs dependence and the symbol $*$ associated with v have been omitted.

The boundary conditions on the plate are

$$u(I, 0) = v(I, 0) = 0 \quad \text{for } I = 0, 1, 2 \ldots N$$

and the asymptotic condition for Blasius flow is

$$u(I, M) = 1.0 \quad \text{for } I = 0, 1, 2 \ldots N.$$

The inlet velocity profiles are

$$u(0, J) = 1.0 \quad \text{and} \quad v(0, J) = 0 \quad \text{for } J = 1, 2 \ldots M.$$

The term $u_I(du_I/dx)$, denoted by SOURCE in the computer program, is zero for Blasius flow.

A *marching scheme* is used to determine the values of $u(I, J)$ from field values at $I - 1$. The terms in equation 5.6.29 are replaced by the following finite difference approximations at the point $(I - 1, J)$,

$$u\frac{\partial u}{\partial x} \simeq u(I-1, J).\left(\frac{u(I, J) - u(I-1, J)}{\delta x}\right) \qquad \text{'forward difference'}$$

$$v*\frac{\partial u}{\partial s} \simeq v(I-1, J).\left(\frac{u(I-1, J+1) - u(I-1, J-1)}{2\delta s}\right) \qquad \text{'central difference'}$$

$$\frac{\partial^2 u}{\partial s^2} \simeq \frac{u(I-1, J+1) - 2u(I-1, J) + u(I-1, J-1)}{\delta s^2} \qquad \text{'central difference'.}$$

This leads to an explicit expression for $u(I, J)$ in terms of the velocities at the points $(I - 1, J - 1)$, $(I - 1, J)$ and $(I - 1, J + 1)$. This is illustrated by the computational molecule shown in Fig. 5.12 where four known field values at the $(I - 1)$th location are used to determine the single field at the Ith location.

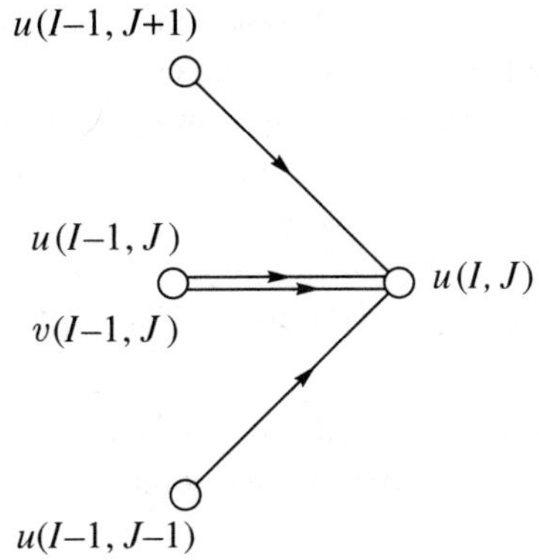

Fig. 5.12 Computational molecule for $u(I, J)$

To complete the marching procedure starting from known inlet fields at $I = 0$ we need to determine $v(I, J)$. Equation 5.6.30′ is represented by central difference approximations at the point $(I - 1/2, J - 1/2)$ using

$$\frac{\partial u}{\partial x} \simeq \frac{1/2[u(I, J) + u(I, J-1)] - 1/2[u(I-1, J) + u(I-1, J-1)]}{\delta x}$$

and

$$\frac{\partial v^*}{\partial s} \simeq \frac{1/2[v(I, J) + v(I-1, J)] - 1/2[v(I, J-1) + v(I-1, J-1)]}{\delta s}.$$

From these an explicit expression can be obtained to determine $v(I, J)$ using other previously calculated velocity fields. The corresponding computational molecule is shown in Fig. 5.13.

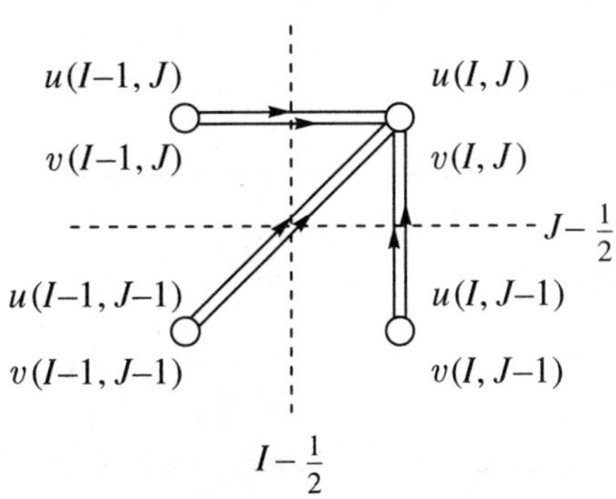

Fig. 5.13 Computational molecule for $v(I, J)$

The order of computation of the velocity field values in the computer program is as follows. For each I, starting at $I = 1$, first $u(I, J)$ and then $v(I, J)$ are calculated for every J value in the range from 1 to $M - 1$, then I is increased by unity and the process repeated until $I = N$. The value of $v(I, M)$ at the outer boundary is not an imposed asymptotic value. (If it were, the problem would be overdetermined since only the first derivatives of v^* appear in the equations.) The value of $v(I, M)$ is calculated for each $I(> 0)$ using the finite difference approximate form of equation 5.6.30′.

The central difference approximations have an accuracy of order of the square of the grid spacing. Thus relatively small errors of order δx^2 and δs^2 are associated with

```
CCC  THIS PROGRAM SOLVES PRANDTL'S BOUNDARY LAYER EQUATIONS
CCC  FOR THE CASE OF A CONSTANT INVISCID VELOCITY (BLASIUS FLOW)
      PARAMETER(N=400,M=32)
      DIMENSION U(0:N,0:M),V(0:N,0:M)
CCC  SETTING THE LENGTH AND HEIGHT ----------------------------------------
      PRINT*,'PLATE LENGTH = ?'
      READ*,XL
      DX=XL/N
      PRINT*,'HEIGHT=?'
      READ*,SMAX
      DS=SMAX/M
CCC  ----------------------------------------------------------------------
CCC  SETTING THE BOUNDARY AND INLET CONDITIONS -----------------------------
      DO 10 I=0,N
      U(I,0)=0.0
      U(I,M)=1.0
      V(I,0)=0.0
   10 CONTINUE
      DO 20 J=1,M
      U(0,J)=1.0
      V(0,J)=0.0
   20 CONTINUE
CCC  ----------------------------------------------------------------------
CCC  SOLVING FOR THE VELOCITIES ---------------------------------------------
      DO 120 I=1,N
      DO 100 J=1,M-1
      SOURCE=0.0
      DUDS=(U(I-1,J+1)-U(I-1,J-1))/2.0/DS
      D2UDS2=(U(I-1,J+1)-2.0*U(I-1,J)+U(I-1,J-1))/DS/DS
      U(I,J)=U(I-1,J)+DX*(SOURCE+D2UDS2-V(I-1,J)*DUDS)/U(I-1,J)
      V(I,J)=V(I,J-1)+V(I-1,J-1)-V(I-1,J)-
     1DS/DX*(U(I,J)+U(I,J-1)-U(I-1,J)-U(I-1,J-1))
  100 CONTINUE
      V(I,M)=V(I,M-1)+V(I-1,M-1)-V(I-1,M)-
     1DS/DX*(U(I,M)+U(I,M-1)-U(I-1,M)-U(I-1,M-1))
  120 CONTINUE
CCC  ----------------------------------------------------------------------
CCC  WRITING THE RESULTS ---------------------------------------------------
      PRINT*,'U VALUES'
      DO 500 J=M,0,-2
      WRITE(1,1000)(U(I,J),I=0,N,40)
  500 CONTINUE
      PRINT*,'V VALUES'
      DO 600 J=M,0,-2
      WRITE(1,1000)(V(I,J),I=0,N,40)
  600 CONTINUE
 1000 FORMAT(1X,11F7.3)
CCC  ----------------------------------------------------------------------
      END
```

FORTRAN program to solve Prandtl's momentum boundary layer equations for a uniform inviscid field U_0

these terms. The forward difference approximation used to represent $\partial u/\partial x$ has only first order accuracy, i.e. the error is of order δx. It is therefore necessary to choose a small value of δx in comparison with δs to achieve comparable accuracy. The FORTRAN program implementing this numerical scheme is listed opposite. The corresponding results shown below are for a grid size of $N = 400$ and $M = 32$.

There is a problem of numerical instability which can occur when using explicit schemes of this type. A description of this effect can be found in texts on the subject of the numerical solution of partial differential equations. For our purposes it is sufficient to use the rule that δx should be of the order of δs^2 to avoid instabilities.

To compare the finite difference solution field with the similarity solution it is convenient to choose the point $x = 0.5$, $s = 1$. The similarity variable, η, is related to x and s as follows

$$\eta = Y\sqrt{\frac{U_0}{2X\nu}} = \frac{Y}{L}\sqrt{\frac{1}{2}\frac{L}{X}\cdot\frac{LU_0}{\nu}} = \frac{s}{\sqrt{2x}}.$$

```
 PLATE LENGTH  =  ?
1.0
 HEIGHT=?
8.0
 U VALUES
  1.000  1.000  1.000  1.000  1.000  1.000  1.000  1.000  1.000  1.000  1.000
  1.000  1.000  1.000  1.000  1.000  1.000  1.000  1.000  1.000  1.000  1.000
  1.000  1.000  1.000  1.000  1.000  1.000  1.000  1.000  1.000  1.000  1.000
  1.000  1.000  1.000  1.000  1.000  1.000  1.000  1.000  1.000  1.000  1.000
  1.000  1.000  1.000  1.000  1.000  1.000  1.000  1.000  1.000  0.999  0.999
  1.000  1.000  1.000  1.000  1.000  1.000  1.000  1.000  0.999  0.998  0.996
  1.000  1.000  1.000  1.000  1.000  1.000  1.000  0.999  0.997  0.994  0.991
  1.000  1.000  1.000  1.000  1.000  0.999  0.998  0.995  0.991  0.986  0.979
  1.000  1.000  1.000  1.000  0.999  0.997  0.993  0.986  0.977  0.967  0.954
  1.000  1.000  1.000  0.999  0.996  0.989  0.979  0.965  0.948  0.930  0.912
  1.000  1.000  1.000  0.996  0.985  0.967  0.945  0.920  0.895  0.869  0.845
  1.000  1.000  0.996  0.979  0.950  0.915  0.878  0.842  0.809  0.778  0.750
  1.000  0.999  0.975  0.925  0.869  0.815  0.767  0.725  0.689  0.657  0.629
  1.000  0.982  0.893  0.798  0.721  0.660  0.611  0.572  0.538  0.510  0.486
  1.000  0.869  0.690  0.581  0.510  0.460  0.421  0.391  0.367  0.346  0.329
  1.000  0.514  0.367  0.301  0.261  0.234  0.213  0.198  0.185  0.174  0.165
  0.000  0.000  0.000  0.000  0.000  0.000  0.000  0.000  0.000  0.000  0.000
 V VALUES
  0.000  2.739  1.924  1.571  1.359  1.215  1.109  1.027  0.960  0.906  0.859
  0.000  2.739  1.924  1.571  1.359  1.215  1.109  1.027  0.960  0.906  0.859
  0.000  2.739  1.924  1.571  1.359  1.215  1.109  1.027  0.960  0.905  0.859
  0.000  2.739  1.924  1.571  1.359  1.215  1.109  1.027  0.960  0.905  0.858
  0.000  2.739  1.924  1.571  1.359  1.215  1.109  1.026  0.959  0.903  0.855
  0.000  2.739  1.924  1.571  1.359  1.215  1.109  1.025  0.957  0.899  0.849
  0.000  2.739  1.924  1.571  1.359  1.214  1.107  1.021  0.950  0.889  0.834
  0.000  2.739  1.924  1.571  1.358  1.212  1.101  1.011  0.933  0.866  0.805
  0.000  2.739  1.924  1.571  1.355  1.203  1.084  0.983  0.897  0.821  0.754
  0.000  2.739  1.924  1.567  1.341  1.174  1.040  0.926  0.829  0.747  0.675
  0.000  2.739  1.921  1.548  1.295  1.104  0.949  0.825  0.722  0.638  0.567
  0.000  2.738  1.898  1.475  1.178  0.962  0.799  0.674  0.577  0.500  0.439
  0.000  2.724  1.786  1.277  0.955  0.741  0.594  0.488  0.410  0.350  0.303
  0.000  2.588  1.442  0.915  0.638  0.474  0.369  0.298  0.247  0.208  0.179
  0.000  1.928  0.822  0.471  0.313  0.226  0.173  0.138  0.114  0.095  0.082
  0.000  0.643  0.229  0.125  0.081  0.058  0.044  0.035  0.029  0.024  0.021
  0.000  0.000  0.000  0.000  0.000  0.000  0.000  0.000  0.000  0.000  0.000
```

Velocities for $U_{INV} = U_0$

Thus when $x = 0.5$ and $s = 1$, η has the value 1. The similarity solution yields $U/U_0 = 0.461$ and

$$\frac{V}{U_0}\sqrt{\frac{U_0 L}{\nu}} = \frac{1}{\sqrt{2x}}\left(\eta\frac{\mathrm{d}f}{\mathrm{d}\eta} - f\right)\bigg|_{\substack{x=0.5\\ \eta=1}} = 0.228.$$

The finite difference solution yields the corresponding values 0.460 and 0.226 respectively. A further convenient comparison is between the values of V for large s (or equivalently large η). At the point $x = 0.5$, the Blasius solution yields $(V_\infty/U_0)\sqrt{Re} = 1.218$ whereas the finite difference solution yields the value 1.215. This excellent

```
CCC  THIS PROGRAM SOLVES PRANDTL'S BOUNDARY LAYER EQUATIONS
CCC  FOR THE INVISCID VELOCITY UINV=1.0+X
      PARAMETER(N=400,M=32)
      DIMENSION U(0:N,0:M),V(0:N,0:M)
CCC  SETTING THE LENGTH AND HEIGHT -----------------------------------
      PRINT*,'PLATE LENGTH = ?'
      READ*,XL
      DX=XL/N
      PRINT*,'HEIGHT=?'
      READ*,SMAX
      DS=SMAX/M
CCC  ----------------------------------------------------------------
CCC  SETTING THE BOUNDARY AND INLET CONDITIONS -----------------------
      DO 10 I-0,N
      U(I,0)=0.0
      U(I,M)=1.0+I*DX
      V(I,0)=0.0
  10  CONTINUE
      DO 20 J=1,M
      U(0,J)=1.0
      V(0,J)=-J*DS
  20  CONTINUE
CCC  ----------------------------------------------------------------
CCC  SOLVING FOR THE VELOCITIES --------------------------------------
      DO 120 I=1,N
      DO 100 J=1,M-1
      SOURCE=1.0+(I-1)*DX
      DUDS=(U(I-1,J+1)-U(I-1,J-1))/2.0/DS
      D2UDS2=(U(I-1,J+1)-2.0*U(I-1,J)+U(I-1,J-1))/DS/DS
      U(I,J)=U(I-1,J)+DX*(SOURCE+D2UDS2-V(I-1,J)*DUDS)/U(I-1,J)
      V(I,J)=V(I,J-1)+V(I-1,J-1)-V(I-1,J)-
     1DS/DX*(U(I,J)+U(I,J-1)-U(I-1,J)-U(I-1,J-1))
 100  CONTINUE
      V(I,M)=V(I,M-1)+V(I-1,M-1)-V(I-1,M)-
     1DS/DX*(U(I,M)+U(I,M-1)-U(I-1,M)-U(I-1,M-1))
 120  CONTINUE
CCC  ----------------------------------------------------------------
CCC  WRITING THE RESULTS ---------------------------------------------
      PRINT*,'U VALUES'
      DO 500 J=M,0,-2
      WRITE(1,1000)(U(I,J),I=0,N,40)
 500  CONTINUE
      PRINT*,'V VALUES'
      DO 600 J=M,0,-2
      WRITE(1,1000)(V(I,J),I=0,N,40)
 600  CONTINUE
1000  FORMAT(1X,11F7.3)
CCC  ----------------------------------------------------------------
      END
```

FORTRAN program to solve Prandtl's momentum boundary layer equations for the inviscid field $U_{\mathrm{INV}} = U_0(1 + X/L)$

agreement for the case of Blasius flow with $U_I = U_0$ allows us to use the finite difference scheme with some confidence for other inviscid flow fields.

Worked example

Modify the computer program to solve for the case of the inviscid flow field

$$U_{INV} = U_0(1 + X/L),\ 0 < X < L,\ Y > 0.$$

Solution

There are three modifications to make to the program.

(i) The asymptotic value of u, namely u_I is changed from 1.0 to $1.0 + x$

(ii) The pressure term $u_I(du_I/dx)$ is no longer zero; it is now $1.0 + x$. (This term is called SOURCE in the program.)

(iii) The inlet values of u and v must be considered. The continuity equation yields

$$\frac{\partial V}{\partial Y} = -\frac{\partial U}{\partial X} = -\frac{U_0}{L}.$$

```
 PLATE LENGTH = ?
1.0
 HEIGHT=?
8.0
 U VALUES
  1.000  1.100  1.200  1.300  1.400  1.500  1.600  1.700  1.800  1.900  2.000
  1.000  1.100  1.200  1.300  1.400  1.500  1.600  1.700  1.800  1.900  2.000
  1.000  1.100  1.200  1.300  1.400  1.500  1.600  1.700  1.800  1.900  2.000
  1.000  1.100  1.200  1.300  1.400  1.500  1.600  1.700  1.800  1.900  2.000
  1.000  1.100  1.200  1.300  1.400  1.500  1.600  1.700  1.800  1.900  2.000
  1.000  1.100  1.200  1.300  1.400  1.500  1.600  1.700  1.800  1.900  2.000
  1.000  1.100  1.200  1.300  1.400  1.500  1.600  1.700  1.800  1.900  2.000
  1.000  1.100  1.200  1.300  1.400  1.500  1.600  1.700  1.800  1.900  2.000
  1.000  1.100  1.200  1.300  1.400  1.500  1.600  1.700  1.800  1.900  2.000
  1.000  1.100  1.200  1.300  1.400  1.500  1.600  1.700  1.800  1.900  2.000
  1.000  1.100  1.200  1.300  1.400  1.500  1.599  1.699  1.799  1.898  1.998
  1.000  1.100  1.200  1.299  1.397  1.496  1.594  1.693  1.792  1.890  1.989
  1.000  1.100  1.196  1.290  1.383  1.476  1.571  1.666  1.761  1.857  1.954
  1.000  1.093  1.166  1.240  1.319  1.402  1.487  1.575  1.663  1.752  1.842
  1.000  1.023  1.029  1.071  1.130  1.196  1.267  1.340  1.414  1.489  1.565
  1.000  0.691  0.659  0.681  0.718  0.760  0.805  0.851  0.899  0.947  0.995
  0.000  0.000  0.000  0.000  0.000  0.000  0.000  0.000  0.000  0.000  0.000
 V VALUES
 -8.000 -6.034 -6.721 -7.069 -7.112 -7.188 -7.283 -7.323 -7.316 -7.293 -7.278
 -7.500 -5.534 -6.221 -6.569 -6.612 -6.688 -6.783 -6.823 -6.816 -6.793 -6.778
 -7.000 -5.034 -5.721 -6.069 -6.112 -6.188 -6.283 -6.323 -6.316 -6.293 -6.278
 -6.500 -4.534 -5.221 -5.569 -5.612 -5.688 -5.783 -5.823 -5.816 -5.793 -5.778
 -6.000 -4.034 -4.721 -5.069 -5.112 -5.188 -5.283 -5.323 -5.316 -5.293 -5.278
 -5.500 -3.534 -4.221 -4.569 -4.612 -4.688 -4.783 -4.823 -4.816 -4.793 -4.778
 -5.000 -3.034 -3.721 -4.069 -4.112 -4.188 -4.283 -4.323 -4.316 -4.293 -4.278
 -4.500 -2.534 -3.221 -3.569 -3.612 -3.688 -3.783 -3.823 -3.816 -3.793 -3.778
 -4.000 -2.034 -2.721 -3.069 -3.112 -3.188 -3.283 -3.323 -3.316 -3.293 -3.278
 -3.500 -1.534 -2.221 -2.569 -2.612 -2.688 -2.783 -2.824 -2.816 -2.793 -2.778
 -3.000 -1.034 -1.721 -2.069 -2.113 -2.188 -2.282 -2.326 -2.320 -2.297 -2.280
 -2.500 -0.534 -1.222 -1.573 -1.618 -1.685 -1.779 -1.834 -1.843 -1.827 -1.804
 -2.000 -0.037 -0.731 -1.098 -1.157 -1.173 -1.215 -1.258 -1.291 -1.312 -1.326
 -1.500  0.425 -0.288 -0.584 -0.726 -0.788 -0.820 -0.839 -0.851 -0.859 -0.865
 -1.000  0.701 -0.103 -0.295 -0.369 -0.403 -0.421 -0.432 -0.439 -0.443 -0.446
 -0.500  0.278 -0.024 -0.085 -0.107 -0.117 -0.122 -0.125 -0.127 -0.128 -0.129
  0.000  0.000  0.000  0.000  0.000  0.000  0.000  0.000  0.000  0.000  0.000
```

Velocities for $U_{INV} = U_0(1 + X/L)$

Integrating yields

$$V(X, Y) - V(X, 0) = -U_0 Y/L,$$

and $V(X, 0) = 0$. Thus in nondimensional form the inlet conditions are:

$u(0, s)$ remains unchanged and is $u(0, s) = 1$

$v^*(0, s)$ is no longer zero it is now equal to $-s$.

These modifications are included in the program listing on p. 206 and the corresponding velocity fields are shown on p. 207. A noticeable feature of the solution is that the accelerating flow field (u_I increasing with x) causes the boundary layer to remain closer to the plate than in the case of constant u_I.

Exercise

Modify the computer program to treat the case of the inviscid velocity field

$$U_{INV} = U_0 \exp\left(\frac{X}{2L}\right) \quad 0 < X < L,\ Y > 0.$$

Thermal boundary layer equation

We continue with our study of flow over a plate occupying the region $0 < x < 1, y = 0$ and consider the temperature variation in the fluid when the incoming flow is at a different temperature than the plate. The governing equation for the two-dimensional steady temperature field is given by 5.6.5′. The first member of the right-hand side of this equation corresponds to heat diffusion in the tangential direction. In the case of convection dominated heat transfer (large Peclet number, Pe) the dominant direction of diffusion is normal to the plate. The tangential diffusion term may be omitted to leading order in an expansion based on the small quantity $1/Pe$. The leading order term in the expansion of the temperature field in the boundary layer satisfies the *thermal boundary layer equation*

$$u\frac{\partial T}{\partial x} + v\frac{\partial T}{\partial y} = \frac{1}{Pe}\frac{\partial^2 T}{\partial y^2}.$$

It is convenient to use the same stretched normal variable, $s = Y\sqrt{Re}/L$, and the scaled normal velocity, $v^* = v\sqrt{Re}$, as were used in the momentum boundary layer equation. The thermal boundary layer equation becomes

$$u\frac{\partial T}{\partial x} + v^*\frac{\partial T}{\partial s} = \frac{Re}{Pe}\frac{\partial^2 T}{\partial s^2} = \frac{1}{Pr}\frac{\partial^2 T}{\partial s^2}, \qquad 5.6.37$$

where the Prandtl number, Pr, has already been introduced as the ratio Pe/Re.

Equation 5.6.37 is a parabolic partial differential equation and requires appropriate boundary conditions. In the case of flow over a plate these are:

the plate temperature:	$T(x, s = 0)$
the far field temperature:	$T(x, s \to \infty)$
the inlet temperature field:	$T(0, s)$.

The equations governing heat and momentum transfer, 5.6.37 and 5.6.29′ respectively, are very similar. Indeed for the case of Blasius flow for a fluid with $Pr = 1$ they are identical. In this case, if the temperature boundary conditions are of the form

$$T(x, 0) = T_w, \quad T(x, s \to \infty) = T(0, s) = T_0,$$

where T_w and T_0 are constants, then the temperature field can be obtained directly from the Blasius velocity field $U = U_0(\mathrm{d}f/\mathrm{d}\eta)$. It is

$$T = (T_0 - T_w)\frac{\mathrm{d}f}{\mathrm{d}\eta} + T_w.$$

The parabolic curves of constant velocity shown in Fig. 5.11 are also curves of constant temperature.

For general plate temperature distributions or inlet fluid temperatures the partial differential equation, 5.6.37, must be solved. It is straightforward to extend the finite difference scheme for solving the momentum equations to include the temperature calculation. The velocity fields are first determined and then the temperature obtained from a 'marching scheme' using a forward difference to approximate the convection term $u(\partial T/\partial x)$.

The example of a thermal boundary layer computer calculation overleaf is for the case of uniform flow $U_I = U_0$ over a flat plate where the incoming fluid has temperature zero,

$$T(x, = 0, s) = 0 = T(x, s \to \infty).$$

The plate temperature is given by

$$T(x, s = 0) = 1 + x$$

and the fluid Prandtl number is 0.71 (see p. 211).

Results are also shown on pages 211 and 212 for the case $Pr = 7.1$ and $Pr = 0.071$. These illustrate the narrowing and widening of the thermal boundary layer as the Prandtl number is changed. (In the case of small Pr, the coefficient $1/Pr$ multiplying the diffusion term in equation 5.6.37 is large. Numerical stability of the marching scheme requires that the value of δx be reduced. This reduction has been effected by reducing the plate length from 1.0 to 0.25 for the case $Pr = 0.071$.)

```
CCC   THIS PROGRAM SOLVES PRANDTL'S BOUNDARY LAYER EQUATIONS
CCC   AND THE THERMAL BOUNDARY LAYER EQUATION
CCC   FOR THE CASE OF A CONSTANT INVISCID VELOCITY (BLASIUS FLOW)
CCC   THE INCOMING FLUID HAS T=0,THE PLATE HAS T=1+X
        PARAMETER(N=400,M=32)
        DIMENSION U(0:N,0:M),V(0:N,0:M),T(0:N,0:M)
CCC   READING THE LENGTH AND HEIGHT ----------------------------------------
        PRINT*,'PLATE LENGTH = ?'
        READ*,XL
        DX=XL/N
        PRINT*,'HEIGHT=?'
        READ*,SMAX
        DS=SMAX/M
CCC   ----------------------------------------------------------------------
CCC   READING THE PRANDTL NUMBER -------------------------------------------
        PRINT*,'PRANDTL NUMBER=?'
        READ*,PRAND
CCC   ----------------------------------------------------------------------
CCC   SETTING THE BOUNDARY AND INLET CONDITIONS ----------------------------
        DO 10 I=0,N
        U(I,0)=0.0
        U(I,M)=1.0
        V(I,0)=0.0
        T(I,0)=1.0+I*DX
        T(I,M)=0.0
   10   CONTINUE
        DO 20 J=1,M
        U(0,J)=1.0
        V(0,J)=0.0
        T(0,J)=0.0
   20   CONTINUE
CCC   ----------------------------------------------------------------------
CCC   SOLVING FOR THE VELOCITIES -------------------------------------------
        DO 120 I=1,N
        DO 100 J=1,M-1
        DUDS=(U(I-1,J+1)-U(I-1,J-1))/2.0/DS
        D2UDS2=(U(I-1,J+1)-2.0*U(I-1,J)+U(I-1,J-1))/DS/DS
        U(I,J)=U(I-1,J)+DX*(D2UDS2-V(I-1,J)*DUDS)/U(I-1,J)
        V(I,J)=V(I,J-1)+V(I-1,J-1)-V(I-1,J)-
       1DS/DX*(U(I,J)+U(I,J-1)-U(I-1,J)-U(I-1,J-1))
  100   CONTINUE
        V(I,M)=V(I,M-1)+V(I-1,M-1)-V(I-1,M)-
       1DS/DX*(U(I,M)+U(I,M-1)-U(I-1,M)-U(I-1,M-1))
  120   CONTINUE
CCC   ----------------------------------------------------------------------
CCC     SOLVING FOR THE TEMPERATURES ---------------------------------------
        DO 220 I=1,N
        DO 220 J=1,M-1
        DTDS=(T(I-1,J+1)-T(I-1,J-1))/2.0/DS
        D2TDS2=(T(I-1,J+1)-2.0*T(I-1,J)+T(I-1,J-1))/DS/DS
        T(I,J)=T(I-1,J)+DX*(D2TDS2/PRAND-V(I-1,J)*DTDS)/U(I-1,J)
  220   CONTINUE
CCC   ----------------------------------------------------------------------
CCC   WRITING THE RESULTS --------------------------------------------------
        PRINT*,'TEMPERATURE VALUES'
        DO 500 J=M,0,-2
        WRITE(1,1000)(T(I,J),I=0,N,40)
  500   CONTINUE
 1000   FORMAT(1X,11F7.3)
CCC   ----------------------------------------------------------------------
        END
```

FORTRAN program to solve the momentum and thermal boundary layer equations

```
 PLATE LENGTH = ?
1.0
 HEIGHT=?
8.0
 PRANDTL NUMBER=?
0.71
 TEMPERATURE VALUES
  0.000  0.000  0.000  0.000  0.000  0.000  0.000  0.000  0.000  0.000  0.000
  0.000  0.000  0.000  0.000  0.000  0.000  0.000  0.000  0.000  0.000  0.000
  0.000  0.000  0.000  0.000  0.000  0.000  0.000  0.000  0.000  0.000  0.001
  0.000  0.000  0.000  0.000  0.000  0.000  0.000  0.000  0.001  0.001  0.003
  0.000  0.000  0.000  0.000  0.000  0.000  0.000  0.001  0.002  0.003  0.006
  0.000  0.000  0.000  0.000  0.000  0.000  0.001  0.002  0.005  0.009  0.014
  0.000  0.000  0.000  0.000  0.000  0.001  0.003  0.006  0.012  0.020  0.029
  0.000  0.000  0.000  0.000  0.001  0.003  0.008  0.016  0.027  0.041  0.058
  0.000  0.000  0.000  0.001  0.004  0.011  0.022  0.038  0.058  0.081  0.106
  0.000  0.000  0.000  0.003  0.013  0.030  0.053  0.081  0.113  0.147  0.184
  0.000  0.000  0.002  0.014  0.038  0.073  0.113  0.156  0.202  0.250  0.299
  0.000  0.000  0.012  0.048  0.099  0.157  0.217  0.277  0.338  0.399  0.459
  0.000  0.004  0.052  0.132  0.217  0.298  0.377  0.453  0.527  0.600  0.672
  0.000  0.038  0.170  0.299  0.410  0.509  0.601  0.689  0.774  0.857  0.939
  0.000  0.193  0.412  0.562  0.683  0.789  0.889  0.984  1.076  1.167  1.257
  0.000  0.574  0.773  0.907  1.022  1.127  1.229  1.327  1.425  1.521  1.616
  1.000  1.100  1.200  1.300  1.400  1.500  1.600  1.700  1.800  1.900  2.000
```

Temperatures for $Pr = 0.71$

```
 PLATE LENGTH = ?
1.0
 HEIGHT=?
8.0
 PRANDTL NUMBER=?
7.1
 TEMPERATURE VALUES
  0.000  0.000  0.000  0.000  0.000  0.000  0.000  0.000  0.000  0.000  0.000
  0.000  0.000  0.000  0.000  0.000  0.000  0.000  0.000  0.000  0.000  0.000
  0.000  0.000  0.000  0.000  0.000  0.000  0.000  0.000  0.000  0.000  0.000
  0.000  0.000  0.000  0.000  0.000  0.000  0.000  0.000  0.000  0.000  0.000
  0.000  0.000  0.000  0.000  0.000  0.000  0.000  0.000  0.000  0.000  0.000
  0.000  0.000  0.000  0.000  0.000  0.000  0.000  0.000  0.000  0.000  0.000
  0.000  0.000  0.000  0.000  0.000  0.000  0.000  0.000  0.000  0.000  0.000
  0.000  0.000  0.000  0.000  0.000  0.000  0.000  0.000  0.000  0.000  0.000
  0.000  0.000  0.000  0.000  0.000  0.000  0.000  0.000  0.000  0.000  0.000
  0.000  0.000  0.000  0.000  0.000  0.000  0.000  0.000  0.000  0.000  0.001
  0.000  0.000  0.000  0.000  0.000  0.000  0.000  0.001  0.001  0.002  0.004
  0.000  0.000  0.000  0.000  0.000  0.001  0.002  0.005  0.008  0.012  0.018
  0.000  0.000  0.000  0.001  0.004  0.010  0.017  0.028  0.041  0.057  0.075
  0.000  0.000  0.005  0.017  0.035  0.060  0.090  0.123  0.160  0.199  0.240
  0.000  0.014  0.059  0.121  0.190  0.260  0.329  0.397  0.465  0.532  0.599
  0.000  0.201  0.389  0.530  0.645  0.748  0.845  0.937  1.027  1.115  1.203
  1.000  1.100  1.200  1.300  1.400  1.500  1.600  1.700  1.800  1.900  2.000
```

Temperatures for $Pr = 7.1$

Mass transfer to a slowly falling particle

This section on boundary layer applications is concluded with a consideration of convection dominated mass transfer. Material dissolved in a liquid is absorbed by a spherical particle which is falling through the liquid under the action of gravity. We will assume that the particle has attained its terminal velocity U_0. The Reynolds

```
 PLATE LENGTH  =  ?
0.25
 HEIGHT=?
8.0
 PRANDTL NUMBER=?
0.071
 TEMPERATURE VALUES
  0.000   0.000   0.000   0.000   0.000   0.000   0.000   0.000   0.000   0.000   0.000
  0.000   0.000   0.000   0.000   0.000   0.000   0.001   0.001   0.003   0.005   0.007
  0.000   0.000   0.000   0.000   0.000   0.001   0.002   0.004   0.007   0.011   0.015
  0.000   0.000   0.000   0.000   0.000   0.001   0.004   0.007   0.013   0.019   0.027
  0.000   0.000   0.000   0.000   0.001   0.003   0.007   0.014   0.022   0.032   0.043
  0.000   0.000   0.000   0.000   0.002   0.007   0.015   0.025   0.038   0.052   0.067
  0.000   0.000   0.000   0.001   0.006   0.015   0.028   0.044   0.061   0.080   0.100
  0.000   0.000   0.000   0.004   0.014   0.030   0.050   0.072   0.096   0.120   0.144
  0.000   0.000   0.002   0.012   0.031   0.056   0.085   0.115   0.145   0.174   0.203
  0.000   0.000   0.006   0.029   0.062   0.099   0.138   0.175   0.211   0.245   0.277
  0.000   0.001   0.020   0.064   0.115   0.165   0.212   0.256   0.296   0.334   0.369
  0.000   0.005   0.056   0.128   0.197   0.258   0.311   0.359   0.402   0.442   0.478
  0.000   0.027   0.134   0.234   0.315   0.382   0.438   0.486   0.530   0.569   0.606
  0.000   0.102   0.273   0.389   0.471   0.536   0.590   0.636   0.677   0.715   0.750
  0.000   0.289   0.485   0.589   0.661   0.717   0.763   0.804   0.841   0.875   0.908
  0.000   0.616   0.754   0.824   0.875   0.916   0.953   0.986   1.017   1.047   1.076
  1.000   1.025   1.050   1.075   1.100   1.125   1.150   1.175   1.200   1.225   1.250
```

Temperatures for $Pr = 0.071$

number for the flow is $U_0 A/\nu$ where A is the radius of the particle and ν is the kinematic viscosity of the fluid. Small particles travelling at low speeds with a value of $U_0 A$ which is typically less than 10^{-6} m^2/s will have low Reynolds numbers. This allows the slow flow approximation of the Navier–Stokes equations where the convection term $(\mathbf{V}\,.\,\nabla)\mathbf{V}$ is omitted from the material derivative.

In the rest frame of the particle the flow is steady. The solution of the slow viscous flow equations for the flow of a fluid past a sphere is well known (see Batchelor[16]). In spherical polar coordinates the two nonzero components of velocity are

$$V_R = -U_0 \cos\theta\left(1 - \frac{3A}{2R} + \frac{A^3}{2R^3}\right), \quad V_\theta = U_0 \sin\theta\left(1 - \frac{3A}{4R} - \frac{A^3}{4R^3}\right).$$

The velocity components are shown in Fig. 5.14. The steady convection diffusion equation governing the concentration, C, is

$$V_R \frac{\partial C}{\partial R} + V_\theta \frac{1}{R}\frac{\partial C}{\partial \theta} = \Gamma \nabla^2 C$$

where Γ is the diffusivity. Using the following nondimensional variables

$r = R/A, \quad u_r = V_R/U_0, \quad u_\theta = V_\theta/U_0, \quad c = C/C_\infty$ this equation becomes

$$u_r \frac{\partial c}{\partial r} + u_\theta \frac{1}{r}\frac{\partial c}{\partial \theta} = \frac{\Gamma}{AU_0}\left[\frac{\partial^2 c}{\partial r^2} + \frac{2}{r}\frac{\partial c}{\partial r} + \frac{1}{r^2 \sin\theta}\frac{\partial}{\partial \theta}\left(\sin\theta \frac{\partial c}{\partial \theta}\right)\right]. \qquad 5.6.38$$

The diffusivity, Γ, is usually far smaller than the kinematic viscosity, ν, with a typical value of 10^{-9} m/s^2. Consequently it is common for the Péclet number, $Pe\ (= AU_0/\Gamma)$ to be large while the Reynolds number (AU_0/ν) remains small. In these circumstances

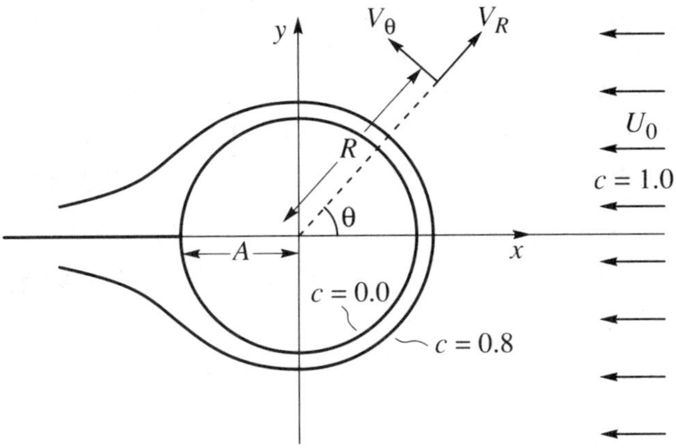

Fig. 5.14 The spherical polar velocity components and the $c = 0.8$ concentration curve

although there is no momentum boundary layer there is a mass transfer boundary layer close to the solid surface. Away from this region the concentration remains at its undisturbed value which we will take as a constant.

Within the boundary layer normal derivatives of the concentration dominate the diffusion term so the last member of the right-hand side of equation 5.6.38 may be neglected. It is convenient to introduce the variable $y\ (=r-1)$ so that the boundary layer occurs near $y = 0$. Expanding the velocities for small values of y yields the following form of 5.6.38:

$$-\frac{3}{2}\cos\theta[y^2 + O(y^3)]\frac{\partial c}{\partial y} + \frac{3}{2}\sin\theta[y + O(y^2)]\frac{\partial c}{\partial \theta}$$

$$= \frac{1}{Pe}\left(\frac{\partial^2 c}{\partial y^2} + 2\frac{\partial c}{\partial y}[1 + O(y)]\right). \qquad 5.6.39$$

The coefficient, $1/Pe$, of the highest derivative is a small parameter. We assume that the boundary layer thickness is characterized by the stretching transformation $s = y/(1/Pe)^t$ where the power, t, is determined from the principle of least degeneracy. Equation 5.6.39 becomes

$$-\frac{3}{2}\cos\theta\left[s^2 + O\left(\frac{1}{Pe^t}\right)\right]\left(\frac{1}{Pe}\right)^{2t-t}\frac{\partial c}{\partial s} + \frac{3}{2}\sin\theta\left[s + O\left(\frac{1}{Pe^t}\right)\right]\left(\frac{1}{Pe}\right)^{t}\frac{\partial c}{\partial \theta}$$

$$= \left(\frac{1}{Pe}\right)^{1-2t}\cdot\frac{\partial^2 c}{\partial s^2} + O\left(\frac{1}{Pe^{1-t}}\right).$$

Thus $t = 1 - 2t$, i.e. $t = 1/3$ and the concentration is governed, to leading order, by the following equation

$$-\frac{3}{2}\cos\theta s^2\frac{\partial c}{\partial s} + \frac{3}{2}\sin\theta s\frac{\partial c}{\partial \theta} = \frac{\partial^2 c}{\partial s^2}. \qquad 5.6.40$$

The boundary layer thickness is of order $1/Pe^{1/3}$, i.e. $y = O(1/Pe^{1/3})$ so within the boundary layer region $R - A = O(A/Pe^{1/3})$. Equation 5.6.40 governs the concentration, subject to the specification of boundary conditions. If we choose $C \to C_\infty$ as $R \to \infty$, i.e. $c = 1$ as $s \to \infty$ and $C = 0$ on $R = A$, i.e. $c(s = 0) = 0$, then a solution of 5.6.40 can be obtained by a similarity transformation (see Levich[17]). The solution is

$$c = \frac{\int_0^z \exp(-4x^3/9)\mathrm{d}x}{\int_0^\infty \exp(-4x^3/9)\mathrm{d}x}, \qquad 5.6.41$$

where

$$z = (3Pe)^{1/3}\frac{R - A}{2R}\frac{\sin\theta}{\left(\dfrac{\theta}{2} - \dfrac{1}{4}\sin 2\theta\right)^{1/3}}. \qquad 5.6.42$$

The curve of concentration $c = 0.8$ is shown in Fig. 5.14 for the case $Pe = 500$. It demonstrates typical features of the concentration distribution. The boundary layer thickness is a minimum on the front surface of the particle. The thickness increases as $\theta \to \pi$. This feature is represented mathematically by the angular dependence of the z variable in equation 5.6.42. In the region near $\theta = \pi$ the boundary layer thickness tends to infinity. This invalidates the approximation in this region. However, the boundary layer solution remains valid over most of the flow region.

In the case of general boundary conditions the governing partial differential equation can only be solved approximately using a numerical scheme for discretized values. The above analysis is helpful in providing essential information about the scale of the required computational grid.

The mass transfer rate, j, from the liquid to the particle is given by the gradient of the concentration at the surface,

$$j = -\Gamma \left.\frac{\partial C}{\partial R}\right|_{R=A}$$

This is proportional to $\Gamma(C_\infty - C_0)$ divided by the boundary layer thickness $A/Pe^{1/3}$,

$$j \propto \frac{\Gamma(C_\infty - C_0)Pe^{1/3}}{A}.$$

There will be an angular dependence as given by the z variable dependence on θ in equation 5.6.42.

The case of a slowly falling liquid drop can be dealt with in a similar fashion. There exists an exact solution of the slow flow equations for the two fluids (see Batchelor[16]). In this case the tangential motion at the droplet boundary enhances mass transfer. The convection diffusion equation has a similar form to 5.6.39 except that the left-hand side is of lower order in y. The terms $y^2(\partial c/\partial y)$ and $y(\partial c/\partial\theta)$ are replaced by

$y(\partial c/\partial y)$ and $(\partial c/\partial \theta)$ respectively. The boundary layer thickness, determined by the principle of least degeneracy, is now of order $A/Pe^{1/2}$. The associated mass transfer rate is enhanced to give

$$j \propto \frac{\Gamma(C_\infty - C_0)Pe^{1/2}}{A},$$

(see Levich[17]).

Chapter 6
The Dominant Balance and WKB Methods

In this chapter asymptotic expansions for the solutions of linear differential equations will be constructed. The behavior of the solutions for large values of the independent variable will be analyzed using the method of dominant balance. Their behavior for large values of a parameter will be analyzed using the WKB method.

The method of dominant balance will be applied to linear homogeneous second order differential equations of the form

$$\frac{\mathrm{d}^2 y}{\mathrm{d}x^2} + p(x)\frac{\mathrm{d}y}{\mathrm{d}x} + q(x)y = 0.$$

Any coefficient of the highest derivative which may initially be present can be divided out and simply changes the coefficients $p(x)$ and $q(x)$. Higher order equations can be analyzed in a similar way to those of second order.

6.1 Frobenius series

The Frobenius method allows the solution of the differential equation near the point $x = x_0$ to be represented as an infinite series in powers of $(x - x_0)$. The condition which must be satisfied for the Frobenius method to be successful is that the point $x = x_0$ be either an ordinary point of the equation when $p(x_0)$ and $q(x_0)$ are finite, or a regular singular point when

$$\lim_{x \to x_0} [(x - x_0)p(x)] \quad \text{and} \quad \lim_{x \to x_0} [(x - x_0)^2 q(x)]$$

are finite. If these conditions are not satisfied then the point $x = x_0$ is an irregular singular point.

We will be concerned with the behavior of solutions as $x \to \infty$. The classification of 'the point at infinity' is obtained by transforming to the new independent variable s, where $s = 1/x$, and investigating the behavior of the coefficients of the equation in the limit $s \to 0$.

Consider as an example the application of the Frobenius technique the differential equation

$$\frac{d^2 y}{dx^2} + \frac{1}{x^4} y = 0, \qquad 6.1.1$$

as $x \to \infty$. The substitution $x = 1/s$ leads to

$$\frac{dy}{dx} = \frac{ds}{dx}\frac{dy}{ds} = -s^2 \frac{dy}{ds},$$

and

$$\frac{d^2 y}{dx^2} = -s^2 \frac{d}{ds}\left(-s^2 \frac{dy}{ds}\right) = s^4 \frac{d^2 y}{ds^2} + 2s^3 \frac{dy}{ds},$$

so that 6.1.1 becomes

$$\frac{d^2 y}{ds^2} + \frac{2}{s}\frac{dy}{ds} + y = 0. \qquad 6.1.2$$

Clearly $s = 0$ is a singularity of the differential equation. However it is a regular singularity so that we are assured of the existence of at least one solution of the standard Frobenius form

$$y = \sum_{n=0}^{\infty} a_n s^{n+\sigma}, \qquad 6.1.3$$

where, to fix the value of σ, we require $a_0 \neq 0$. The second solution can sometimes involve logarithmic terms. In fact for the example 6.1.2 two independent solutions are generated from the form 6.1.3.

Differentiating the series and substituting into equation 6.1.2 leads to the following:

$$a_0[\sigma(\sigma - 1) + 2\sigma]s^{\sigma-2}$$
$$+ a_1[(\sigma + 1)\sigma + 2(\sigma + 1)]s^{\sigma-1}$$
$$+ \sum_{n=2}^{\infty} \{a_n[(n + \sigma)(n + \sigma - 1) + 2(n + \sigma)] + a_{n-2}\} s^{n+\sigma-2} = 0.$$

Equating coefficients of powers of s to zero yields

$$s^{\sigma-2} \quad : \quad \sigma(\sigma + 1) = 0 \qquad 6.1.4$$

$$s^{\sigma-1} \quad : \quad a_1(\sigma + 1)(\sigma + 2) = 0 \qquad 6.1.5$$

$$s^{n+\sigma-2} \quad : a_n(n + \sigma)(n + \sigma + 1) + a_{n-2} = 0 \quad \text{for } n \geqslant 2. \qquad 6.1.6$$

There are two solutions of equation 6.1.4 namely $\sigma = 0$ and $\sigma = -1$. Consider first the case $\sigma = 0$. Equation 6.1.5 requires that a_1 be zero. Then the recurrence relation

6.1.6 yields the result $a_n = 0$ for all odd n while for n even

$$a_2 = -\frac{a_0}{2.3}, \quad a_4 = -\frac{a_2}{4.5} = \frac{a_0}{5!}, \quad a_6 = -\frac{a_4}{6.7} = -\frac{a_0}{7!},$$

etc.

Thus the solution corresponding to $\sigma = 0$ is

$$y_a = a_0\left(1 - \frac{s^2}{3!} + \frac{s^4}{5!} - \frac{s^6}{7!} + \cdots\right). \qquad 6.1.7$$

In the case $\sigma = -1$ equation 6.1.5 is satisfied for all values of a_1. The recurrence relation 6.1.6 becomes

$$b_n = -\frac{b_{n-2}}{n(n-1)} \quad \text{for } n \geqslant 2,$$

where the b_n are used to distinguish between the first and second solutions. We have

$$b_2 = -\frac{b_0}{2.1}, \quad b_3 = -\frac{b_1}{3.2}, \quad b_4 = -\frac{b_2}{4.3} = \frac{b_0}{4!}, \quad b_5 = -\frac{b_3}{5.4} = \frac{b_1}{5!},$$

which generates the following solution

$$y_b = \frac{b_0}{s}\left(1 - \frac{s^2}{2!} + \frac{s^4}{4!} - \cdots\right) + \frac{b_1}{s}\left(s - \frac{s^3}{3!} + \frac{s^5}{5!} - \cdots\right). \qquad 6.1.8$$

The second solution, y_b, contains the first solution, y_a, as a special case obtained by setting $b_0 = 0$ and $b_1 = a_0$. The general solution is therefore given by the second solution. The corresponding series in terms of the original variable is obtained by replacing s by $1/x$. It is possible to sum the series in this particular case because within the expression 6.1.8 can be recognized the expansions for sine s and cosine s. Thus the solution of 6.1.1 has been found to be expressible in closed form using elementary functions and is

$$y = Ax\cos(1/x) + Bx\sin(1/x). \qquad 6.1.9$$

It is merely a convenient occurrence that in this case the series obtained were those of elementary functions. Usually we are left with an infinite series form of solution.

The failure of the Frobenius series

An example of a case where the Frobenius method fails is provided by the differential equation

$$\frac{d^2y}{dx^2} - \frac{1}{x}y = 0, \qquad 6.1.10$$

as $x \to \infty$.

The substitution $s = 1/x$ leads to the equation

$$\frac{d^2 y}{ds^2} + \frac{2}{s}\frac{dy}{ds} - \frac{1}{s^3} y = 0. \tag{6.1.11}$$

There is an irregular singular point at $s = 0$. To see the failure of the Frobenius method we attempt to construct a solution of the form 6.1.3. Then substituting into equation 6.1.11 leads to the following:

$$\sum_{n=0}^{\infty} a_n[(n+\sigma)(n+\sigma-1) + 2(n+\sigma)]s^{n+\sigma-2} - \sum_{n=0}^{\infty} a_n s^{n+\sigma-3} = 0.$$

Equating coefficients of powers of s to zero yields

$$s^{\sigma-3} \quad : \quad a_0 = 0.$$

$$s^{n+\sigma-3} \quad : a_n - a_{n-1}(n+\sigma-1)(n+\sigma) = 0 \quad \text{for } n \geqslant 1.$$

Thus all of the coefficients, a_n, are zero.

We will show in the next section that the solution of 6.1.10 behaves as follows:

$$y = Ax^{1/4}e^{2\sqrt{x}}[1 + O(1/\sqrt{x})] + Bx^{1/4}e^{-2\sqrt{x}}[1 + O(1/\sqrt{x})] \quad \text{as } x \to \infty. \tag{6.1.12}$$

The presence of the square root in the exponential function shows that this cannot be represented in the Frobenius series form $\sum_{n=0}^{\infty} a_n(1/x)^{n+\sigma}$.

6.2 The method of dominant balance

The method of dominant balance will be described by constructing the approximation 6.1.12 for the solution of 6.1.10. We may choose to express the solution in the form $y = \exp(f)$.

Then

$$\frac{dy}{dx} = \frac{df}{dx}.e^f \quad \text{and} \quad \frac{d^2 y}{dx^2} = \left[\frac{d^2 f}{dx^2} + \left(\frac{df}{dx}\right)^2\right]e^f,$$

so that on substituting into 6.1.10 we obtain

$$\frac{d^2 f}{dx^2} + \left(\frac{df}{dx}\right)^2 - \frac{1}{x} = 0. \tag{6.2.1}$$

It is more difficult to obtain the exact solution of this equation than that of the original equation 6.1.10. However, an approximate solution of 6.2.1 for large x can easily be obtained as follows. The function f is required to behave in such a way that either d^2f/dx^2 or $(df/dx)^2$ or both balance the term $1/x$ on the left-hand side of 6.2.1. If we assume

$$\frac{d^2f}{dx^2} = O\left(\frac{1}{x}\right) \quad \text{as } x \to \infty \text{ then } \quad \frac{df}{dx} = O[\ln(x)],$$

so that the dominant term on the left-hand side of 6.2.1 would be of order $[\ln(x)]^2$ with no other term to balance it. On the other hand, if we assume

$$\left(\frac{df}{dx}\right)^2 = O\left(\frac{1}{x}\right) \quad \text{as } x \to \infty \text{ then } \quad \frac{df}{dx} = O\left(\frac{1}{\sqrt{x}}\right),$$

so that

$$\frac{d^2f}{dx^2} = O\left(\frac{1}{x^{3/2}}\right) \quad \text{as } x \to \infty .$$

For this choice 6.2.1 can be satisfied to leading order, i.e. to $O(1/x)$ with an error of order $1/x^{3/2}$.

Thus the leading order expression for f is obtained by solving the first order equation

$$\left(\frac{df}{dx}\right)^2 - \frac{1}{x} = 0,$$

i.e.

$$\frac{df}{dx} = \pm\frac{1}{\sqrt{x}},$$

so that $f = \pm 2\sqrt{x}$. No constant of integration is added to f because this becomes a multiplicative constant for $y(=e^f)$ and an arbitrary multiplicative constant is included at the end of the solution procedure.

The full solution is not simply $e^{\pm 2\sqrt{x}}$, of course, but may again be expressed in exponential form

$$y = e^{\pm 2\sqrt{x}} e^g, \tag{6.2.2}$$

where the leading behavior $e^{\pm 2\sqrt{x}}$ has been factored out. The function g is found as follows. Differentiating 6.2.2 yields

$$\frac{dy}{dx} = \left(\pm\frac{1}{\sqrt{x}} + \frac{dg}{dx}\right)\exp(\pm 2\sqrt{x} + g),$$

and

$$\frac{d^2y}{dx^2} = \left[\mp\frac{1}{2x^{3/2}} + \frac{d^2g}{dx^2} + \left(\pm\frac{1}{\sqrt{x}} + \frac{dg}{dx}\right)^2\right]\exp(\pm 2\sqrt{x} + g).$$

Substituting into equation 6.1.10 leads to the expression

$$\frac{d^2g}{dx^2} + \left(\frac{dg}{dx}\right)^2 \pm \frac{2}{\sqrt{x}}\frac{dg}{dx} + \frac{1}{x} \mp \frac{1}{2x^{3/2}} - \frac{1}{x} = 0. \qquad 6.2.3$$

The dominant terms on the left-hand side of 6.2.3 are $O(1/x)$ and cancel exactly leaving a driving term $\mp 1/2x^{3/2}$. The function g is determined to leading order by requiring that this driving term be balanced by either

$$\frac{d^2g}{dx^2}, \quad \left(\frac{dg}{dx}\right)^2, \quad \pm\frac{2}{\sqrt{x}}\frac{dg}{dx},$$

or a combination of them.

If we assume that

$$\frac{d^2g}{dx^2} = O\left(\frac{1}{x^{3/2}}\right) \quad \text{as } x \to \infty,$$

then $\dfrac{dg}{dx} = O\left(\dfrac{1}{\sqrt{x}}\right)$ so that the terms $\left(\dfrac{dg}{dx}\right)^2$ and $\pm\dfrac{2}{\sqrt{x}}\dfrac{dg}{dx}$ are $O(1/x)$.

These are of higher order than the driving term, so this possibility is rejected. Next assume

$$\left(\frac{dg}{dx}\right)^2 = O\left(\frac{1}{x^{3/2}}\right) \quad \text{as } x \to \infty$$

then $\dfrac{dg}{dx} = O\left(\dfrac{1}{x^{3/4}}\right)$ so that the term $\pm\dfrac{2}{\sqrt{x}}\dfrac{dg}{dx}$ is $O\left(\dfrac{1}{x^{5/4}}\right)$ which is again of higher order than the driving term.

The remaining possibility is to assume that

$$\pm\frac{2}{\sqrt{x}}\frac{dg}{dx} = O\left(\frac{1}{x^{3/2}}\right) \quad \text{as } x \to \infty.$$

Then $\dfrac{dg}{dx} = O(1/x)$ so that the terms $\dfrac{d^2g}{dx^2}$ and $\left(\dfrac{dg}{dx}\right)^2$ have order $1/x^2$ and are thus of lower order than the driving term. This is therefore the correct approximation and the equation

$$\pm\frac{2}{\sqrt{x}}\frac{dg}{dx} \mp \frac{1}{2x^{3/2}} = 0,$$

determines g to leading order. Integrating yields $g = \frac{1}{4}\ln|x|$ where again the constant of integration is omitted since it merely contributes a multiplicative constant for the solution y. We choose to consider x large and positive so that the modulus sign may be omitted. The solution form 6.2.2 may be expressed as

$$y = \exp(\pm 2\sqrt{x} + \tfrac{1}{4}\ln x).\exp(h),$$

with the leading behavior factored out.

Differentiating yields

$$\frac{dy}{dx} = \left(\pm\frac{1}{\sqrt{x}} + \frac{1}{4x} + \frac{dh}{dx}\right)\exp(\pm 2\sqrt{x} + \tfrac{1}{4}\ln x + h)$$

and

$$\frac{d^2y}{dx^2} = \left[\mp\frac{1}{2x^{3/2}} - \frac{1}{4x^2} + \frac{d^2h}{dx^2} + \left(\pm\frac{1}{\sqrt{x}} + \frac{1}{4x} + \frac{dh}{dx}\right)^2\right]$$
$$\times \exp(\pm 2\sqrt{x} + \tfrac{1}{4}\ln x + h).$$

Substituting into equation 6.1.10 leads to the expression

$$\mp\frac{1}{2x^{3/2}} - \frac{1}{4x^2} + \frac{d^2h}{dx^2} + \frac{1}{x} + \frac{1}{16x^2} + \left(\frac{dh}{dx}\right)^2 \pm \frac{1}{2\sqrt{x}\,.\,x}$$
$$\pm\frac{2}{\sqrt{x}}\frac{dh}{dx} + \frac{1}{2x}\frac{dh}{dx} - \frac{1}{x} = 0. \qquad 6.2.4$$

The driving term on the left-hand side is $-3/16x^2$. If we assume that the term $\pm\dfrac{2}{\sqrt{x}}\dfrac{dh}{dx}$ is of this order then $\dfrac{dh}{dx} = O\left(\dfrac{1}{x^{3/2}}\right)$ so that $h = O\left(\dfrac{1}{\sqrt{x}}\right)$. Then the remaining terms on the left-hand side of 6.2.4 behave as follows,

$$\frac{d^2h}{dx} = O\left(\frac{1}{x^{5/2}}\right) \quad \text{as} \quad x \to \infty$$

$$\left(\frac{dh}{dx}\right)^2 = O\left(\frac{1}{x^3}\right) \quad \text{as} \quad x \to \infty$$

$$\frac{1}{2x}\frac{dh}{dx} = O\left(\frac{1}{x^{5/2}}\right) \quad \text{as} \quad x \to \infty .$$

These terms are of lower order than the driving term, so the leading approximation for h is given by solution of the equation

$$\pm\frac{2}{\sqrt{x}}\frac{dh}{dx} - \frac{3}{16x^2} = 0, \quad \text{i.e.} \quad h = \mp\frac{3}{16\sqrt{x}}.$$

Thus the approximation which has been constructed has the form

$$y(x) \sim \exp[f(x) + g(x) + h(x) + \cdots] \qquad 6.2.5$$

where f is of dominant order, followed by g and then h, i.e.

$$g(x) = o[f(x)] \quad \text{as } x \to \infty$$

$$h(x) = o[g(x)] \quad \text{as } x \to \infty .$$

The tilde symbol in the expression 6.2.5 is used to allow for the case where the series is divergent but asymptotic.

For the example considered we have the solution

$$y \sim A \exp\left(2\sqrt{x} + \frac{1}{4}\ln x - \frac{3}{16\sqrt{x}} + \cdots\right)$$
$$+ B\exp\left(-2\sqrt{x} + \frac{1}{4}\ln x + \frac{3}{16\sqrt{x}} + \cdots\right).$$

Further terms in the argument of the exponentials can be found by continuing the dominant balance procedure. We will be content in this example to truncate the series at the first term which tends to zero as $x \to \infty$. The coefficient associated with the last term has the behavior,

$$\exp\left(\mp \frac{3}{16\sqrt{x}}\right) = 1 + O\left(\frac{1}{\sqrt{x}}\right) \quad \text{as } x \to \infty .$$

Thus we may write the solution in the form

$$y = Ax^{1/4}\exp(2\sqrt{x})\left[1 + O\left(\frac{1}{\sqrt{x}}\right)\right]$$
$$+ Bx^{1/4}\exp(-2\sqrt{x})\left[1 + O\left(\frac{1}{\sqrt{x}}\right)\right] \quad \text{as } x \to \infty . \qquad 6.2.6$$

The relative error associated with the remainder,

$$\frac{y_{\text{exact}} - Ax^{1/4}\exp(2\sqrt{x}) - Bx^{1/4}\exp(-2\sqrt{x})}{y_{\text{exact}}}$$

is of order $1/\sqrt{x}$ and so tends to zero as $x \to \infty$.

It would not be sufficient to truncate the procedure after having obtained the leading approximation, $f = \pm 2\sqrt{x}$, because the relative error associated with the approximation $y \simeq A\exp(2\sqrt{x}) + B\exp(-2\sqrt{x})$ is of order 1 as $x \to \infty$.

The technique described above for obtaining the approximation 6.2.6 for the solution of equation 6.1.10 illustrates the method of dominant balance.

Removal of the first derivative

Before proceeding further with our study of the technique it is useful to take advantage of the fact that second order linear equations of the form

$$\frac{d^2y}{dx^2} + p(x)\frac{dy}{dx} + q(x)y = 0, \qquad 6.2.7$$

can be reduced, by a change of dependent variable, to a form with $p(x) = 0$. The transformation is $Y = y\exp(\frac{1}{2}\int p\,dx)$. This is verified as follows:

$$\frac{dy}{dx} = \left(\frac{dY}{dx} - \frac{1}{2}pY\right)\exp\left(-\frac{1}{2}\int p\,dx\right)$$

and

$$\frac{d^2y}{dx^2} = \left(\frac{d^2Y}{dx^2} - p\frac{dY}{dx} - \frac{1}{2}\frac{dp}{dx}Y + \frac{1}{4}p^2Y\right)\exp\left(-\frac{1}{2}\int p\,dx\right).$$

Substituting into equation 6.2.7 and canceling the exponential leads to the equation

$$\frac{d^2Y}{dx^2} - p\frac{dY}{dx} - \frac{1}{2}\frac{dp}{dx}Y + \frac{1}{4}p^2Y + p\frac{dY}{dx} - \frac{1}{2}p^2Y + qY = 0,$$

i.e.

$$\frac{d^2Y}{dx^2} + \left(q - \frac{1}{2}\frac{dp}{dx} - \frac{1}{4}p^2\right)Y = 0,$$

where the first derivative is absent.

Thus without loss of generality we may restrict our study to the form

$$\frac{d^2y}{dx} + q(x)y = 0. \qquad 6.2.8$$

Construction of an asymptotic expansion

The condition imposed on the point at infinity such that a Frobenius solution exists is obtained by considering the behavior as $s \to 0$ where $s = 1/x$. Then 6.2.8 becomes

$$\frac{d^2y}{ds^2} + \frac{2}{s}\frac{dy}{dx} + \frac{1}{s^4}q\left(\frac{1}{s}\right)y = 0,$$

so that for $s = 0$ to be a regular singular point we require $\lim_{s\to 0}\left[\frac{1}{s^2}q\left(\frac{1}{s}\right)\right]$ to be finite. In terms of the independent variable x the condition for the existence of a Frobenius solution is that $\lim_{x \to \infty} [x^2 q(x)]$ be finite, i.e.

$$q(x) = O\left(\frac{1}{x^2}\right) \quad \text{as } x \to \infty .$$

If the behavior of $q(x)$ dominates $1/x^2$ as $x \to \infty$ then a Frobenius solution does not exist. It is for such cases that the method of dominant balance is vital in providing a solution procedure where more straightforward techniques fail.

In Murray's[18] text a systematic approach is described for the application of the method of dominant balance. Following his notation, the solution $y(x)$ of 6.2.8 is assumed to have an expansion

$$y \sim \exp\left(\sum_{n=0}^{\infty} \phi_n(x)\right) \quad \text{as } x \to \infty ,$$

where the functions $\phi_n(x)$ form an asymptotic sequence, $\phi_{n+1}(x) = o[\phi_n(x)]$ as $x \to \infty$. The differentiated sequences are assumed to also be asymptotic (this is the case for all the examples which we consider). The derivatives

$$\frac{dy}{dx} \sim \left(\sum_{n=0}^{\infty} \phi_n'(x)\right) \exp\left(\sum_{n=0}^{\infty} \phi_n(x)\right),$$

and

$$\frac{d^2 y}{dx^2} \sim \left[\sum_{n=0}^{\infty} \phi_n''(x) + \left(\sum_{n=0}^{\infty} \phi_n'(x)\right)^2\right] \exp\left(\sum_{n=0}^{\infty} \phi_n(x)\right),$$

when substituted into the equation 6.2.8 yield the expression

$$\sum_{n=0}^{\infty} \phi_n''(x) + \left(\sum_{n=0}^{\infty} \phi_n'(x)\right)^2 + q(x) \sim 0. \qquad 6.2.9$$

The function $q(x)$ determines the sequence $\phi_n(x)$ by dominant balance arguments. These are best described by studying various examples.

Consider first the solution of the equation

$$\frac{d^2 y}{dx^2} - \left(a_0 + \frac{a_1}{x} + \frac{a_2}{x^2}\right) y = 0. \qquad 6.2.10$$

We let $\phi \sim \exp(\phi_0 + \phi_1 + \phi_2 + \cdots)$ and the condition 6.2.9 for determining the ϕ_n leads to

$$\phi_0'' + \phi_1'' + \phi_2'' + \cdots + (\phi_0')^2 + (\phi_1')^2 + (\phi_2')^2 + \cdots + 2\phi_0'\phi_1'$$

$$+ 2\phi_0'\phi_2' + 2\phi_1'\phi_2' + \cdots \sim a_0 + \frac{a_1}{x} + \frac{a_2}{x^2}. \qquad 6.2.11$$

We require

$$\phi_{n+1} = o(\phi_n) \quad \text{as } x \to \infty$$

$$\phi'_{n+1} = o(\phi'_n) \quad \text{as } x \to \infty$$

$$\phi''_{n+1} = o(\phi''_n) \quad \text{as } x \to \infty .$$

Thus the possible dominant terms on the left-hand side of 6.2.11 are ϕ''_0 and $(\phi'_0)^2$. One or other (or both) are to equal the dominant driving term a_0. If we choose $\phi''_0 = a_0$ then $\phi'_0 = a_0 x$ which creates an unbalanced dominant term $(\phi'_0)^2 = a_0^2 x^2$. The alternative choice $(\phi'_0)^2 = a_0$ leads to $\phi'_0 = \pm\sqrt{a_0}$ and $\phi''_0 = 0$ which is consistent with the dominant balance requirement.

It is useful to digress here and consider the occasions when ϕ''_0 is of equal order or higher order than $(\phi'_0)^2$. Suppose the dominant term on the right-hand side of 6.2.11 is of order x^n. For ϕ''_0 and $(\phi'_0)^2$ to both be of order x^n we have $\phi_0 = O(x^{n+2})$ and $\phi'_0 = O(x^{n+1})$ such that $x^n = O(x^{2(n+1)})$ i.e., $n = -2$. Thus if the right-hand side of 6.2.11 is of order $1/x^2$ then both ϕ''_0 and $(\phi'_0)^2$ will be included in the dominant order equation. If the right-hand side is of higher order than $1/x^2$ then the dominant member of the left-hand side is $(\phi'_0)^2$ while if the right-hand side is of lower order than $1/x^2$ then the dominant member of the left-hand side is ϕ''_0.

Returning to the example, the right-hand side of 6.2.11 is of order one so $(\phi'_0)^2 = a_0$ and $\phi_0 = \pm\sqrt{a_0}\,x$. The remaining possible dominant members of the left-hand side are ϕ''_1 and $2\phi'_0\phi'_1$. The driving term is a_1/x, so we consider separately the possibilities

$$\text{(i)}\ \phi''_1 = \frac{a_1}{x} \qquad\qquad \text{(ii)}\ 2\phi'_0\phi'_1 = \frac{a_1}{x}.$$

The former yields $\phi'_1 = a_1 \ln|x|$. Hence $\phi_1 = O(x \ln x)$ which is contrary to the requirement that $\phi_1 = o(\phi_0)$ as $x \to \infty$.

The second possibility yields

$$\phi'_1 = \pm\frac{a_1}{2\sqrt{a_0}\,x}, \quad \phi_1 = \pm\frac{a_1}{2\sqrt{a_0}} \ln x,$$

so that $\phi_1 = o(\phi_0)$ as required. Thus the second possibility is adopted.

It is necessary to keep track of the terms which have been used in the expression 6.2.11. So far $(\phi'_0)^2$ has been equated to a_0 and $2\phi'_0\phi'_1$ equated to a_1/x. The term ϕ''_0 is equal to zero so we are left with

$$\phi''_1 + \phi''_2 + \cdots + (\phi'_1)^2 + (\phi'_2)^2 + \cdots + 2\phi'_0\phi'_2 + 2\phi'_1\phi'_2 + \cdots \sim \frac{a_2}{x^2}.$$

The terms ϕ''_1 and $(\phi'_1)^2$ are known and become driving terms. They are both of order $(1/x^2)$ which coincides with the order of the right-hand side term a_2/x^2. The term

$2\phi'_0\phi'_2$ will dominate $\phi'_1\phi'_2$ so the choice is between

$$\text{(i) } \phi''_2 = \mathrm{O}\left(\frac{1}{x^2}\right) \quad \text{and} \quad \text{(ii) } 2\phi'_0\phi'_2 = \mathrm{O}\left(\frac{1}{x^2}\right).$$

The former yields $\phi_2 = \mathrm{O}(\ln x)$ and violates the requirement that $\phi_2 = \mathrm{o}(\phi_1)$. The latter yields $\phi_2 = \mathrm{O}(1/x)$ which is acceptable. Thus

$$\phi'_2 = \pm\frac{1}{2\sqrt{a_0}}\left(\frac{a_2}{x^2} \pm \frac{a_1}{2\sqrt{a_0}}\frac{1}{x^2} - \frac{a_1^2}{4a_0x^2}\right),$$

from which the function ϕ_2 is obtained. This process may be continued to obtain as many ϕ_n as required. There are two solutions of the form

$$y \sim \exp(\phi_0 + \phi_1 + \phi_2 + \cdots),$$

associated with the choice of sign of ϕ_0. Each solution may be multiplied by an arbitrary constant. Thus the general solution of equation 6.2.10 has the following behavior for large x,

$$y = A\exp\left[\sqrt{a_0}\,x + \frac{a_1}{2\sqrt{a_0}}\ln x + \mathrm{O}\left(\frac{1}{x}\right)\right]$$
$$+ B\exp\left[-\sqrt{a_0}\,x - \frac{a_1}{2\sqrt{a_0}}\ln x + \mathrm{O}\left(\frac{1}{x}\right)\right].$$

This simplifies to yield the expression

$$y = Ax^{a_1/2\sqrt{a_0}}\exp(\sqrt{a_0}\,x)\cdot\left[1 + \mathrm{O}\left(\frac{1}{x}\right)\right]$$
$$+ Bx^{-a_1/2\sqrt{a_0}}\exp(-\sqrt{a_0}\,x)\cdot\left[1 + \mathrm{O}\left(\frac{1}{x}\right)\right] \quad \text{as } x \to \infty.$$

If a_0 is negative the exponentials in the above expression are best expressed as trigonometric functions. The next example demonstrates this type of solution. Consider the behavior of the solution of the equation

$$\frac{d^2y}{dx^2} + \left(1 + \frac{c}{x^2}\right)y = 0, \tag{6.2.12}$$

for large values of x, where c is a constant.

Let $y \sim \exp(\phi_0 + \phi_1 + \phi_2 + \phi_3 + \cdots)$, then

$$\phi''_0 + \phi''_1 + \phi''_2 + \cdots + (\phi'_0)^2 + (\phi'_1)^2 + (\phi'_2)^2 + \cdots$$
$$+ 2\phi'_0\phi'_1 + 2\phi'_0\phi'_2 + 2\phi'_0\phi'_3 + 2\phi'_1\phi'_2 + \cdots \sim -1 - \frac{c}{x^2}. \tag{6.2.13}$$

The dominant equation is O(1) so

$$(\phi'_0)^2 = -1, \quad \phi_0 = \pm \mathrm{i}x,$$

and $\phi''_0 = 0$. The terms ϕ''_0 and $(\phi'_0)^2$ can be removed from the left-hand side of 6.2.13 while the term -1 is removed from the right-hand side. This leaves the driving term $-c/x^2$. The candidates on the left-hand side to have order $1/x^2$ are either ϕ''_1 or $2\phi'_0\phi'_1$. If $\phi''_1 = \mathrm{O}(1/x^2)$ then $\phi'_1 = \mathrm{O}(1/x)$ which causes the term $2\phi'_0\phi'_1$ to be of order $1/x$ with no other term to balance it. If $2\phi'_0\phi'_1 = \mathrm{O}(1/x^2)$ then $\phi'_1 = \mathrm{O}(1/x^2)$ so that ϕ''_1 is of lower order and can be balanced.

Thus

$$2\phi'_0\phi'_1 = -\frac{c}{x^2},$$

so that

$$\phi'_1 = \pm\frac{\mathrm{i}c}{2x^2} \quad \text{and} \quad \phi_1 = \mp\frac{\mathrm{i}c}{2x}.$$

The terms ϕ''_1 and $(\phi'_1)^2$ become driving terms. The dominant term is $\phi''_1(= \mp\,\mathrm{i}c/x^3)$. The only term which can be of this order is $2\phi'_0\phi'_2$, therefore

$$2\phi'_0\phi'_2 = -\phi''_1 = \pm\frac{\mathrm{i}c}{x^3},$$

so that

$$\phi'_2 = \frac{c}{2x^3} \quad \text{and} \quad \phi_2 = -\frac{c}{4x^2}.$$

The terms $(\phi'_1)^2$ and ϕ''_2 in 6.2.13 are each of order $1/x^4$ and are balanced by the term $2\phi'_0\phi'_3$. Therefore ϕ_3 is of order $1/x^3$. Working to order ϕ_2, i.e. $\mathrm{O}(1/x^2)$ we have obtained the following approximation for large x,

$$y = A\exp\left[\mathrm{i}x - \frac{\mathrm{i}c}{2x} - \frac{c}{4x^2} + \mathrm{O}\left(\frac{1}{x^3}\right)\right] + B\exp\left[-\mathrm{i}x + \frac{\mathrm{i}c}{2x} - \frac{c}{4x^2} + \mathrm{O}\left(\frac{1}{x^3}\right)\right].$$

This may be expressed in the form

$$\begin{aligned} y = {} & A\exp(-c/4x^2)\exp[\mathrm{i}(x - c/2x)].[1 + \mathrm{O}(1/x^3)] \\ & + B\exp(-c/4x^2)\exp[-\mathrm{i}(x - c/2x)].[1 + \mathrm{O}(1/x^3)], \end{aligned}$$

or more conveniently as

$$\begin{aligned} y = {} & a\exp(-c/4x^2)\cos(x - c/2x).[1 + \mathrm{O}(1/x^3)] \\ & + b\exp(-c/4x^2)\sin(x - c/2x).[1 + \mathrm{O}(1/x^3)] \quad \text{as } x \to \infty. \end{aligned} \qquad 6.2.14$$

Bessel's equation

The approximation 6.2.14 can be applied directly to the solution of Bessel's equation of order n,

$$\frac{\mathrm{d}^2 w}{\mathrm{d}r^2} + \frac{1}{r}\frac{\mathrm{d}w}{\mathrm{d}r} + \left(1 - \frac{n^2}{r^2}\right)w = 0.$$

The coefficient of the first derivative is $1/r$ so, from the argument following equation 6.2.7, we define the new independent variable

$$y = w\exp\left(\frac{1}{2}\int\frac{\mathrm{d}r}{r}\right) = w\sqrt{r},$$

where y satisfies the equation

$$\frac{\mathrm{d}^2 y}{\mathrm{d}r^2} + \left(1 - \frac{n^2}{r^2} + \frac{1}{2r^2} - \frac{1}{4r^2}\right)y = 0.$$

This is of the form 6.2.12 with $c = 1/4 - n^2$.

If we restrict our study to Bessel's equation of order zero then $c = 1/4$ and 6.2.14 yields

$$w = \frac{1}{\sqrt{r}}\exp(-1/16r^2)\,[a\cos(r - 1/8r) + b\sin(r - 1/8r)]\,.\left[1 + \mathrm{O}\left(\frac{1}{r^3}\right)\right]. \qquad 6.2.15$$

In Section 7.7 an approximation for large values of r is obtained from the integral definition of Bessel functions of the first kind,

$$J_n(r) = \frac{1}{\pi}\int_0^{\pi}\cos(r\sin\theta - n\theta)\,\mathrm{d}\theta \simeq \sqrt{\frac{2}{\pi r}}\cos(r - n\pi/2 - \pi/4) \quad \text{as } r \to \infty\,.$$

In the case $n = 0$ this becomes

$$J_0(r) \simeq \frac{1}{\sqrt{\pi r}}(\cos r + \sin r) \quad \text{as } r \to \infty\,.$$

The corresponding dominant part of 6.2.15 is

$$w = \frac{1}{\sqrt{r}}(a\cos r + b\sin r)\left[1 + \mathrm{O}\left(\frac{1}{r}\right)\right],$$

which shows that the particular solution of Bessel's equation corresponding to the zeroth order Bessel function of the first kind has $a = b = 1/\sqrt{\pi}$.

Airy's equation

Another important equation is Airy's equation

$$\frac{d^2y}{dx^2} - xy = 0. \qquad 6.2.16$$

In our subsequent considerations we will require asymptotic approximations of the solution for both large positive and large negative values of x. Consider first the case of large positive x.

Let $y \sim \exp\{\phi_0 + \phi_1 + \phi_2 + \cdots\}$ then equation 6.2.16 yields the expression

$$\phi_0'' + \phi_1'' + \cdots + (\phi_0')^2 + (\phi_1')^2 + \cdots + 2\phi_0'\phi_1' + 2\phi_0'\phi_2' + \cdots \sim x.$$

The dominant equation is

$$(\phi_0')^2 = x, \quad \text{i.e.} \quad \phi_0' = \pm\sqrt{x} \text{ with solution } \phi_0 = \pm\tfrac{2}{3}x^{3/2}.$$

The next dominant balance equation is

$$\phi_0'' + 2\phi_0'\phi_1' = 0,$$

so that

$$\phi_1' = -\frac{1}{2}\frac{\phi_0''}{\phi_0'} = -\frac{1}{4x} \text{ with solution } \phi_1 = -\frac{1}{4}\ln x.$$

Both ϕ_1'' and $(\phi_1')^2$ are $O(1/x^2)$ terms so that the next dominant balance equation is

$$\phi_1'' + (\phi_1')^2 + 2\phi_0'\phi_2' = 0.$$

Thus $\phi_2' = \mp 5/32x^{5/2}$ so $\phi_2 = O(1/x^{3/2})$.

The leading approximation for the solution of Airy's equation as $x \to +\infty$ is

$$y = \frac{a}{x^{1/4}}\exp\left(\frac{2}{3}x^{3/2}\right)\left[1 + O\left(\frac{1}{x^{3/2}}\right)\right]$$

$$+ \frac{b}{x^{1/4}}\exp\left(-\frac{2}{3}x^{3/2}\right)\left[1 + O\left(\frac{1}{x^{3/2}}\right)\right]. \qquad 6.2.17$$

In the case of x large and negative it is convenient to introduce the new independent variable $z = -x$ and consider the behavior of the equation

$$\frac{d^2y}{dz^2} + zy = 0,$$

for z large and positive. We obtain the dominant balance expression:

$$\phi_0'' + \phi_1'' + \cdots + (\phi_0')^2 + (\phi_1')^2 + \cdots + 2\phi_0'\phi_1' + 2\phi_0'\phi_2' + \cdots \sim -z,$$

where primes now indicate derivatives with respect to z. The dominant balance yields

$$\phi_0' = \pm\, i\sqrt{z}, \quad \text{i.e.} \quad \phi_0 = \pm\, i\tfrac{2}{3}z^{3/2}.$$

The next dominant balance equation is

$$\phi_0'' + 2\phi_0'\phi_1' = 0,$$

with solution $\phi_1 = -\frac{1}{4}\ln z$. Again ϕ_2 is found to be of order $z^{-3/2}$ so

$$y = \exp[\pm\, i\tfrac{2}{3}\, z^{3/2} - \tfrac{1}{4}\ln z + O(1/z^{3/2})].$$

In terms of the original variable this becomes

$$y = \frac{c}{|x|^{1/4}}\cos\left(\frac{2}{3}|x|^{3/2}\right)\cdot\left[1 + O\left(\frac{1}{|x|^{3/2}}\right)\right]$$
$$+ \frac{d}{|x|^{1/4}}\sin\left(\frac{2}{3}|x|^{3/2}\right)\cdot\left[1 + O\left(\frac{1}{|x|^{3/2}}\right)\right] \quad \text{as } x \to -\infty. \qquad 6.2.18$$

The method of dominant balance provides valuable information about the behavior of solutions for large values of the independent variable. However, the coefficients in the general solution cannot be obtained from conditions specified at arbitrary positions. In the case of Airy's equation we would like to link the values of a and b with those of c and d in equations 6.2.17 and 6.2.18. The method of dominant balance cannot help because the two approximations are confined to the regions of large positive and large negative x.

We will show in Section 7.7 that the solution of Airy's equation may be written in the form

$$y = \alpha Ai(x) + \beta Bi(x), \qquad 6.2.19$$

where the Airy functions $Ai(x)$ and $Bi(x)$ have the following asymptotic behavior,

$$Ai(x) \simeq \frac{1}{2\sqrt{\pi}}\frac{1}{x^{1/4}}\exp\left(-\frac{2}{3}x^{3/2}\right) \quad \text{as } x \to \infty, \qquad 6.2.20$$

$$Bi(x) \simeq \frac{1}{\sqrt{\pi}}\frac{1}{x^{1/4}}\exp\left(\frac{2}{3}x^{3/2}\right) \quad \text{as } x \to \infty, \qquad 6.2.21$$

$$Ai(x) \simeq \frac{1}{\sqrt{\pi}}\frac{1}{|x|^{1/4}}\sin\left(\frac{2}{3}|x|^{3/2} + \frac{\pi}{4}\right) \quad \text{as } x \to -\infty, \qquad 6.2.22$$

$$Bi(x) \simeq \frac{1}{\sqrt{\pi}}\frac{1}{|x|^{1/4}}\cos\left(\frac{2}{3}|x|^{3/2} + \frac{\pi}{4}\right) \text{as } x \to -\infty. \qquad 6.2.23$$

Thus comparing 6.2.17 with 6.2.19 using 6.2.20 and 6.2.21 yields

$$a = \beta/\sqrt{\pi} \quad \text{and} \quad b = \alpha/2\sqrt{\pi}.$$

ing 6.2.18 with 6.2.19 using 6.2.22 and 6.2.23 yields

$$\beta)/\sqrt{2\pi} \quad \text{and} \quad d = (\alpha - \beta)/\sqrt{2\pi}.$$

Eliminating α and β provides the connection between the solutions 6.2.17 and 6.2.18. The process of connecting solutions of this type is used later in this chapter when turning points in the WKB method are considered.

Exercises

Obtain asymptotic expansions of the solutions of the following equations for large positive x,

(i) $\dfrac{d^2y}{dx^2} - \left(1 + \dfrac{1}{x}\right)y = 0.$

(ii) $\dfrac{d^2y}{dx^2} + (1 + x)y = 0.$

(iii) $\dfrac{d^2y}{dx^2} - \left(\dfrac{1}{x} - \dfrac{1}{x^2}\right)y = 0.$

(iv) $\dfrac{d^2y}{dx^2} - x^2y = 0.$

(v) $\dfrac{d^2y}{dx^2} + x^2y = 0.$

(vi) $\dfrac{d^2y}{dx^2} - (2 + x^2)y = 0.$

[Work to $O(\phi_2)$ in each case.]

Selected answers

(i) $y = A\sqrt{x}\,e^x\left[1 - \dfrac{1}{8x} + O\left(\dfrac{1}{x^2}\right)\right] + \dfrac{B}{\sqrt{x}}e^{-x}\left[1 - \dfrac{3}{8x} + O\left(\dfrac{1}{x^2}\right)\right].$

(ii) $y = \dfrac{A}{x^{1/4}}\cos\left(\dfrac{2}{3}x^{3/2} + x^{1/2}\right)[1 + O(1/\sqrt{x})] + \dfrac{B}{x^{1/4}}\sin\left(\dfrac{2}{3}x^{3/2} + x^{1/2}\right)[1 + O(1/\sqrt{x})].$

(iv) $y = \dfrac{A}{\sqrt{x}}\exp(x^2/2)\left[1 + \dfrac{3}{16x^2} + O\left(\dfrac{1}{x^4}\right)\right] + \dfrac{B}{\sqrt{x}}\exp(-x^2/2)\left[1 - \dfrac{3}{16x^2} + O\left(\dfrac{1}{x^4}\right)\right].$

6.3 The WKB method

In this section we consider differential equations of the form

$$\frac{d^2y}{dx^2} + q(x;\lambda)y = 0, \tag{6.3.1}$$

where λ is a large parameter. A special case known as Liouville's equation arises when $q = \lambda^2 f(x)$,

$$\frac{d^2y}{dx^2} + \lambda^2 f(x)y = 0. \tag{6.3.2}$$

A physical example of the occurrence of this equation is provided by the one-dimensional wave equation

$$\frac{\partial^2 Y}{\partial t^2} = [c(x)]^2 \frac{\partial^2 Y}{\partial x^2}. \tag{6.3.3}$$

Here the wave speed $c(x)$ is a function of position. This variation could arise in many ways, one cause might be the dependence of the density of the medium on position. Waves with frequency ω have the form

$$Y(x, t) = y(x)\cos(\omega t + \alpha),$$

where α is a phase angle. Substituting this expression into equation 6.3.3 yields the equation

$$\frac{d^2y}{dx^2} + \frac{\omega^2}{[c(x)]^2} y = 0. \tag{6.3.4}$$

This is Liouville's equation 6.3.2 with λ equal to the angular frequency ω.

The physicists Wentzel, Kramers and Brillouin independently studied an extension of Liouville's equation to the case where

$$q = f_0(x) + \lambda f_1(x) + \lambda^2 f_2(x).$$

The solution technique which takes advantage of λ being large is known as the WKB method in recognition of the contributions of Wentzel, Kramers and Brillouin.

Construction of an asymptotic expansion

The method described in this section is a modification of the original WKB procedure. The technique is that advocated by Murray[18] in his text. It is similar to the dominant balance method in that the solution $y(x;\lambda)$ is expressed as an exponential of an asymptotic expansion. In this case the asymptotic sequence involved is a set of functions of λ which we denote by $\delta_n(\lambda)$. Each term in the expansion has a coefficient function $\phi_n(x)$ so that

$$y(x;\lambda) \sim \exp\left(\sum_{n=0}^{\infty} \delta_n(\lambda)\phi_n(x)\right) \quad \text{as } \lambda \to \infty,$$

where

$$\delta_{n+1}(\lambda) = o[\delta_n(\lambda)] \quad \text{as } \lambda \to \infty.$$

The form of the asymptotic sequence is not assumed at the outset. Instead the functions $\delta_n(\lambda)$ are determined by the term $q(x;\lambda)$ in the equation 6.3.1 using a dominant balance approach which is similar to that of the previous section. However, in this case the functions considered do not involve derivatives but merely algebraic combinations of the $\delta_n(\lambda)$. The resulting analysis associated with large values of a parameter turns out to be rather easier to perform than that associated with large values of the independent variable.

For our first example we consider a Liouville equation

$$\frac{d^2y}{dx^2} + \lambda^2(1+x^2)^2 y = 0. \tag{6.3.5}$$

Let

$$y \sim \exp\left(\sum_{n=0}^{\infty} \delta_n(\lambda)\phi_n(x)\right) \quad \text{as } \lambda \to \infty, \tag{6.3.6}$$

then

$$\frac{dy}{dx} \sim \left(\sum_{n=0}^{\infty} \delta_n(\lambda)\phi_n'(x)\right)\exp\left(\sum_{n=0}^{\infty} \delta_n(\lambda)\phi_n(x)\right),$$

$$\frac{d^2y}{dx^2} \sim \left[\sum_{n=0}^{\infty} \delta_n(\lambda)\phi_n''(x) + \left(\sum_{n=0}^{\infty} \delta_n(\lambda)\phi_n'(x)\right)^2\right]\exp\left(\sum_{n=0}^{\infty} \delta_n(\lambda)\phi_n(x)\right).$$

Substituting into the differential equation and canceling the exponential yields the expression

$$\delta_0(\lambda)\phi_0'' + \delta_1(\lambda)\phi_1'' + \cdots + [\delta_0(\lambda)]^2(\phi_0')^2 + [\delta_1(\lambda)]^2(\phi_1')^2 + \cdots$$
$$+ 2\delta_0(\lambda)\delta_1(\lambda)\phi_0'\phi_1' + 2\delta_0(\lambda)\delta_2(\lambda)\phi_0'\phi_2' + \cdots \sim -\lambda^2(1+x^2)^2. \tag{6.3.7}$$

The dominant term in this expression is of order λ^2. This will correspond to the term $[\delta_0(\lambda)]^2(\phi_0')^2$ so that $\delta_0 = \lambda$ and

$$\phi_0' = \pm i(1+x^2).$$

Hence

$$\phi_0 = \pm i(x + x^3/3).$$

(The simplest choice for the δ_n is always adopted with the complexity reserved for the ϕ_n.)

The term $\delta_0(\lambda)\phi_0''$ provides the next dominant order, namely λ. This must be balanced by $2\delta_0(\lambda)\delta_1(\lambda)\phi_0'\phi_1'$ so $\delta_1 = 1$ and

$$\phi_0'' + 2\phi_0'\phi_1' = 0.$$

Thus

$$\phi_1' = -\frac{1}{2}\frac{\phi_0''}{\phi_0'},$$

so that

$$\phi_1 = -\tfrac{1}{2}\ln|\phi_0'| = -\tfrac{1}{2}\ln(1 + x^2).$$

The next equation is of O(1),

$$\delta_1\phi_1'' + \delta_1^2(\phi_1')^2 + 2\delta_0\delta_2\phi_0'\phi_2' = 0,$$

which yields $\delta_2 = 1/\lambda$. We choose to terminate the expansion for this example at this stage without evaluating ϕ_2. The approximation 6.3.6 becomes

$$y = \exp\left\{\pm \mathrm{i}\lambda\left(x + \frac{x^3}{3}\right) - \frac{1}{2}\ln(1 + x^2) + \mathrm{O}\left(\frac{1}{\lambda}\right)\right\}.$$

Expressing the exponential as trigonometric functions and including arbitrary multiplicative constants leads to the general solution

$$y = \frac{a}{\sqrt{1 + x^2}}\cos\left\{\lambda\left(x + \frac{x^3}{3}\right)\right\}.\left[1 + \mathrm{O}\left(\frac{1}{\lambda}\right)\right]$$
$$+ \frac{b}{\sqrt{1 + x^2}}\sin\left\{\lambda\left(x + \frac{x^3}{3}\right)\right\}.\left[1 + \mathrm{O}\left(\frac{1}{\lambda}\right)\right]. \qquad 6.3.8$$

This is valid for all x and allows the constants a and b to be determined by extra conditions imposed at any point. Thus, for example, y and $\mathrm{d}y/\mathrm{d}x$ may be specified at $x = 0$. Then, to leading order,

$$y(0) = a \quad \text{and} \quad \frac{\mathrm{d}y}{\mathrm{d}x}(0) = \lambda b.$$

The ability of the WKB approximation to satisfy initial conditions contrasts with the method of dominant balance where the approximation is valid only for large values of the independent variable.

Consider next the equation

$$\frac{\mathrm{d}^2y}{\mathrm{d}x^2} - \lambda^2\sin^2 x.y = 0. \qquad 6.3.9$$

The expansion $y \sim \exp\left(\sum_{n=0}^{\infty} \delta_n(\lambda)\phi_n(x)\right)$, leads to the expression

$$\delta_0\phi_0'' + \delta_1\phi_1'' + \cdots + \delta_0^2(\phi_0')^2 + \delta_1^2(\phi_1')^2 + \cdots$$
$$+ 2\delta_0\delta_1\phi_0'\phi_1' + 2\delta_0\delta_2\phi_0'\phi_2' + \cdots \sim \lambda^2 \sin^2 x.$$

The dominant equation is

$$\delta_0^2(\phi_0')^2 = \lambda^2 \sin^2 x,$$

yielding $\delta_0 = \lambda$ and

$$\phi_0' = \pm \sin x,$$

so that

$$\phi_0 = \mp \cos x.$$

The next dominant equation is

$$\delta_0\phi_0'' + 2\delta_0\delta_1\phi_0'\phi_1' = 0,$$

therefore $\delta_1 = 1$ and

$$\phi_1' = -\frac{1}{2}\frac{\phi_0''}{\phi_0'}.$$

Integrating leads to

$$\phi_1 = -\tfrac{1}{2}\ln|\phi_0'| = -\tfrac{1}{2}\ln|\sin x|.$$

The next dominant equation is of O(1),

$$\delta_1\phi_1'' + \delta_1^2(\phi_1')^2 + 2\delta_0\delta_2\phi_0'\phi_2' = 0,$$

and yields $\delta_2 = 1/\lambda$.

Thus the general solution of 6.3.7 is

$$y = \frac{a}{\sqrt{|\sin x|}}\exp(-\lambda\cos x)\left[1 + \mathrm{O}\left(\frac{1}{\lambda}\right)\right]$$
$$+ \frac{b}{\sqrt{|\sin x|}}\exp(\lambda\cos x)\left[1 + \mathrm{O}\left(\frac{1}{\lambda}\right)\right]. \qquad 6.3.10$$

This solution is not valid when $\sin x = 0$. The approximation breaks down then because the coefficient function $q(x; \lambda)$ in 6.3.9 is $-\lambda^2 \sin^2 x$ and no matter how large the parameter λ may become the term is zero when $x = \pm n\pi$ ($n = 0, 1, 2, \ldots$).

Points where the coefficient of the dominant λ term in $q(x; \lambda)$ vanish are called turning points. The WKB approximation breaks down in the region of turning points. A special study of such regions is required where locally valid solutions are matched to

WKB solutions in an overlap region where both are valid. This idea will be described in Section 6.5. Until then we must be content to restrict the range of validity of the WKB approximations to regions away from turning points.

For our next example consider the behavior of the solution of the equation

$$\frac{d^2y}{dx^2} - \left(\lambda^2 x^4 + \lambda x^3 + \frac{x^2}{4}\right) y = 0, \tag{6.3.11}$$

for large values of λ. We will construct the WKB approximation away from the turning point at $x = 0$.

Substituting the expansion 6.3.6 into 6.3.11 yields the expression

$$\delta_0 \phi_0'' + \delta_1 \phi_1'' + \cdots + \delta_0^2 (\phi_0')^2 + \delta_1^2 (\phi_1')^2 + \cdots$$
$$+ 2\delta_0 \delta_1 \phi_0' \phi_1' + 2\delta_0 \delta_2 \phi_0' \phi_2' + \cdots \sim \lambda^2 x^4 + \lambda x^3 + \frac{x^2}{4}.$$

The dominant equation is $O(\lambda^2)$,

$$\delta_0^2 (\phi_0')^2 = \lambda^2 x^4,$$

yielding $\delta_0 = \lambda$ and $\phi_0 = \pm \frac{1}{3} x^3$.

The $O(\lambda)$ equation is

$$\delta_0 \phi_0'' + 2\delta_0 \delta_1 \phi_0' \phi_1' = \lambda x^3,$$

yielding $\delta_1 = 1$ and $\phi_1' = -\dfrac{1}{x} \pm \dfrac{x}{2}$, i.e. $\phi_1 = -\ln|x| \pm \dfrac{x^2}{4}$.

The $O(1)$ equation is

$$\delta_1 \phi_1'' + \delta_1^2 (\phi_1')^2 + 2\delta_0 \delta_2 \phi_0' \phi_2' = \frac{x^2}{4},$$

yielding $\delta_2 = 1/\lambda$ and $\phi_2 = -\dfrac{1}{4x} \pm \dfrac{1}{3x^3}$.

The next equation will yield $\delta_3 = O(1/\lambda^2)$. The following WKB approximation for the solution of 6.3.11 has thus been obtained,

$$y = \frac{A}{|x|} \exp\left\{\lambda \frac{x^3}{3} + \frac{x^2}{4} - \frac{1}{\lambda}\left(\frac{1}{4x} - \frac{1}{3x^3}\right) + O\left(\frac{1}{\lambda^2}\right)\right\}$$
$$+ \frac{B}{|x|} \exp\left\{-\lambda \frac{x^3}{3} - \frac{x^2}{4} - \frac{1}{\lambda}\left(\frac{1}{4x} + \frac{1}{3x^3}\right) + O\left(\frac{1}{\lambda^2}\right)\right\},$$

for x values away from $x = 0$.

Exercises

Obtain the WKB approximations for large λ of the solutions of the following equations away from the turning points:

(i) $\dfrac{d^2y}{dx^2} - \lambda^2 x^2 y = 0$ — Consider $x > 0$ and $x < 0$, work to $O(1/\lambda)$.

(ii) $\dfrac{d^2y}{dx^2} + (\lambda^2 x^2 + x)y = 0$ — Consider $x > 0$ and $x < 0$, work to $O(1/\lambda)$.

(iii) $\dfrac{d^2y}{dx^2} - \left(\lambda^2 x^2 + \dfrac{\lambda}{x}\right)y = 0$ — Consider $x > 0$ and $x < 0$, work to $O(1/\lambda)$.

(iv) $\dfrac{d^2y}{dx^2} + \lambda^2(x-1)y = 0$ — Consider $x > 1$ and $x < 1$, work to $O(1)$.

(v) $\dfrac{d^2y}{dx^2} + \left(\lambda^2 x - \dfrac{1}{8x^2}\right)y = 0$ — Consider $x > 0$, work to $O(1/\lambda)$.

Selected answers

(ii)
$$y = \frac{a}{\sqrt{|x|}}\cos\left[\frac{\lambda x^2}{2} + \frac{1}{2\lambda}\left(x - \frac{3}{8x^2}\right)\right].\left[1 + O\left(\frac{1}{\lambda^2}\right)\right]$$
$$+ \frac{b}{\sqrt{|x|}}\sin\left[\frac{\lambda x^2}{2} + \frac{1}{2\lambda}\left(x - \frac{3}{8x^2}\right)\right].\left[1 + O\left(\frac{1}{\lambda^2}\right)\right].$$

(iii)
$$y = \frac{a}{\sqrt{|x|}}\exp\left\{\frac{\lambda x^2}{2} - \frac{1}{2x} + \frac{1}{\lambda}\left(\frac{1}{32x^4} - \frac{1}{4x^3} + \frac{3}{16x^2}\right)\right\}.\left[1 + O\left(\frac{1}{\lambda^2}\right)\right]$$
$$+ \frac{b}{\sqrt{|x|}}\exp\left\{-\frac{\lambda x^2}{2} + \frac{1}{2x} - \frac{1}{\lambda}\left(\frac{1}{32x^4} + \frac{1}{4x^3} + \frac{3}{16x^2}\right)\right\}.\left[1 + O\left(\frac{1}{\lambda^2}\right)\right].$$

(iv)
$$\left.\begin{aligned} y = {}&\frac{a}{(1-x)^{1/4}}\exp\left(\frac{2}{3}\lambda(1-x)^{3/2}\right).\left[1 + O\left(\frac{1}{\lambda}\right)\right] \\ &+ \frac{b}{(1-x)^{1/4}}\exp\left(-\frac{2}{3}\lambda(1-x)^{3/2}\right).\left[1 + O\left(\frac{1}{\lambda}\right)\right] \end{aligned}\right\} \text{ for } x < 1$$

$$\left.\begin{aligned} y = {}&\frac{a}{(x-1)^{1/4}}\cos\left(\frac{2}{3}\lambda(x-1)^{3/2}\right).\left[1 + O\left(\frac{1}{\lambda}\right)\right] \\ &+ \frac{b}{(x-1)^{1/4}}\sin\left(\frac{2}{3}\lambda(x-1)^{3/2}\right).\left[1 + O\left(\frac{1}{\lambda}\right)\right] \end{aligned}\right\} \text{ for } x > 1.$$

6.4 Eigenvalue problems

Eigenvalue problems arise in many fields of science. They occur in the description of wave motion in classical physics where, for example, the possible wavelengths of sound in pipes or of vibrating strings are determined by solving an eigenvalue problem. An example of the occurrence of eigenvalues in modern physics is as the energy levels in Schrödinger's wave equation.

The WKB method provides a powerful approximation technique for determining the large eigenvalues of boundary value problems. As our first example we will reconsider equation 6.3.5 of the previous section,

$$\frac{d^2y}{dx^2} + \lambda^2(1 + x^2)^2 y = 0. \qquad \text{6.3.5 (repeated)}$$

The WKB approximation for the solution is given by equation 6.3.8. The constants in the general solution can be determined by prescribing initial values at one point or boundary values at two points. Initial value problems impose no constraint on the values of the parameter λ whereas boundary value problems in general restrict the parameter to a discrete set of values. Suppose we wish to obtain solutions of 6.3.5 in the region $0 < x < 1$ subject to the boundary conditions $y(0) = y(1) = 0$. The trivial solution $y = 0$ allows any value of λ. Nontrivial solutions exist only for certain values of λ. These are called eigenvalues. The solution 6.3.8 is

$$\begin{aligned} y = {} & \frac{a}{\sqrt{1+x^2}}\cos\left\{\lambda\left(x + \frac{x^3}{3}\right)\right\}\cdot\left[1 + O\left(\frac{1}{\lambda}\right)\right] \\ & + \frac{b}{\sqrt{1+x^2}}\sin\left\{\lambda\left(x + \frac{x^3}{3}\right)\right\}\cdot\left[1 + O\left(\frac{1}{\lambda}\right)\right]. \end{aligned} \qquad \text{6.3.8 (repeated)}$$

Imposing the condition $y(0) = 0$ yields the leading approximation $a = 0$. Imposing the condition $y(1) = 0$ and avoiding the trivial solution $b = 0$ yields the condition

$$\sin\left[\lambda\left(1 + \frac{1}{3}\right)\right] \simeq 0 \quad \text{for large } \lambda\text{, i.e. } 4\lambda/3 \simeq n\pi \text{ with } n \text{ integer.}$$

Thus the large eigenvalues are given by $\lambda_n \simeq 3n\pi/4$ with n large and integer.

Derivative boundary conditions

Next consider the same equation but now subject to a derivative boundary condition,

$$\frac{dy}{dx}(0) = 0,$$

along with the previous condition $y(1) = 0$.

Differentiating the solution, 6.3.8, leads to an $O(\lambda)$ contribution from the trigonometric functions and an $O(1)$ contribution from the coefficient $(1 + x^2)^{1/2}$,

$$\frac{dy}{dx} = -a\lambda\sqrt{1+x^2}\sin\left\{\lambda\left(x + \frac{x^3}{3}\right)\right\} + b\lambda\sqrt{1+x^2}\cos\left\{\lambda\left(x + \frac{x^3}{3}\right)\right\} + O(1).$$

Thus to leading order

$$\frac{dy}{dx}(0) = \lambda b, \text{ so } b = 0.$$

Imposing the boundary condition at $x = 1$ yields the condition

$$\cos(4\lambda/3) \simeq 0.$$

Thus the large eigenvalues for the equation with the derivative boundary condition are given by the WKB approximation $\lambda_n \simeq (3\pi/8)(2n + 1)$ with n large and integer.

Worked example

Obtain approximations for the large eigenvalues of the boundary value problem

$$\frac{d^2y}{dx^2} + \lambda^2 x^4 y = 0, \quad 1 < x < 2, \tag{6.4.1}$$

with $y(1) = y(2) = 0$.

Solution

Let $y \sim \exp\left(\sum_{n=0}^{\infty} \delta_n(\lambda)\phi_n(x)\right)$,

then

$$\frac{d^2y}{dx^2} \sim (\delta_0^2\phi_0'^2 + 2\delta_0\delta_1\phi_0'\phi_1' + \cdots + \delta_0\phi_0'' + \cdots)y,$$

and substituting into 6.4.1 leads to the expression

$$\delta_0^2\phi_0'^2 + 2\delta_0\delta_1\phi_0'\phi_1' + \cdots + \delta_0\phi_0'' + \cdots \sim -\lambda^2 x^4.$$

The $O(\lambda^2)$ equation is

$$\delta_0^2\phi_0'^2 = -\lambda^2 x^4,$$

with solution $\delta_0 = \lambda$ and $\phi_0' = \pm\, ix^2$, thus $\phi_0 = \pm\, ix^3/3$.

The next equation is $O(\lambda)$, (determined by the term $\delta_0\phi_0''$),

$$2\delta_0\delta_1\phi_0'\phi_1' + \delta_0\phi_0'' = 0.$$

Therefore $\delta_1 = 1$ and $\phi_1' = -\dfrac{1}{2}\dfrac{\phi_0''}{\phi_0'}$, so

$$\phi_1 = -\tfrac{1}{2}\ln|\phi_0'| = -\ln x.$$

The next order equation will yield the result $\delta_2 = 1/\lambda$. Thus the general s large λ is

$$y \simeq \frac{a}{x}\cos(\lambda x^3/3) + \frac{b}{x}\sin(\lambda x^3/3).$$

The boundary conditions yield, to leading order,

$$a\cos(\lambda/3) + b\sin(\lambda/3) = 0$$

$$a\cos(8\lambda/3) + b\sin(8\lambda/3) = 0.$$

This requires

$$\tan(8\lambda/3) = \tan(\lambda/3), \text{ i.e.}$$

$$8\lambda/3 = \lambda/3 + n\pi.$$

Thus the large eigenvalue approximation is $\lambda_n = 3n\pi/7$ with n large and integer.

Exercises

Obtain approximations for the large eigenvalues of the following boundary value problems:

(i) $\dfrac{d^2y}{dx^2} + \lambda^2(2+x)^2 y = 0 \quad -1 < x < 1$

$y(-1) = y(1) = 0.$

(ii) $\dfrac{d^2y}{dx^2} + \lambda^2(1+x)^2 y = 0 \quad 0 < x < 1$

(a) $y(0) = 0, \ \dfrac{dy}{dx}(1) = 0$ (b) $\dfrac{dy}{dx}(0) = 0, \ \dfrac{dy}{dx}(1) = 0.$

(iii) $\dfrac{d^2y}{dx^2} + \lambda^2 xy = 0 \quad 1 < x < 4$

$y(1) = y(4) = 0.$

(iv) $\dfrac{d^2y}{dx^2} + \lambda^2 e^{4x} y = 0 \quad 0 < x < 1$

$y(0) = y(1) = 0.$

Answers

(i) $\lambda \simeq n\pi/4$ for n large and integer.
(ii) (a) $\lambda \simeq (2n+1)\pi/3$, (b) $\lambda \simeq 2n\pi/3$ for n large and integer.
(iii) $\lambda \simeq 3n\pi/14$ for n large and integer.
(iv) $\lambda \simeq 2n\pi/(e^2 - 1)$ for n large and integer.

6.5 Turning points

The WKB approximation breaks down in regions of nonuniformity of the expansion in the solution form

$$y \sim \exp\left(\sum_{n=0}^{\infty} \delta_n(\lambda)\phi_n(x)\right).$$

Consider the expansion

$$\delta_0(\lambda)\phi_0(x) + \delta_1(\lambda)\phi_1(x) + \delta_2(\lambda)\phi_2(x) + \cdots. \qquad 6.5.1$$

If ϕ_0 is zero at $x = x_0$ then the term $\delta_0(\lambda)\phi_0(x_0)$ is no longer dominant no matter how large λ becomes and the WKB approximation breaks down in the region near $x = x_0$. This is called a turning point.

One of the simplest examples of the breakdown of the WKB approximation is provided by the equation

$$\frac{d^2y}{dx^2} + \lambda^2 xy = 0. \qquad 6.5.2$$

If x is positive the WKB solution is oscillatory,

$$y \simeq \frac{a}{x^{1/4}}\cos\left(\frac{2}{3}\lambda x^{3/2}\right) + \frac{b}{x^{1/4}}\sin\left(\frac{2}{3}\lambda x^{3/2}\right). \qquad 6.5.3$$

If x is negative the WKB solution involves exponentials,

$$y \simeq \frac{c}{|x|^{1/4}}\exp\left(\frac{2}{3}\lambda|x|^{3/2}\right) + \frac{d}{|x|^{1/4}}\exp\left(-\frac{2}{3}\lambda|x|^{3/2}\right). \qquad 6.5.4$$

Both approximations are invalid as x approaches the turning point $x = 0$. The behavior of the solution is radically different either side of this point.

The constants c and d must be related to a and b because the general solution of 6.5.2 involves only two arbitrary constants. If initial conditions are provided at some point x_i where $x_i > 0$ then a and b can be determined directly. In order to determine c and d the connection between the constants is required. Similarly if initial conditions are provided at some point x_i where $x_i < 0$ then in order to obtain the approximate solution for $x > 0$ the connection between the constants is required.

It is quite common for boundary value problems to involve regions which span turning points. This again requires the connection between the constants in the WKB approximations.

Matching exponential and trigonometric solutions

Equation 6.5.2 is of the form of Airy's equation

$$\frac{d^2y}{dz^2} - zy = 0, \tag{6.5.5}$$

where $z = -\lambda^{2/3}x$.

The general solution of 6.5.5 is

$$y = \alpha Ai(z) + \beta Bi(z) = \alpha Ai(-\lambda^{2/3}x) + \beta\, Bi(-\lambda^{2/3}x). \tag{6.5.6}$$

The behavior for large $|z|$ is summarized in equations 6.2.20–6.2.23. The behavior of the solution of 6.5.2 with x positive and λ large is given by 6.5.6 with $z(=-\lambda^{2/3}x)$ large and negative, thus

$$\begin{aligned} y \simeq{}& \alpha\frac{1}{\sqrt{\pi}}\frac{1}{(\lambda^{2/3}x)^{1/4}}\sin\left(\frac{2}{3}\lambda(x)^{3/2}+\frac{\pi}{4}\right)\\ &+\beta\frac{1}{\sqrt{\pi}}\frac{1}{(\lambda^{2/3}x)^{1/4}}\cos\left(\frac{2}{3}\lambda(x)^{3/2}+\frac{\pi}{4}\right) \quad \text{for } x>0. \end{aligned} \tag{6.5.7}$$

Similarly the behavior of the solution of 6.5.2 with x negative and λ large is

$$\begin{aligned} y \simeq{}& \frac{\alpha}{2\sqrt{\pi}}\frac{1}{|\lambda^{2/3}x|^{1/4}}\exp\left(-\frac{2}{3}\lambda|x|^{3/2}\right)\\ &+\frac{\beta}{\sqrt{\pi}}\frac{1}{|\lambda^{2/3}x|^{1/4}}\exp\left(\frac{2}{3}\lambda|x|^{3/2}\right) \quad \text{for } x<0. \end{aligned} \tag{6.5.8}$$

Comparing 6.5.7 and 6.5.3 yields

$$a = \frac{1}{\sqrt{2\pi}}\frac{1}{\lambda^{1/6}}(\alpha+\beta), \quad \text{and} \quad b = \frac{1}{\sqrt{2\pi}}\frac{1}{\lambda^{1/6}}(\alpha-\beta). \tag{6.5.9}$$

Comparing 6.5.8 and 6.5.4 yields

$$c = \frac{\beta}{\sqrt{\pi}}\frac{1}{\lambda^{1/6}} \quad \text{and} \quad d = \frac{\alpha}{2\sqrt{\pi}}\frac{1}{\lambda^{1/6}}. \tag{6.5.10}$$

Eliminating α and β yields the connection formulae

$$2d + c = \sqrt{2}a \quad \text{and} \quad 2d - c = \sqrt{2}b. \tag{6.5.11}$$

As an example of a boundary value problem which spans the turning point we will impose the boundary conditions $y(-\infty) = 0$ and $y(1) = 0$ on the solution of equation 6.5.2. Then 6.5.4 requires that $c = 0$ and from 6.5.11 we have $a = b = \sqrt{2}\,d$. Thus the large eigenvalue solutions of 6.5.2 have the form

$$y = \begin{cases} \dfrac{d}{|x|^{1/4}} \exp\left(-\dfrac{2}{3}\lambda |x|^{3/2}\right) & \text{for } x < 0 \\[2ex] \dfrac{\sqrt{2}d}{|x|^{1/4}} \left[\cos\left(\dfrac{2}{3}\lambda x^{3/2}\right) + \sin\left(\dfrac{2}{3}\lambda x^{3/2}\right)\right] & \text{for } x > 0 \end{cases},$$

away from the turning point $x = 0$.

The boundary condition $y(1) = 0$ requires

$$\cos\left(\frac{2}{3}\lambda\right) + \sin\left(\frac{2}{3}\lambda\right) = 0, \quad \text{i.e.} \quad \tan\left(\frac{2}{3}\lambda\right) = -1.$$

Therefore the large eigenvalues are given approximately by the equation

$$\lambda_n \simeq \frac{3}{2}\left(\frac{3\pi}{4} + n\pi\right) \quad \text{with } n \text{ a large integer.}$$

In this particular example the equation 6.5.2 can be recognized at the outset as a form of Airy's equation, whose solutions are already known, so that the WKB technique is not required. However the above analysis is helpful because it can be extended to equations of the form

$$\frac{d^2y}{dx^2} + \lambda^2 f(x) y = 0, \tag{6.5.12}$$

where $f(x)$ has a simple zero at $x = x_0$. (This means that f may be written in the form $f(x) = (x - x_0)g(x)$ where $g(x_0) \neq 0$. The case of zeros of orders other than unity can be treated by similar techniques but will not be considered in this book.) The location $x = x_0$ of the turning point can, by the simple change of variable $x' = x - x_0$, be transformed to the point $x' = 0$. Therefore without loss of generality we may consider the equation

$$\frac{d^2y}{dx^2} + \lambda^2 x g(x) y = 0, \tag{6.5.13}$$

with $g(0) \neq 0$. We shall restrict our study to functions $g(x)$ which are nonzero over the range of interest and deal with the case of g positive. The corresponding case of g negative is a straightforward modification of the procedure to be described.

WKB expansions can be obtained for the solution of 6.5.13 away from the turning point, i.e. in the ranges $x > 0$ and $x < 0$. The resulting four constants can be matched to the Airy function solutions of 6.5.13 which are valid for small x. Then to leading order $g(x)$ may be replaced by $g(0)$ so that the small x Airy function solutions of 6.5.13 correspond to those of 6.5.2 with λ replaced by $\lambda\sqrt{g(0)}$, i.e.

$$y = \alpha Ai\{-[\lambda\sqrt{g(0)}]^{2/3} x\} + \beta Bi\{-[\lambda\sqrt{g(0)}]^{2/3} x\}. \tag{6.5.14}$$

As an example consider the equation

$$\frac{d^2y}{dx^2} + \lambda^2 x(2+x)^2 y = 0, \quad \text{for } x > -2. \qquad 6.5.15$$

Here $g(x) = (2+x)^2$ and the turning point is at $x = 0$. The WKB expansions, valid away from the turning point, are constructed as follows. We let

$$y \sim \exp\{\delta_0(\lambda)\phi_0(\lambda) + \delta_1(\lambda)\phi_1(\lambda) + \cdots\},$$

and obtain the expressions

$$\delta_0 = \lambda, \quad \phi_0' = \pm\sqrt{-x}(2+x)$$

$$\delta_1 = 1, \quad \phi_1' = -\frac{1}{2}\frac{\phi_0''}{\phi_1'}, \quad \text{so } \phi_1 = \ln[\sqrt{|x|}(2+x)].$$

If $x > 0$ then

$$\phi_0' = \pm i\sqrt{x}(2+x) \quad \text{and} \quad \phi_0 = \pm i\left(\frac{4}{3}x^{3/2} + \frac{2}{5}x^{5/2}\right).$$

For the case $x < 0$ it is helpful to introduce the new variable $z = -x$ then

$$\frac{d\phi_0}{dz} = -\frac{d\phi_0}{dx} = \mp\sqrt{z}(2-z),$$

and on integrating we obtain

$$\phi_0 = \mp\left(\frac{4}{3}z^{3/2} - \frac{2}{5}z^{5/2}\right) = \mp\left(\frac{4}{3}|x|^{3/2} - \frac{2}{5}|x|^{5/2}\right).$$

Thus the WKB expansions valid away from the turning point $x = 0$ are

$$y \simeq \begin{cases} \dfrac{1}{x^{1/4}\sqrt{2+x}}\left\{a\cos\left[\lambda\left(\dfrac{4}{3}x^{3/2} + \dfrac{2}{5}x^{5/2}\right)\right] + b\sin\left[\lambda\left(\dfrac{4}{3}x^{3/2} + \dfrac{2}{5}x^{5/2}\right)\right]\right\}, & x > 0 \\ \dfrac{1}{|x|^{1/4}\sqrt{2+x}}\left\{c\exp\left[\lambda\left(\dfrac{4}{3}|x|^{3/2} - \dfrac{2}{5}|x|^{5/2}\right)\right]\right. & \\ \qquad \left. + d\exp\left[-\lambda\left(\dfrac{4}{3}|x|^{3/2} - \dfrac{2}{5}|x|^{5/2}\right)\right]\right\}, & x < 0. \end{cases} \qquad 6.5.16$$

They can be matched to the Airy function solution 6.5.14 by expressing the Airy functions in their asymptotic form for large values of $\lambda^{2/3}|x|$. Matching requires that the terms $\lambda|x|^{5/2}$ in 6.5.16 are small. A suitable overlap region is $x = O(1/\sqrt{\lambda})$ as $\lambda \to \infty$ then $\lambda^{2/3}|x| = O(\lambda^{1/6})$ as $\lambda \to \infty$, while $\lambda|x|^{5/2} = O(\lambda^{-1/4})$ as $\lambda \to \infty$. To leading order in this region the terms involving $x^{5/2}$ which occur in equation 6.5.16

may be neglected and the coefficient $\sqrt{2+x}$ replaced by $\sqrt{2}$. Then matching 6.5.16 to the Airy function solution expansions yields,

$$\frac{\alpha+\beta}{\sqrt{2\pi}(2\lambda)^{1/6}}=\frac{a}{\sqrt{2}},\quad \frac{\alpha-\beta}{\sqrt{2\pi}(2\lambda)^{1/6}}=\frac{b}{\sqrt{2}}, \tag{6.5.17}$$

$$\frac{\alpha}{2\sqrt{\pi}(2\lambda)^{1/6}}=\frac{d}{\sqrt{2}},\quad \frac{\beta}{\sqrt{\pi}(2\lambda)^{1/6}}=\frac{c}{\sqrt{2}}. \tag{6.5.18}$$

Eliminating α and β from these expressions provides the connection formulae between the constants in the WKB expansions of equation 6.5.16, namely

$$2d+c=\sqrt{2}a \quad\text{and}\quad 2d-c=\sqrt{2}b. \tag{6.5.19}$$

Notice that these are the same as 6.5.11 and they do not involve the value of $g(0)$.

The full approximation consists of the expansions 6.5.16 valid for $|x| > O(1/\lambda^{2/3})$ as $\lambda \to \infty$ and the Airy function solution 6.5.14 in the region x equal to zero. When initial conditions are specified at some point x_i they determine a and b if $x_i > 0$ or c and d if $x_i < 0$. In either case the other pair of constants along with α and β are determined from 6.5.17, 18 and 19. If conditions are specified at $x = 0$ then we require the results,

$$A_i(0)=B_i(0)/\sqrt{3}=1\bigg/\left[3^{2/3}\Gamma\left(\frac{2}{3}\right)\right]=0.3550 \text{ (to 4 significant figures.)}$$

(see Abramowitz and Stegun[1b]).

We conclude this chapter with an example involving a boundary condition imposed at a turning point. We seek an approximation for the large eigenvalues of the boundary value problem 6.5.15 with the boundary conditions $y(0) = y(1) = 0$. Then

$$y(0)=\alpha A_i(0)+\beta B_i(0)=0,\quad \text{i.e.}\quad \alpha=-\sqrt{3}\beta$$

and

$$y(1)=\frac{1}{\sqrt{3}}\left[a\cos\left(\frac{26\lambda}{15}\right)+b\sin\left(\frac{26\lambda}{15}\right)\right]=0,\quad \text{i.e.}\quad \frac{a}{b}=-\tan\left(\frac{26\lambda}{15}\right).$$

The equations 6.5.17 yield

$$\frac{a}{b}=\frac{\alpha+\beta}{\alpha-\beta},$$

so that

$$\frac{a}{b}=\frac{\sqrt{3}-1}{1+\sqrt{3}}=-\tan\left(\frac{26\lambda}{15}\right).$$

Therefore the large eigenvalues are approximated by the equation

$$\lambda_n \simeq \frac{15}{26}\left[\tan^{-1}\left(\frac{1-\sqrt{3}}{1+\sqrt{3}}\right) + n\pi\right],$$

where n is a large integer.

Exercise

Obtain an approximate expression for the large eigenvalues of the following boundary value problem,

$$\frac{d^2y}{dx^2} + \lambda^2 x(1+x)^4 y = 0$$

$$y(0) = y(1) = 0.$$

Chapter 7
The Asymptotic Approximation of Integrals

In this chapter we will consider methods of constructing asymptotic approximations of integrals. Specifically we will study integrals of the form

$$\int_a^b e^{ph(t)}.f(t)\,dt, \quad \int_a^b \cos[ph(t)].f(t)\,dt \quad \text{and} \quad \int_a^b \sin[ph(t)].f(t)\,dt,$$

where h and f are given functions and p is a large parameter.

To motivate this study consider the one-dimensional diffusion convection equation

$$-\nu\frac{d^2\phi}{dX^2} + U(X)\frac{d\phi}{dX} = F(X),$$

where ϕ may represent various physical properties of the medium such as temperature, chemical concentration or fluid momentum. The property ϕ is transported by the medium with speed $U(X)$ and diffuses through the medium whose diffusivity is given by ν. The right-hand side, $F(x)$, represents the source of the quantity ϕ.

Multiplying throughout by the integrating factor $\exp(-UX/\nu)$ allows the equation to be written in the following form,

$$-\nu\frac{d}{dX}\left(e^{-UX/\nu}.\frac{d\phi}{dX}\right) = e^{UX/\nu}.F.$$

Thus on integrating we have

$$\frac{d\phi}{dX} = -\frac{1}{\nu}e^{UX/\nu}\int e^{-UX/\nu}F\,dX.$$

After introducing the nondimensional variables $t = X/L$ and $h(t) = -UX/U_0L$, L and U_0 are a characteristic length and velocity, the integral in the above expression takes the form

$$\int e^{ph(t)}f(t)\,dt,$$

where $f(t) = F(X)$ and $p = U_0L/\nu$.

The parameter p is the nondimensional combination U_0L/v which provides a measure of the strength of convection relative to diffusion. In the description of heat transfer p is the Péclet number while in fluid flow it is the Reynolds number. It is common for convection to dominate diffusion so that these nondimensional numbers are often large. Thus we are led to consider integrals of the above form for large values of p. Such integrals are called *Laplace integrals* and their asymptotic approximation for large p is obtained using a technique known as *Laplace's method.*

Another important class of integrals has the exponential replaced by sinusoidal functions

$$\int_a^b \cos[ph(t)].f(t)\mathrm{d}t \quad \text{and} \quad \int_a^b \sin[ph(t)].f(t)\mathrm{d}t.$$

We will consider the stationary phase method of obtaining asymptotic approximations of such integrals.

We first consider a restricted class of Laplace integral obtained by choosing $h(t) = -t$,

$$\int_a^b \mathrm{e}^{-pt}f(t)\mathrm{d}t.$$

We will derive a method of obtaining an expansion for large values of p involving inverse powers of p. This is known as Watson's lemma.

In the above classes of integrals there is a region (or regions) in the interval $a \leqslant t \leqslant b$ which provide the dominant contribution to the integration. This allows the simplification of replacing both $f(x)$ and $h(x)$ by functions representing their local behavior in the region of dominant contribution. The resulting expressions can be evaluated by straightforward exact techniques and in this way an asymptotic approximation is constructed.

There are, of course, many numerical quadrature techniques which can be used to provide a direct approximation of the original integrals. However, in the case of large values of the parameter a knowledge of the region of dominant contribution is required so that sufficient quadrature points can be located in this region. Furthermore, the use of a numerical method of integration provides an approximation only for the one particular parameter value which happens to have been chosen. In contrast the asymptotic approach provides an analytic expression for the dependence of the integral on the parameter in a continuous range.

7.1 The region of dominant contribution

The crucial idea in the asymptotic approximation of integrals involves identifying the region(s) of dominant contribution and replacing functions in the integrand by simpler functions which represent their local behavior in this region. To introduce this idea we consider the following integral for large values of p,

$$\int_0^\infty e^{-pt} \cos \pi t \, dt. \tag{7.1.1}$$

The integrand consists of the product of the functions e^{-pt} and $\cos \pi t$. These along with their product are shown for the cases $p = 1$, 5 and 10 in Fig. 7.1.

It can be seen from Fig. 7.1 that, as p increases, the region of dominant contribution approaches the left-hand limit of integration. The behavior of $\cos \pi t$ for large values of t is unimportant because the rapidly decaying exponential function which multiplies $\cos \pi t$ annihilates its contribution except near $t = 0$. This suggests replacing

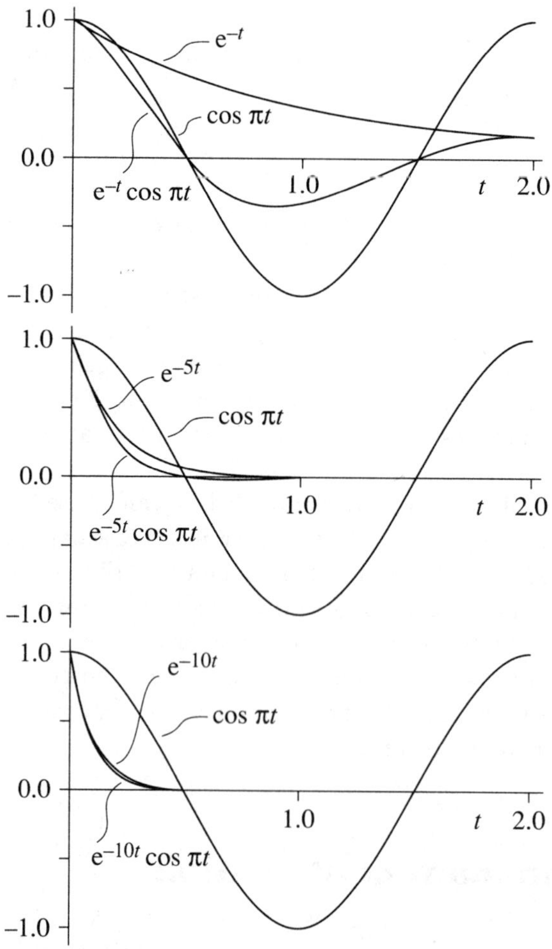

Fig. 7.1 The functions, e^{-pt}, $\cos \pi t$ and their product as p increases

$\cos \pi t$ by its value at $t = 0$, i.e.

$$I = \int_0^\infty \mathrm{e}^{-pt} \cos \pi t \, \mathrm{d}t \simeq \int_0^\infty \mathrm{e}^{-pt} \cos \pi 0 \, \mathrm{d}t = \int_0^\infty \mathrm{e}^{-pt} \, \mathrm{d}t = \frac{1}{p}.$$

This approximation can be improved by replacing $\cos \pi t$ with its Maclaurin expansion,

$$I = \int_0^\infty \mathrm{e}^{-pt}\left(1 - \frac{\pi^2 t^2}{2} + \frac{\pi^4 t^4}{4!} - \cdots\right)\mathrm{d}t. \qquad 7.1.2$$

To proceed with the expansion in 7.1.2 we need to evaluate integrals of the form

$$I_n = \int_0^\infty t^n \mathrm{e}^{-pt} \, \mathrm{d}t.$$

Integrating by parts yields

$$I_n = \left[\frac{t^n \mathrm{e}^{-pt}}{-p}\right]_0^\infty + \frac{n}{p}\int_0^\infty t^{n-1} \mathrm{e}^{-pt} \, \mathrm{d}t.$$

The first member of the right-hand side is zero at the upper limit since the exponential function annihilates the algebraic term as $t \to \infty$. The lower limit is zero for $n \geqslant 1$. Thus

$$I_n = \frac{n}{p} I_{n-1} \quad \text{for } n \geqslant 1.$$

For the case $n = 0$, $I_0 = 1/p$ so that,

$$I_1 = \frac{1}{p} I_0 = \frac{1}{p^2}$$

$$I_2 = \frac{2}{p} I_1 = \frac{2.1}{p^3}$$

$$I_3 = \frac{3}{p} I_2 = \frac{3.2.1}{p^4},$$

and in general $I_n = n!/p^{n+1}$.

Returning to 7.1.2 we have

$$I = I_0 - \frac{\pi^2}{2} I_2 + \frac{\pi^4}{4!} I_4 - \cdots = \frac{1}{p} - \frac{\pi^2}{p^3} + \frac{\pi^4}{p^5} - \cdots. \qquad 7.1.3$$

We will postpone a consideration of the remainder term in such series until Watson's lemma is derived in Section 7.2.

It is a common occurrence for the above expansion procedure to lead to a diverging series (although not in this case provided $p > \pi$). However, we will show that under rather general conditions the expansion generated by termwise integration in the above fashion is an asymptotic expansion. This result will be anticipated in the following examples and the equal to symbol will be replaced by a tilde symbol.

The integral 7.1.1 happens to be of a type which can be evaluated exactly by repeated integration by parts, i.e.

$$I = \int_0^\infty e^{-pt}\cos \pi t\, dt = \left[\frac{e^{-pt}\cos \pi t}{-p}\right]_0^\infty - \frac{\pi}{p}\int_0^\infty e^{-pt}\sin \pi t\, dt$$

$$= \frac{1}{p} - \frac{\pi}{p}\left\{\left[\frac{e^{-pt}\sin \pi t}{-p}\right]_0^\infty + \frac{\pi}{p}\int_0^\infty e^{-pt}\cos \pi t\, dt\right\}$$

$$= \frac{1}{p} - \frac{\pi^2}{p^2}I.$$

Thus

$$I = \frac{1}{p\left(1 + \dfrac{\pi^2}{p^2}\right)},$$

and the expansion 7.1.3 is recovered using the binomial expansion (provided $p > \pi$).

The following integral cannot be evaluated by exact techniques,

$$I = \int_0^\infty e^{-pt}\cos t^2\, dt.$$

An asymptotic expansion for large p is obtained by arguing that, just as in the previous example, the region of dominant contribution is near $t = 0$. Then on replacing $\cos t^2$ by its Maclaurin expansion we have

$$I \sim \int_0^\infty e^{-pt}\left(1 - \frac{t^4}{2!} + \frac{t^8}{4!} - \cdots\right) dt$$

$$\sim \frac{1}{p}\int_0^\infty e^{-s} ds - \frac{1}{p^5}\frac{1}{2!}\int_0^\infty s^4 e^{-s} ds + \frac{1}{p^9}\frac{1}{4!}\int_0^\infty s^8 e^{-s} ds - \cdots$$

$$\sim \frac{1}{p} - \frac{4!}{2!}\frac{1}{p^5} + \frac{8!}{4!}\frac{1}{p^9} - \cdots. \qquad 7.1.4$$

The series represented by 7.1.4 diverges for all p. A careful consideration of the behavior of the remainder bound would show that the series is an asymptotic series. The corresponding general result will be established when Watson's lemma is derived.

The previous two examples have involved the integration range from zero to infinity. Suppose instead that we are interested in the behavior of the following integral for large p,

$$I = \int_0^T e^{-pt} f(t)\mathrm{d}t, \quad (T > 0). \tag{7.1.5}$$

This can be expressed as the difference of two integrals,

$$I = \underbrace{\int_0^\infty e^{-pt} f(t)\mathrm{d}t}_{I_1} - \underbrace{\int_T^\infty e^{-pt} f(t)\mathrm{d}t}_{I_2}.$$

If we let $t' = t - T$ in the second integral and define a function g by the equation $g(t') = f(t' + T)$ then

$$I_2 = e^{-pT} \int_0^\infty e^{-pt'} g(t')\mathrm{d}t'.$$

The multiplicative coefficient e^{-pT} is transcendentally small as $p \to \infty$. Thus, provided $g(t')$ is a function which does not increase too rapidly, the integral I_2 is a transcendentally small term (T.S.T.). We will see in the next section that a sufficient condition for g and hence f is that it is exponentially bounded for large t, i.e. $|f(t)| < Ke^{ct}$ for some fixed constants K and c.

The above consideration shows that integrals of the form 7.1.5 have the same asymptotic expansions in terms of inverse powers of p as those integrals with the upper limit T replaced by ∞, the difference between them being transcendentally small,

$$\int_0^T e^{-pt} f(t)\mathrm{d}t = \int_0^\infty e^{-pt} f(t)\mathrm{d}t + \text{T.S.T.} \tag{7.1.6}$$

Next consider the case when the lower limit of integration is greater than zero:

$$I = \int_{T_1}^{T_2} e^{-pt} f(t)\mathrm{d}t \quad 0 < T_1 < T_2. \tag{7.1.7}$$

Let $t' = t - T_1$ and $g(t') = f(t' + T_1)$. Then

$$I = e^{-pT_1} \int_0^{T_2 - T_1} e^{-pt'} g(t')\mathrm{d}t',$$

and, from 7.1.6, the upper limit of integration may be replaced by infinity with a transcendentally small error. Thus the integral 7.1.7 takes the standard form with the integration range being from zero to infinity,

$$I = \int_{T_1}^{T_2} e^{-pt} f(t)\mathrm{d}t = \exp(-pT_1)\left(\int_0^\infty e^{-pt'} g(t')\mathrm{d}t' + \text{T.S.T.}\right).$$

The following integral provides an example of this idea,

$$I = \int_1^2 e^{-pt} \frac{1}{\sqrt{1+t^3}} dt.$$

Let $s = t - 1$, then $t^3 = (1+s)^3 = 1 + 3s + 3s^2 + s^3$ so that

$$\begin{aligned} I &= \int_0^1 e^{-p(s+1)} \cdot \frac{1}{\sqrt{2+3s+3s^2+s^3}} ds \\ &= \frac{e^{-p}}{\sqrt{2}} \int_0^1 e^{-ps} \left(1 + \frac{3}{2}s + \frac{3}{2}s^2 + \frac{s^3}{2}\right)^{-1/2} ds \\ &= \frac{e^{-p}}{\sqrt{2}} \int_0^1 e^{-ps} \left(1 - \frac{3}{4}s + O(s^2)\right) ds \\ &= \frac{e^{-p}}{\sqrt{2}} \left[\int_0^\infty e^{-ps} \left(1 - \frac{3}{4}s + O(s^2)\right) ds + \text{T.S.T.}\right] \\ &= \frac{e^{-p}}{\sqrt{2}} \left[\frac{1}{p} - \frac{3}{4}\frac{1}{p^2} + O\left(\frac{1}{p^3}\right) + \text{T.S.T.}\right]. \end{aligned}$$

The integrals considered so far have generated asymptotic expansions involving inverse integer powers of p. In the following example fractional powers occur in the expansion

$$I = \int_0^\infty e^{-pt} \frac{1}{\sqrt{t}} \cdot \frac{1}{1+t} dt. \qquad 7.1.8$$

The integrand is singular at $t = 0$. However, the integral exists because upon integration the function behaves like $\sqrt{t}$ sufficiently near $t = 0$. (The general condition, associated with the small t behavior, for the existence of an integral is that the integrand be less singular than $1/t$ as $t \to 0$.)

For large values of p the dominant region of contribution to the integral 7.1.8 will be near $t = 0$. We are led to expand the factor $1/(1+t)$ about $t = 0$ to obtain the following:

$$I \sim \int_0^\infty e^{-pt} \left(\frac{1}{\sqrt{t}} - \frac{t}{\sqrt{t}} + \frac{t^2}{\sqrt{t}} - \cdots\right) dt.$$

On replacing pt by the variable s we have

$$I \sim \frac{1}{p^{1/2}} \int_0^\infty s^{-1/2} e^{-s} ds - \frac{1}{p^{3/2}} \int_0^\infty s^{1/2} e^{-s} ds + \frac{1}{p^{5/2}} \int_0^\infty s^{3/2} e^{-s} ds - \cdots. \qquad 7.1.9$$

Digression: the gamma function

The integrals in the expansion 7.1.9 are particular examples of the gamma function $\Gamma(x)$ defined by the general expression

$$\Gamma(x) = \int_0^\infty s^{x-1} e^{-s} ds \quad \text{for } x > 0.$$

The fundamental property of the gamma function is:

$\Gamma(x+1) = x\Gamma(x)$ Fundamental property of Γ.

This is easily verified by integration by parts,

$$\Gamma(x+1) = \int_0^\infty s^x e^{-s} ds = \underbrace{\Big[-s^x e^{-s}\Big]_0^\infty}_{\text{zero for } x>0} + x \underbrace{\int_0^\infty s^{x-1} e^{-s} ds.}_{\Gamma(x)}$$

When $x = 1$ we have

$$\Gamma(1) = \int_0^\infty e^{-s} ds = 1.$$

Then, from the fundamental property,

$$\Gamma(2) = 1\Gamma(1) = 1$$

$$\Gamma(3) = 2\Gamma(2) = 2!$$

$$\Gamma(4) = 3\Gamma(3) = 3!$$

Clearly in general for integer n

$$\Gamma(n+1) = n!$$

(with the convention $0! = 1$).

The gamma function generalizes the idea of the factorial function to noninteger values of the argument x. Values of $\Gamma(x)$ are tabulated in books of special functions (see for example Abramowitz and Stegun[1c]). It is only necessary to provide values of Γ over an interval of unit length of the independent variable. Other values are obtained using the fundamental property of Γ. Tables of values of $\Gamma(x)$ are usually provided for values of x in the range $1 \leqslant x < 2$. Thus for example, given that $\Gamma(1.1) = 0.95135$ we may obtain $\Gamma(0.1)$ and $\Gamma(3.1)$ as follows:

$$\Gamma(1.1) = 0.1\Gamma(0.1).$$

Hence $\Gamma(0.1) = \Gamma(1.1)/0.1 = 9.5135$

$$\Gamma(3.1) = 2.1\Gamma(2.1) = 2.1 \times 1.1\Gamma(1.1).$$

Hence $\Gamma(3.1) = 2.1 \times 1.1 \times 0{\cdot}95135 = 2.19762.$

There is a special result which occurs quite often, it is the case $\Gamma(x)$ when x is half integer:

$$\Gamma(\tfrac{1}{2}) = \int_0^\infty \frac{e^{-s}}{\sqrt{s}}\, ds.$$

Let $s = x^2$ then

$$\Gamma[\tfrac{1}{2}] = 2\int_0^\infty \exp(-x^2)\, dx = \int_{-\infty}^\infty \exp(-x^2)\, dx$$

$$= \sqrt{\int_{-\infty}^\infty \exp(-x^2)\, dx \,.\int_{-\infty}^\infty \exp(-y^2)\, dy}.$$

If x and y are identified as rectangular axes of the two-dimensional plane then the product integral within the square root becomes a double integral over the entire plane.

$$\int_{-\infty}^\infty \int_{-\infty}^\infty \exp[-(x^2 + y^2)]\, dx\, dy.$$

This can be evaluated using plane polar coordinates r, θ. The integral becomes

$$\int_0^{2\pi} \int_0^\infty \exp(-r^2) r\, dr\, d\theta = \pi.$$

Thus we have established the result that $\Gamma(\frac{1}{2}) = \sqrt{\pi}$. ■

Returning to the expansion 7.1.9 we may identify the integrals as gamma functions so that

$$\int_0^\infty e^{-pt} \frac{1}{\sqrt{t}} \cdot \frac{1}{1+t}\, dt \sim \frac{1}{p^{1/2}}\Gamma(\tfrac{1}{2}) - \frac{1}{p^{3/2}}\Gamma(\tfrac{3}{2}) + \frac{1}{p^{5/2}}\Gamma(\tfrac{5}{2})$$

$$\sim \frac{\sqrt{\pi}}{p^{1/2}} - \frac{1}{p^{3/2}}\cdot\frac{1}{2}\sqrt{\pi} + \frac{1}{p^{5/2}}\cdot\frac{3}{2}\cdot\frac{1}{2}\sqrt{\pi} - \cdots.$$

This example along with our digression on the gamma function leads us naturally to a consideration of Watson's lemma.[19]

7.2 Watson's lemma

Let

$$I(p) = \int_0^T e^{-pt} t^{\lambda} g(t) dt, \quad (T > 0) \qquad 7.2.1$$

where

(i) $\lambda > -1$ 7.2.2

(ii) $g(t)$ is exponentially bounded in the interval $0 \leqslant t \leqslant T$ 7.2.3

(iii) $g(t)$ possesses a Maclaurin series 7.2.4

then

$$I(p) \sim \sum_{n=0}^{\infty} \frac{d^n g}{dt^n}(0) \frac{1}{n!} \frac{\Gamma(\lambda + n + 1)}{p^{\lambda+n+1}} \quad \text{as } p \to \infty . \qquad 7.2.5$$

The condition (i) is the extent to which the integrand may be singular near $t = 0$ while ensuring the existence of the integral.

The condition (ii), namely $|g(t)| < Ke^{ct}$ in $0 \leqslant t \leqslant T$ for fixed constants K and c ensures that e^{-pt} dominates the integrand away from the lower limit when p is large. It will allow the upper limit of the integration range, T, to be changed to another non-zero value with a transcendentally small error. The condition means for example that, in the case $T = \infty$, we may have $g = e^t, e^{10t}$, etc., but it does not allow $g = \exp(t^2)$. This latter type of behavior must be excluded to allow e^{-pt} to annihilate $g(t)$ away from $t = 0$ for large p. The requirement of an exponential bound serves to eliminate singular forms of $g(t)$ such as $g = 1/(1 - t)$ for $T \geqslant 1$ (although as we shall see later, a weaker singularity is permissible).

The condition (iii) allows $g(t)$ to be replaced by an expression which describes its behavior in the region of dominant contribution ($t = 0$). We will see that the radius of convergence, R, of the Maclaurin expansion is unimportant. In particular, the size of R relative to the upper limit of integration, T, is irrelevant.

A 'quick derivation' of Watson's lemma is provided by the following stages.

I Replace the upper limit of integration, T, by ∞ with a transcendentally small error

$$I(p) = \int_0^{\infty} e^{-pt} t^{\lambda} g(t) dt + \text{T.S.T.}$$

II Replace $g(t)$ with its Maclaurin expansion,

$$I(p) = \int_0^{\infty} e^{-pt} t^{\lambda} \sum_{n=0}^{\infty} \frac{d^n g}{dt^n}(0) \frac{t^n}{n!} dt$$

III Interchange the order of summing and integrating,

$$I(p) = \sum_{n=0}^{\infty} \frac{d^n g}{dt^n}(0) \frac{1}{n!} \int_0^{\infty} t^{\lambda+n} e^{-pt} dt$$

IV Let $s = pt$,

$$I(p) = \sum_{n=0}^{\infty} \frac{d^n g}{dt^n}(0) \frac{1}{n!} \frac{1}{p^{\lambda+n+1}} \int_0^{\infty} s^{\lambda+n} e^{-s} ds$$

V Identify the integrals as gamma functions to obtain the result

$$I(p) = \sum_{n=0}^{\infty} \frac{d^n g}{dt^n}(0) \frac{1}{n!} \frac{\Gamma(\lambda + n + 1)}{p^{\lambda+n+1}}.$$

The above derivation is flawed but provides a final result which is correct provided the equal to symbol $(=)$ is replaced by the asymptotically equivalent to symbol $(\sim)$ with the associated interpretation of the summation as explained in Section 2.2. There are two procedures in the above derivation which may be invalid. Firstly the Maclaurin expansion may not have an infinite radius of convergence so that step II is not necessarily valid. Secondly the operations of infinite summation and integration may not always commute so that step III may be invalid.

A more careful proof of Watson's lemma will now be presented. In outline the procedure is as follows. We first reduce the interval of integration from $0 < t < T$ to $0 < t < T^*$ where T^* is finite and chosen to lie within the interval of convergence of the Maclaurin series for $g(t)$. The difference between the two integrals will be shown to be transcendentally small,

$$I(p) = \int_0^{T} e^{-pt} t^{\lambda} g(t) dt = \int_0^{T^*} e^{-pt} t^{\lambda} g(t) dt + \text{T.S.T.} \quad 7.2.6$$

The second step is to replace $g(t)$ by its truncated Maclaurin expansion with remainder, i.e.

$$g(t) = \sum_{n=0}^{N} \frac{d^n g}{dt^n}(0) \frac{t^n}{n!} + R_N(t). \quad 7.2.7$$

Substituting into 7.2.6 yields

$$I(p) = \int_0^{T^*} e^{-pt} t^{\lambda} \sum_{n=0}^{N} \frac{d^n g}{dt^n}(0) \frac{t^n}{n!} dt + \text{integrated remainder}. \quad 7.2.8$$

The integrated remainder will be shown to be of order the first neglected term. This establishes the asymptotic nature of the expansion.

The next step is to interchange the process of integration and summation – this is a valid process because of the finite number of terms involved,

$$I(p) = \sum_{n=0}^{N} \frac{d^n g}{dt^n}(0) \frac{1}{n!} \int_0^{T^*} t^{\lambda+n} e^{-pt} dt + \text{integrated remainder}. \quad 7.2.9$$

At this stage the upper limit of integration T^* is replaced by infinity with an associated transcendentally small error. Then since

$$\int_0^\infty t^{\lambda+n}\mathrm{e}^{-pt}\mathrm{d}t = \frac{1}{p^{\lambda+n+1}}\int_0^\infty s^{\lambda+n}\mathrm{e}^{-s}\mathrm{d}s = \frac{\Gamma(\lambda+n+1)}{p^{\lambda+n+1}},$$

the result 7.2.5 follows.

To complete the proof we must establish equation 7.2.6 and consider the behavior of the remainder term in equation 7.2.8. Consider first the expression 7.2.6. We start from the equation

$$\begin{aligned} I(p) &= \int_0^T \mathrm{e}^{-pt}t^\lambda g(t)\mathrm{d}t \\ &= \int_0^{T^*} \mathrm{e}^{-pt}t^\lambda g(t)\mathrm{d}t + \int_{T^*}^T \mathrm{e}^{-pt}t^\lambda g(t)\mathrm{d}t, \end{aligned}$$

and consider the second integral

$$I_2 = \int_{T^*}^T \mathrm{e}^{-pt}t^\lambda g(t)\mathrm{d}t.$$

The terms e^{-pt} and t^λ in the integrand are positive and $g(t)$ is exponentially bounded, $|g(t)| < K\mathrm{e}^{ct}$, so that I_2 has the bound

$$|I_2| \leqslant K\int_{T^*}^T \mathrm{e}^{-(p-c)t}t^\lambda\mathrm{d}t.$$

We prove that this bound is transcendentally small as follows. If λ is negative then t^λ is bounded by $(T^*)^\lambda$ in the range $T^* < t < T$ so that

$$|I_2| \leqslant K(T^*)^\lambda\int_{T^*}^T \mathrm{e}^{-(p-c)t}\mathrm{d}t \leqslant K(T^*)^\lambda.\frac{\mathrm{e}^{-(p-c)T^*} - \mathrm{e}^{-(p-c)T}}{p-c}.$$

This bound is of order e^{-pT^*}/p as $p \to \infty$. If λ is positive t^λ can be bounded by T^λ and the same argument as above used provided that T is finite. However T may be infinite so we must establish the transcendentally small nature of the bound for $|I_2|$ in another way. A bound for t^λ where λ is positive is $t^\lambda \leqslant (\mathrm{e}^t)^\lambda$ so that

$$|I_2| \leqslant K\int_{T^*}^T \mathrm{e}^{-pt}\mathrm{e}^{\lambda t}\mathrm{e}^{ct}\mathrm{d}t \leqslant K\frac{\mathrm{e}^{-(p-\lambda-c)T^*} - \mathrm{e}^{-(p-\lambda-c)T}}{p-\lambda-c}.$$

Again this bound is of order e^{-pT^*}/p as $p \to \infty$.

Thus equation 7.2.6 has been established. This allows us to consider the integral

$$\int_0^{T^*} \mathrm{e}^{-pt}t^\lambda g(t)\mathrm{d}t, \qquad 7.2.10$$

where T^* is less than the radius of convergence of the function $g(t)$. Then throughout the integration range we may replace $g(t)$ with the truncated series and remainder,

$$g(t) = \sum_{n=0}^{N} \frac{d^n g}{dt^n}(0)\frac{t^n}{n!} + R_N,$$

where the remainder has the bound

$$|R_N| \leqslant L t^{N+1}/(N+1)!$$

The value of the constant L is irrelevant. All we require is that such a constant exists. From the consideration of Maclaurin series in Chapter 2 we know that it is sufficient for L to exceed the maximum value of the quantity $\left|\frac{d^{N+1}}{dt^{N+1}}g(t)\right|$ over the range $0 < t < T^*$.

The integral 7.2.10 is equal to

$$\sum_{n=0}^{N} \frac{d^n g}{dt^n}(0)\frac{1}{n!}\int_0^{T^*} t^{\lambda+n} e^{-pt}\,dt + IR,$$

where the integrated remainder, IR, has the following bound,

$$|IR| \leqslant L\int_0^{T^*} t^{\lambda+N+1} e^{-pt}\,dt/(N+1)!$$

The inequality is unchanged if T^* is replaced by infinity since the integrand is everywhere nonnegative. This and the substitution $s = pt$ leads to the bound

$$IR \leqslant L\frac{\Gamma(\lambda+N+2)}{(N+1)!\,p^{\lambda+N+2}}.$$

Thus the integrated remainder is of order $1/p^{\lambda+N+2}$ as $p \to \infty$ which establishes the asymptotic nature of the series and completes the proof of Watson's lemma.

Worked examples

Obtain the first three terms in the asymptotic expansion of the following for large p,

(i) $\int_0^4 e^{-pt} t^2 \sqrt{1+t}\,dt$ (ii) $\int_0^3 e^{-pt}\frac{1}{\sqrt{t}}\ln(1+t^2)dt.$

Solutions

(i) The conditions of Watson's lemma are met so we may write

$$I = \int_0^4 e^{-pt} t^2 \sqrt{1+t}\,dt \sim \int_0^\infty e^{-pt} t^2 \left(1 + \tfrac{1}{2}t - \tfrac{1}{8}t^2 + \cdots\right)dt.$$

Notice here that the radius of convergence of the Maclaurin series for $\sqrt{1+t}$ is unity. This is exceeded by the upper limit, namely $T = 4$, but this is not a difficulty. All we require is that

$\sqrt{1+t}$ is exponentially bounded which it certainly is. Then the various manipulations associated with the proof of Watson's lemma are valid and allow the last expression for I to be used. The evaluation in terms of gamma functions is straightforward. We let $s = pt$ to obtain

$$I \sim \frac{1}{p^3}\int_0^\infty s^2 e^{-s} ds + \frac{1}{2}\frac{1}{p^4}\int_0^\infty s^3 e^{-s} ds - \frac{1}{8}\frac{1}{p^5}\int_0^\infty s^4 e^{-s} ds + \cdots$$

$$\sim \frac{\Gamma(3)}{p^3} + \frac{1}{2}\frac{\Gamma(4)}{p^4} - \frac{1}{8}\frac{\Gamma(5)}{p^5} + \cdots \quad \text{as } p \to \infty .$$

(ii)

$$I = \int_0^3 e^{-pt}\frac{1}{\sqrt{t}}\ln(1+t^2)dt \sim \int_0^\infty e^{-pt}\frac{1}{\sqrt{t}}\left(t^2 - \frac{t^4}{2} + \frac{t^6}{3} - \cdots\right)dt$$

$$\sim \frac{\Gamma(5/2)}{p^{5/2}} - \frac{1}{2}\frac{\Gamma(9/2)}{p^{9/2}} + \frac{1}{3}\frac{\Gamma(13/2)}{p^{13/2}} - \cdots \quad \text{as } p \to \infty .$$

An extension of the permissible range of the value of λ in Watson's lemma is possible if some of the earlier terms in the Maclaurin series of $g(t)$ are absent. Thus in the second worked example the function $g(t) = \ln(1+t^2)$ has a Maclaurin series starting with the term t^2 which allows λ to be any value greater than -3. So if we choose $\lambda = -2.5$ we have,

$$I = \int_0^3 e^{-pt}\frac{1}{t^2\sqrt{t}}\ln(1+t^2)dt$$

$$\sim \int_0^\infty e^{-pt}\frac{1}{t^2\sqrt{t}}\left(t^2 - \frac{t^4}{2} + \frac{t^6}{3} - \cdots\right)dt$$

$$\sim \frac{\Gamma(1/2)}{p^{1/2}} - \frac{1}{2}\frac{\Gamma(5/2)}{p^{5/2}} + \frac{1}{3}\frac{\Gamma(9/2)}{p^{9/2}} - \cdots \quad \text{as } p \to \infty .$$

Exercises

1 Obtain the first two nonzero terms in the asymptotic expansion for large p of the following integrals:

(a) $\int_0^1 e^{-pt}\sin\sqrt{t}\,dt$, (b) $\int_0^3 e^{-pt} t\sin(t^2)dt$,

(c) $\int_0^1 e^{-pt}\frac{\sin(t^2)}{t^2}dt$, (d) $\int_0^\infty e^{-pt}\frac{1}{1+t}dt$,

(e) $\int_0^1 e^{-pt}\frac{1}{\sqrt{1+t^{3/2}}}dt$, (f) $\int_0^\pi e^{-pt}\frac{\cos t}{\sqrt{t}}dt$,

(g) $\int_0^{10} e^{-pt}\frac{\ln(1+\sqrt{t})}{t}dt$, (h) $\int_0^1 \exp(-pt^2)dt$.

2 The integral

$$I_n(r) = \frac{1}{\pi}\int_0^{\pi} e^{r\cos\theta}.\cos(n\theta)\,d\theta$$

is a modified Bessel function of order n.
Prove that

$$I_n(r) \sim \frac{e^r}{\sqrt{2\pi r}}\left(1 - \frac{(n^2 - \frac{1}{4})}{2r} + \cdots\right) \quad \text{as } r \to \infty\,.$$

[*Hint*: Let $t = 1 - \cos\theta$, expand $\cos n\theta$ in powers of θ and express θ in powers of t.]

7.3 Extensions of Watson's lemma

Watson's lemma may be extended to integrals of the Laplace form

$$\int_a^b e^{ph(t)} f(t)\,dt, \qquad 7.3.1$$

by the substitution $s = -h(t)$. Consider the following example:

$$I = \int_0^{\infty} \exp(-pt^2)\sin t\,dt \quad \text{as } p \to \infty\,.$$

Substituting $s = t^2$ yields

$$I = \int_0^{\infty} e^{-ps}\sin(\sqrt{s})\frac{ds}{2\sqrt{s}}. \qquad 7.3.2$$

Watson's lemma may be applied to 7.3.2 and leads to the following:

$$I \sim \frac{1}{2}\int_0^{\infty}\frac{e^{-ps}}{\sqrt{s}}\left(\sqrt{s} - \frac{1}{3!}s\sqrt{s} + \frac{1}{5!}s^2\sqrt{s} - \cdots\right)ds$$

$$\sim \frac{1}{2}.\frac{1}{p} - \frac{1}{2}.\frac{1}{3!}\frac{\Gamma(2)}{p^2} + \frac{1}{2}.\frac{1}{5!}\frac{\Gamma(3)}{p^3} - \cdots \quad \text{as } p \to \infty\,.$$

Consider next the integral

$$I = \int_0^1 \exp(-pt^3)\,dt \quad \text{as } p \to \infty\,. \qquad 7.3.3$$

Let $s = t^3$ so that 7.3.3 becomes

$$I = \int_0^1 e^{-ps}\frac{1}{3s^{2/3}}\,ds.$$

Watson's lemma will contribute a single-term expansion with a transcendenta... small remainder.

$$I = \int_0^\infty e^{-ps} \frac{1}{3s^{2/3}} ds + \text{T.S.T.} = \frac{1}{3} \frac{\Gamma(1/3)}{p^{1/3}} + \text{T.S.T.} \quad \text{as } p \to \infty .$$

Watson's lemma can be applied to integrals which do not appear to have the form 7.3.1. Consider the complementary error function for large values of its argument,

$$\operatorname{erfc} p = \frac{2}{\sqrt{\pi}} \int_p^\infty \exp(-t^2) dt \quad \text{as } p \to \infty . \qquad 7.3.4$$

Here the large parameter appears as the lower limit of integration. An integral of the required form for the application of Watson's lemma can be obtained by the substitution $s = t - p$ which makes the lower limit of the integral zero. Equation 7.3.4 becomes

$$\operatorname{erfc} p = \frac{2}{\sqrt{\pi}} \int_0^\infty \exp(-p^2 - 2ps - s^2) ds$$

$$= \frac{2}{\sqrt{\pi}} \exp(-p^2) \int_0^\infty \exp(-2ps) . \exp(-s^2) ds.$$

Then on expanding $\exp(-s^2)$ we obtain

$$\operatorname{erfc} p \sim \frac{2}{\sqrt{\pi}} \exp(-p^2) \int_0^\infty \exp(-2ps) \left(1 - s^2 + \frac{s^4}{2!} - \cdots \right) ds$$

$$\sim \frac{2}{\sqrt{\pi}} \exp(-p^2) \left(\frac{1}{2p} - \frac{\Gamma(3)}{(2p)^3} + \frac{\Gamma(5)}{2!(2p)^5} - \cdots \right) \quad \text{as } p \to \infty . \qquad 7.3.5$$

(This is equivalent to the expansion obtained in Chapter 1, equation 1.4.14, by repeated integration by parts.)

Exercises

Obtain the first two nonzero terms in the asymptotic expansion for large p of the following integrals

(i) $\displaystyle\int_0^2 \exp(-pt^2) \frac{1}{1+t} dt,$ (ii) $\displaystyle\int_0^1 \exp(-pt^2) . e^t \, dt,$

(iii) $\displaystyle\int_0^1 \exp(-pt^4) . \cos t \, dt,$ (iv) $\displaystyle\int_p^\infty \exp(-t^3) \, dt.$

Further extensions

So far in this section the examples chosen for the functions $h(t)$ in the integral 7.3.1 have been powers of t. More general functions can be dealt with using the substitution $s = -h(t)$. Consider the following example

$$I = \int_0^{\pi/4} e^{-p \sin t} dt \quad \text{as } p \to \infty. \tag{7.3.6}$$

The substitution $s = \sin t$ leads to

$$I = \int_0^{1/\sqrt{2}} e^{-ps} \frac{1}{\cos t} ds.$$

In the integration range $\cos t$ is positive so that $\cos t = +\sqrt{1 - \sin^2 t} = \sqrt{1 - s^2}$. This yields the following expression

$$I = \int_0^{1/\sqrt{2}} e^{-ps} \frac{1}{\sqrt{1 - s^2}} ds. \tag{7.3.7}$$

All the conditions of Watson's lemma are met so that

$$I \sim \int_0^{\infty} e^{-ps} \left(1 + \frac{s^2}{2} + \frac{3}{8} s^4 + \cdots \right) ds$$

$$\sim \frac{1}{p} + \frac{1}{2} \frac{\Gamma(3)}{p^3} + \frac{3}{8} \cdot \frac{\Gamma(5)}{p^5} + \cdots \quad \text{as } p \to \infty.$$

Suppose that the integral 7.3.6 has the upper limit $\pi/4$ replaced by $\pi/2$. Then 7.3.7 becomes

$$I = \int_0^{\pi/2} e^{-p \sin t} dt = \int_0^{1} e^{-ps} \frac{1}{\sqrt{1 - s^2}} ds.$$

The conditions stated at the start of Section 7.2 for the application of Watson's Lemma are not satisfied since the function $1/\sqrt{1 - s^2}$ is not exponentially bounded in the interval $0 \leqslant t \leqslant 1$. However, the singularity at $s = 1$ of the integrand is of strength $1/\sqrt{1 - s}$ which ensures that the integral exists. Indeed the contribution to the integral from the region near $s = 1$ is transcendentally small. This is most directly demonstrated by splitting the original integral into two,

$$I = \underbrace{\int_0^{\pi/4} e^{-p \sin t} dt}_{I_1} + \underbrace{\int_{\pi/4}^{\pi/2} e^{-p \sin t} dt}_{I_2}.$$

The contribution from the second integral is bounded by replacing $\sin t$ with its minimum value over the integration range namely $\sin \pi/4$. Thus

$$I_2 \leqslant \int_{\pi/4}^{\pi/2} e^{-p/\sqrt{2}}\,dt \leqslant \pi/4 e^{-p/\sqrt{2}},$$

which proves that I_2 is transcendentally small. We have already shown that $I_1 = O(1/p)$ as $p \to \infty$ so I_1 alone need be considered.

The exponential bound condition which was used to prove Watson's lemma is in fact not necessary unless the upper limit of integration is infinite. It is then required to ensure the existence of the integral. The minimum requirements of the function $f(t)$ such that Watson's lemma is valid are associated with ensuring both the existence of the integral and the exponentially small contribution of the integrand away from the region near $t = 0$. It is beyond the scope of this book to explore the minimum requirements further.

In the following example the straightforward substitution $s = -h(t)$ is insufficient,

$$I = \int_0^1 e^{-p/(1+t)}\,dt.$$

On setting $s = 1/(1+t)$ the integral becomes

$$I = \int_{1/2}^1 e^{-ps}.\frac{1}{s^2}\,ds.$$

The region of maximum contribution for large p will be near the lower limit $s = 1/2$. The further substitution $x = s - 1/2$ is required to produce the form to which Watson's lemma is applicable

$$I = e^{-p/2}\int_0^{1/2} e^{-px}\frac{1}{(x+1/2)^2}\,dx.$$

Then on using the binomial expansion

$$\frac{1}{(x+1/2)^2} = 4(1+2x)^{-2} = 4(1 - 4x + 12x^2 \cdots),$$

we have

$$I \sim e^{-p/2}\int_0^{\infty} e^{-px}\,4(1 - 4x + 12x^2 \cdots)dx$$

$$\sim e^{-p/2}\left(\frac{4}{p} - 16\frac{\Gamma(2)}{p^2} + 48\frac{\Gamma(3)}{p^3} - \cdots\right) \quad \text{as } p \to \infty.$$

Instead of transforming the original integral into a form to which Watson's lemma is directly applicable it is sometimes more convenient to deal with the integral directly using Laplace's method. This will now be described.

7.4 Laplace's method

Consider the integral

$$I = \int_0^\pi e^{p \sin t}\,dt. \qquad 7.4.1$$

The dominant contribution to the integral for large values of p will be from the region near $t = \pi/2$ where $\sin t$ takes its maximum value. If we compare, for example, the value of the integrand at $t = \pi/4$ and $t = \pi/2$ we have

$$e^{p\sin \pi/4} = e^{p/\sqrt{2}} \quad \text{and} \quad e^{p\sin(\pi/2)} = e^p.$$

This demonstrates that the contribution from the region near $t = \pi/2$ is exponentially greater than that from other regions.

On introducing the variable s $(= t - \pi/2)$ we have

$$\sin t = \sin(s + \pi/2) = \cos s = 1 - \frac{s^2}{2} + O(s^4). \qquad 7.4.2$$

Since the integral 7.4.1 has its dominant contribution from the region near $t = \pi/2$, and hence near $s = 0$, it is reasonable to assume that $\cos s$ may be replaced by its truncated Maclaurin series. Then 7.4.1 becomes

$$I = \int_{-\pi/2}^{\pi/2} \exp\{p[1 - s^2/2 + O(s^4)]\}ds = e^p \int_{-\pi/2}^{\pi/2} \exp(-ps^2/2)[1 + O(ps^4)]ds.$$

Let $x = s\sqrt{p/2}$ so that

$$I = \sqrt{\frac{2}{p}}\,e^p \int_{(-\pi/2)\sqrt{p/2}}^{(\pi/2)\sqrt{p/2}} \exp(-x^2)\left[1 + O\left(\frac{x^4}{p}\right)\right]dx. \qquad 7.4.3$$

The limits of integration $\pm\dfrac{\pi}{2}\sqrt{\dfrac{p}{2}}$ may be replaced by $\pm\infty$ with a transcendentally small error. Thus 7.4.3 becomes

$$I = \sqrt{\frac{2}{p}}\,e^p \int_{-\infty}^{\infty} \exp(-x^2)dx + O\left(\frac{e^p}{p\sqrt{p}}\right) \quad \text{as } p \to \infty.$$

The integral on the right-hand side of this expression commonly occurs in Laplace's method, its value, namely $\sqrt{\pi}$, was obtained in Section 7.1. The asymptotic approximation for the integral 7.4.1 is therefore

$$I = \sqrt{\frac{2\pi}{p}}\,e^p + O\left(\frac{e^p}{p\sqrt{p}}\right) \quad \text{as } p \to \infty. \qquad 7.4.4$$

In this approximation procedure it is essential to preserve the exponential decay away from the region of dominant contribution. It would be quite wrong to replace

$\cos s$ by its value at $s = 0$ and attempt to approximate the integral as follows:

$$I = \int_{-\pi/2}^{\pi/2} e^{p\cos s}\, ds \doteq \int_{-\pi/2}^{\pi/2} e^{p}\, ds = \pi e^{p} \text{ (WRONG)}.$$

The above process is erroneous because the integrand is assigned the value $e^{p\cos 0}$ for all values of s while in fact, for values of s away from zero, the integrand has a value which is exponentially smaller than that when $s = 0$.

The integral 7.4.1 can be cast into a form suitable for the application of Watson's lemma by the substitution $z = 1 - \sin t$,

$$I = 2\int_{0}^{\pi/2} e^{p\sin t}\, dt = 2\int_{1}^{0} e^{p} e^{-pz}\left(-\frac{dz}{\cos t}\right).$$

Throughout the integration range $0 < t < \pi/2$, we have

$$\cos t = \sqrt{1-(\sin t)^2} = \sqrt{1-(1-z)^2} = \sqrt{2z - z^2} = \sqrt{2z}\sqrt{1 - z/2}$$

so that

$$I = \sqrt{2}e^{p}\int_{0}^{1} e^{-pz}\frac{1}{\sqrt{z}}\frac{1}{\sqrt{1-z/2}}\, dz. \tag{7.4.5}$$

Watson's lemma may be applied to yield

$$\begin{aligned} I &\sim \sqrt{2}e^{p}\int_{0}^{\infty} e^{-pz}\frac{1}{\sqrt{z}}\left(1 + \frac{z}{4} + \cdots\right) dz \\ &\sim \sqrt{2}e^{p}\left(\frac{\Gamma(1/2)}{p^{1/2}} + \frac{1}{4}\frac{\Gamma(3/2)}{p^{3/2}} + \cdots\right) \\ &\sim \sqrt{\frac{2\pi}{p}}\, e^{p} + \frac{1}{8p}\sqrt{\frac{2\pi}{p}}\, e^{p} + \cdots \quad \text{as } p \to \infty. \end{aligned}$$

This verifies the approximation 7.4.4 obtained by Laplace's method.

Consider next the integrals of the general Laplace form,

$$I = \int_{t_1}^{t_2} f(t) e^{ph(t)}\, dt \quad \text{as } p \to \infty. \tag{7.4.6}$$

We seek the leading asymptotic approximation for large p. The region of dominant contribution will be near $t = t_0$ where $h(t)$ takes its maximum value. The function $f(t)$ may be replaced by $f(t_0)$. However, $h(t)$ cannot simply be replaced by $h(t_0)$ because the exponential decay away from the region of dominant contribution must be preserved. The Taylor expansion of $h(t)$ about $t = t_0$ is

$$h(t) = h(t_0) + (t - t_0)h'(t_0) + \frac{(t-t_0)^2}{2!}h''(t_0) + \cdots, \tag{7.4.7}$$

where the primes denote derivatives. The leading approximation to 7.4.6 is obtained by replacing $h(t)$ with $h(t_0)$ plus the first nonzero term in the expansion 7.4.7.

There are two different forms of approximation depending on whether $h(t)$ possesses a turning point maximum ($h'(t_0) = 0$) in the integration range or whether $h(t)$ increases or decreases monotonically having a maximum either at $t = t_1$ with $h(t_1) < 0$ or at $t = t_2$ with $h(t_2) > 0$. These forms of behavior are illustrated in Fig. 7.2.

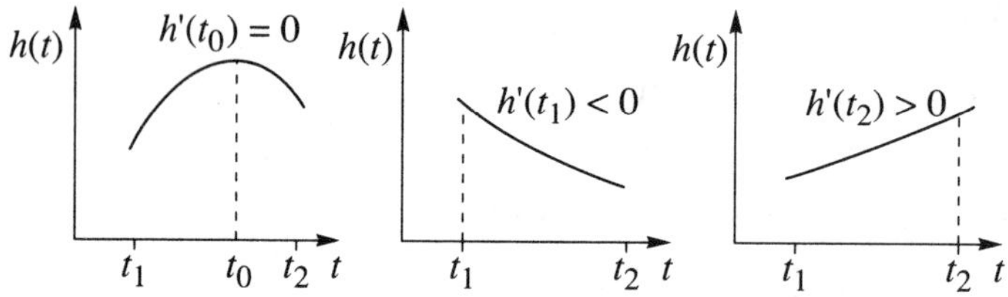

Fig. 7.2 Turning point maximum and end point maxima of $h(t)$

Turning point maxima

In the case of a turning point maximum there is the possibility of its occurring either at $t = t_1$ or $t = t_2$. In this case the approximation to the integral takes half the value that it would have taken had the maximum occurred at an interior point.

We first deal with the case of a turning point maximum occurring at an interior point $t_1 < t_0 < t_2$, $h'(t_0) = 0$. We will assume that $h''(t_0) < 0$. (The case $h''(t_0) = h'''(t_0) = 0$, $h''''(t_0) < 0$ can be treated by a straightforward extension of the technique.) Substituting $f(t) \simeq f(t_0)$ and $h(t) \simeq h(t_0) + ((t - t_0)^2/2)\, h''(t_0)$ leads to the approximation

$$I \simeq f(t_0)\exp[ph(t_0)] \int_{t_1}^{t_2} \exp\{p[(t - t_0)^2/2]h''(t_0)\}\, dt.$$

Then on letting $s = (t - t_0)\sqrt{-ph''(t_0)/2}$, noting that the quantity $-ph''(t_0)$ is positive, we obtain the approximation

$$I \simeq f(t_0)\exp[ph(t_0)] \sqrt{\frac{-2}{ph''(t_0)}} \int_{(t_1 - t_0)\sqrt{-ph''(t_0)/2}}^{(t_2 - t_0)\sqrt{-ph''(t_0)/2}} \exp(-s^2)\, ds. \qquad 7.4.8$$

The limits of the integral 7.4.8 can be replaced by $\pm\infty$ with an exponentially small error so that the integral may be approximated by the value $\sqrt{\pi}$. Thus we obtain the leading order approximation

$$I \simeq f(t_0)\exp[ph(t_0)] \sqrt{\frac{2\pi}{p|h''(t_0)|}} \quad \text{as } p \to \infty. \qquad 7.4.9$$

In the case of the turning point maximum occurring at an end point $t_0 = t_1$ or $t_0 = t_2$ then either the lower or upper limit of integration in the expression 7.4.8 will be zero. The asymptotic approximation of the integral is then $\sqrt{\pi}/2$ so that the approximation 7.4.9 is replaced by half its value.

Worked examples

Obtain leading order approximations of the following integrals

(i) $\displaystyle\int_{-2}^{0} \exp(t).\exp[p(3t^2 + 2t^3)]\mathrm{d}t$ (ii) $\displaystyle\int_{0}^{1} \sqrt{1+t}\,\exp[p(2t - t^2)]\mathrm{d}t$ as $p \to \infty$.

Solutions

(i) $h = 3t^2 + 2t^3$, $h' = 6t + 6t^2$, $h'' = 6 + 12t$

The gradient of h is zero when $t = 0$ and $t = -1$

$h''(0) = 6$ $\quad\therefore t = 0$ is a minimum

$h''(-1) = -6$ $\quad\therefore t = -1$ is a maximum

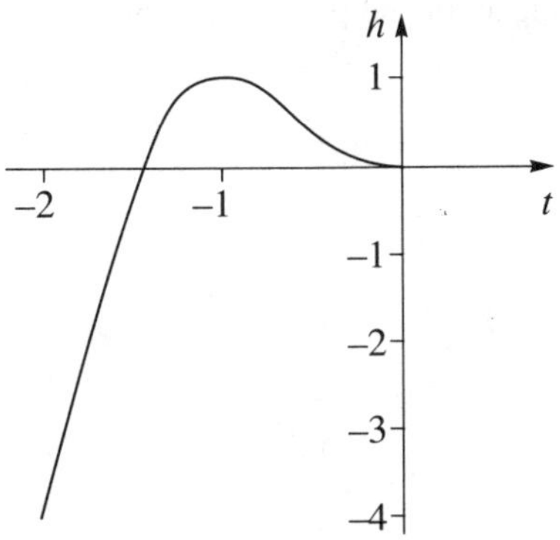

The function $h(t) = 3t^2 + 2t^3$ in the range $-2 < t < 0$.

To expand $h(t)$ about its maximum at $t = -1$ we let $s = t + 1$ and expand in powers of s. Then

$$\begin{aligned} h &= 3(s-1)^2 + 2(s-1)^3 \\ &= 3(s^2 - 2s + 1) + 2(s^3 - 3s^2 + 3s - 1) \\ &= 1 - 3s^2 + 2s^3, \end{aligned}$$

thus

$$\begin{aligned} I &= \int_{-2}^{0} \exp(t).\exp[p(3t^2 + 2t^3)]\mathrm{d}t \\ &= \mathrm{e}^{-1}\int_{-1}^{1} \exp[p(1 - 3s^2 + 2s^3)]\mathrm{d}s \\ &\simeq \mathrm{e}^{-1}\mathrm{e}^{p}\int_{-\infty}^{\infty} \exp(-3ps^2)\mathrm{d}s = \mathrm{e}^{p-1}\sqrt{\frac{\pi}{3p}}. \end{aligned}$$

(ii) $h = 2t - t^2$, $h' = 2 - 2t$, $h'' = -2$.
The gradient of h is zero when $t = 1$ and $h''(1) = -2$ so this is a maximum.

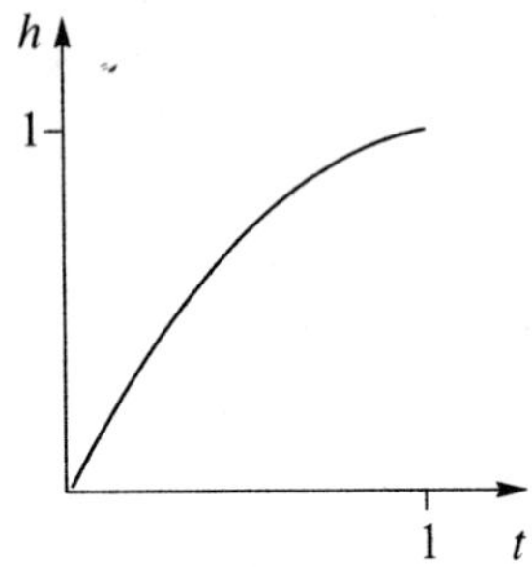

The function $h(t) = 2t - t^2$ in the range $0 < t < 1$.

In this case $t = 1$ is a turning point maximum occurring at an end point of the integration range.
Let $s = t - 1$ then $h = 2(s + 1) - (s + 1)^2 = 1 - s^2$.
Thus

$$\begin{aligned} I &= \int_0^1 \sqrt{1 + t}\, \exp[p(2t - t^2)]\,dt \\ &= \int_{-1}^{0} \sqrt{2 + s}\, \exp[p(1 - s^2)]\,ds \\ &\simeq \sqrt{2}e^p \int_{-\infty}^{0} \exp(-ps^2)\,ds = \sqrt{2}e^p \frac{1}{2}\sqrt{\frac{\pi}{p}}. \end{aligned}$$

End point maxima

In the case when the maximum value of $h(t)$ occurs at an end point of the integration range $t_1 < t < t_2$ with $h'(t) \neq 0$ the function is replaced by

$$h(t_0) + (t - t_0)h'(t_0),$$

with $t_0 = t_1$ or t_2 depending on where the maximum occurs. Suppose that the maximum occurs at $t = t_1$ then 7.4.6 becomes

$$I \simeq \int_{t_1}^{t_2} f(t_1) \exp\{p[h(t_1) + (t - t_1)h'(t_1)]\}\,dt.$$

On substituting $s = t - t_1$ we have

$$I \simeq f(t_1)\exp[ph(t_1)] \int_0^{t_2 - t_1} \exp[psh'(t_1)]\mathrm{d}s$$

$$= \frac{f(t_1)\exp[ph(t_1)]}{ph'(t_1)} \Big[\exp[psh'(t_1)] \Big]_0^{t_2 - t_1}$$

$$= \frac{f(t_1)\exp[ph(t_1)]}{ph'(t_1)} \{\exp[ph'(t_1)(t_2 - t_1)] - 1\}.$$

The first member of the bracketed term is exponentially small as $p \to \infty$ (since $h'(t_1)$ is negative and $t_2 - t_1$ is positive). Thus

$$I \simeq \frac{f(t_1)\exp[ph(t_1)]}{p|h'(t_1)|} \quad \text{as } p \to \infty.$$

In the case when the end point maximum occurs at the right-hand limit, t_2, the corresponding approximation is

$$I \simeq \frac{f(t_2)\exp[ph(t_2)]}{ph'(t_2)} \quad \text{as } p \to \infty.$$

Worked examples

Obtain leading order approximations of the following integrals

(i) $\displaystyle\int_1^2 e^{p/(1+t)}.\sqrt{3+t}\,\mathrm{d}t$ (ii) $\displaystyle\int_0^1 \exp(t).\exp[p(3t^2 + 2t^3)]\mathrm{d}t$ as $p \to \infty$.

Solutions

(i) $h = 1/(1+t)$ and $h' = -1/(1+t)^2$ which is negative throughout the integration range. Thus there is an end point maximum at the lower limit $t = 1$.

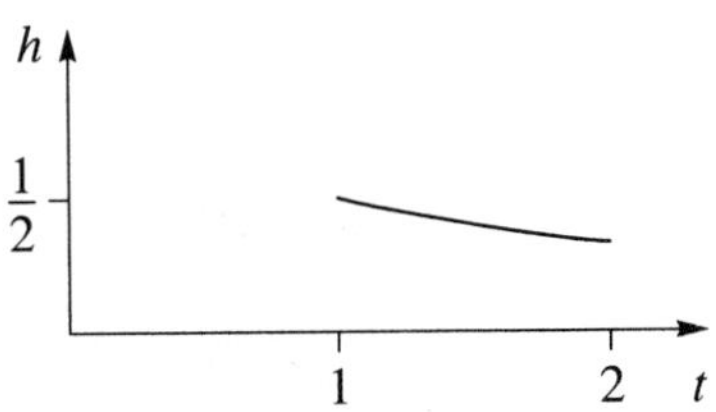

The function $h(t) = 1/(1+t)$ in the range $1 < t < 2$.

Let $s = t - 1$ and expand in powers of s,

$$h = \frac{1}{1+t} = \frac{1}{2+s} = \frac{1}{2}\left(1 - \frac{s}{2} + \cdots\right).$$

Then

$$I = \int_1^2 e^{p/(1+t)} \sqrt{3+t}\,dt$$

$$\simeq \sqrt{3+1}\, e^{p/2} \int_0^{\infty} e^{-ps/4}\,ds$$

$$= 2e^{p/2} \frac{4}{p} \quad \text{as } p \to \infty.$$

(ii) Notice that the integrand is the same as the earlier example used to illustrate the case of an interior turning point maximum. However the integration range is now from 0 to 1 and over this interval the function $h = 3t^2 + 2t^3$ has an end point maximum at $t = 1$. Let $s = t - 1$ then

$$h = 3t^2 + 2t^3 = 3(s+1)^2 + 2(s+1)^3 = 5 + 12s + \cdots,$$

and

$$I = \int_0^1 e^t . \exp[p(3t^2 + 2t^3)]dt \simeq e^1 \int_{-\infty}^0 e^{5p + 12sp}\,ds$$

$$= e^{1+5p} . \frac{1}{12p} \quad \text{as } p \to \infty.$$

In those cases when $h(t)$ has multiple maxima in the integration range the interval $t_1 < t < t_2$ should be divided into subintervals in which $h(t)$ has a single maximum. The location of the points of division is unimportant provided they do not coincide with maxima. Consider, for example, the following integral

$$I = \int_{-2}^{2} \exp[p(t^3/3 - t)] \ln(1 + t^2)dt \quad \text{as } p \to \infty.$$

Here $h = t^3/3 - t$, $h' = t^2 - 1$, $h'' = 2t$, so that a turning point maximum occurs at $t = -1$ and there is an end point maximum at $t = 2$.

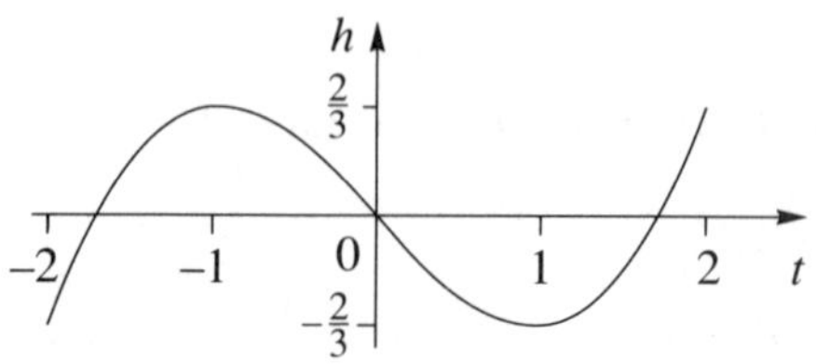

The function $h(t) = t^3/3 - t$ in the range $-2 < t < 2$.

The values of $h(t)$ at the internal maximum and the end point maximum are the same, $h(-1) = h(2) = 2/3$. Thus they share the same exponential coefficient $e^{2p/3}$. The difference between their contributions is algebraic, the turning point maximum has an algebraic coefficient of $O(1/\sqrt{p})$ while the end point maximum has a coefficient

$O(1/p)$. To see this it is convenient to divide the integration range at $t = 0$,

$$I = \underbrace{\int_{-2}^{0} \ldots \mathrm{d}t}_{I_1} + \underbrace{\int_{0}^{2} \ldots \mathrm{d}t}_{I_2}.$$

Then

$$I_1 = \int_{-2}^{0} \exp[p(t^3/3 - t)] \ln(1 + t^2)\mathrm{d}t,$$

and we let $s = t + 1$ so that

$$t^3/3 - t = \tfrac{1}{3}(s^3 - 3s^2 + 3s - 1 - 3s + 3) = 2/3 - s^2 + \cdots.$$

Therefore

$$I_1 \simeq e^{2p/3} \ln 2 \int_{-\infty}^{\infty} \exp(-ps^2)\mathrm{d}s = \sqrt{\frac{\pi}{p}}\, e^{2p/3} \ln 2.$$

For the second integral,

$$I_2 = \int_{0}^{2} \exp[p(t^3/3 - t)] \ln(1 + t^2)\mathrm{d}t,$$

we let $s = t - 2$ so that

$$t^3/3 - t = \tfrac{1}{3}(s^3 + 6s^2 + 12s + 8 - 3s - 6) = \tfrac{2}{3} + 3s + \cdots.$$

Therefore

$$I_2 \simeq e^{2p/3} \ln 5 \int_{-\infty}^{0} e^{3ps}\,\mathrm{d}s = \frac{e^{2p/3}}{3p} \ln 5.$$

Thus I_1 provides the dominant contribution and the corresponding leading approximation

$$I \simeq \sqrt{\frac{\pi}{p}}\, e^{2p/3} \ln 2 \quad \text{as } p \to \infty.$$

It would be a mistake to include the contribution from I_2 because lower order terms in the expansion of I_1 have not been considered and they may well be of the same order as I_2.

Before leaving this example it is instructive to note that had the integration range been from -2 to 3 then

$$I_2 = \int_{0}^{3} \exp[p(t^3/3 - t)] \ln(1 + t^2)\mathrm{d}t = O(e^{6p}/p) \quad \text{as } p \to \infty.$$

In this case the contribution from the end point maximum dominates the integral.

The gamma function and Stirling's approximation

The gamma function provides an interesting nonstandard application of Laplace's method. Consider the integral

$$\Gamma(p+1) = \int_0^\infty e^{-t} t^p \, dt,$$

for large values of p.

We may write t^p in the exponential form

$$t^p = (e^{\ln t})^p = e^{p \ln t}.$$

An attempt at a straightforward application of Laplace's method yields an incorrect result. We have

$$\Gamma(p+1) = \int_0^\infty e^{p \ln t} e^{-t} \, dt, \qquad 7.4.10$$

which suggests identifying the function $h(t)$ of Laplace's method as $\ln t$. This has an 'end point' maximum as $t \to \infty$ of $\ln \infty$. However, the multiplicative function e^{-t} tends to zero as $t \to \infty$!

The correct procedure is to investigate the location of the maximum of the full argument of the exponential in the integrand, $\exp(p \ln t - t)$, i.e.

$$\frac{d}{dt}(p \ln t - t) = \frac{p}{t} - 1.$$

Thus the maximum occurs at $t = p$. On introducing a new variable s by the equation $t = ps$ we recover a standard form for the application of Laplace's method.

$$\Gamma(p+1) = \int_0^\infty e^{p \ln p} e^{p(\ln s - s)} p \, ds = p \, . \, p^p \int_0^\infty e^{p(\ln s - s)} \, ds.$$

The function $h(s) = \ln s - s$ has derivative $1/s - 1$ which is zero at $s = 1$. Let $s = 1 + r$ then

$$h = \ln(1+r) - 1 - r = r - \frac{r^2}{2} + \cdots - 1 - r \simeq -1 - \frac{r^2}{2}$$

so that

$$\Gamma(p+1) \simeq p^{p+1} e^{-p} \int_{-\infty}^{\infty} \exp(-pr^2/2) \, dr \simeq p^{p+1} e^{-p} \sqrt{\frac{2}{p}} \sqrt{\pi} \quad \text{as } p \to \infty.$$

In the case when p is a large integer, N, we obtain Stirling's approximation for $N!$,

$$N! \simeq N^{N+1} e^{-N} \sqrt{\frac{2\pi}{N}}.$$

Consider the value $N = 10$ to test the accuracy of this approximation,

$$10! = 3628800$$

while $10^{11}e^{-10}\sqrt{\frac{2\pi}{10}} = 3.59 \times 10^6$ (to 3 significant figures). Thus for $N = 10$ the approximation is accurate to within about 1 per cent.

Another example with a similar behavior is provided by the integral

$$I = \int_0^\infty e^{-tp}e^{-1/t}\,dt. \qquad 7.4.11$$

A straightforward application of Laplace's method, or indeed of Watson's lemma, suggests that the maximum contribution occurs at $t = 0$. However the coefficient function $e^{-1/t}$ is zero at $t = 0$. Furthermore all the derivatives of $e^{-1/t}$ are zero, so Watson's lemma provides the result

$$I = 0 + 0 \times \left(\frac{1}{p}\right) + 0 \times \left(\frac{1}{p^2}\right) + 0 \times \left(\frac{1}{p^3}\right) + \cdots.$$

In fact the integral has a nonzero asymptotic approximation but it is smaller than any power of $1/p$. We consider the integrand in the form $e^{-tp-1/t}$ and locate the maximum of the argument $-tp - 1/t$, i.e.

$$\frac{d}{dt}\left(-tp - \frac{1}{t}\right) = -p + \frac{1}{t^2}.$$

Thus the maximum occurs at $t = 1/\sqrt{p}$. We recover a form suitable for the application of Laplace's technique by the substitution $t = s/\sqrt{p}$. Then

$$I = \frac{1}{\sqrt{p}}\int_0^\infty e^{-\sqrt{p}(s+1/s)}\,ds,$$

and the maximum contribution occurs at $s = 1$.

Let $s = 1 + r$ then

$$s + \frac{1}{s} = 1 + r + (1 - r + r^2 - \cdots) \simeq 2 + r^2,$$

so that

$$I \simeq \frac{1}{\sqrt{p}}e^{-2\sqrt{p}}\int_{-\infty}^\infty \exp(-\sqrt{p}.r^2)\,dr \simeq \frac{\sqrt{\pi}}{p^{3/4}}e^{-2\sqrt{p}} \quad \text{as } p \to \infty.$$

Exercises

Obtain asymptotic approximations for large p of the following integrals

(i) $\int_{-3}^{6} \exp(-pt^2)\sqrt{1+t^2}\,dt.$

(ii) $\int_{0}^{\pi/2} \exp[p(\sin t + \cos t)]\sqrt{t}\,dt.$

(iii) $\int_{0}^{2} \frac{\exp[p(3t^2 - 2t^3)]}{\sqrt{1+t^2}}\,dt.$

(iv) $\int_{-1}^{2} \frac{\exp[p(3t^2 - 2t^3)]}{\sqrt{1+t^2}}\,dt.$

(v) $\int_{-1}^{0} \exp(-p\cosh t).\cos t\,dt.$

(vi) $\int_{\pi/4}^{\pi/2} e^{-p\cosh t}.\cos t\,dt.$

(vii) $\int_{0}^{\infty} \exp[p(2t - t^2)].\ln(1+t^2)dt.$

(viii) $\int_{-\infty}^{\infty} \exp\{p[t - \exp(t)]\}dt.$

(ix) $\int_{0}^{1} \exp[p(t - t^2)]dt.$

(x) $\int_{0}^{1} \exp[p(t^2 - t)]dt.$

(xi) $\int_{0}^{\pi/4} \exp(p\cos t).e^{-t}\,dt.$

(xii) $\int_{-1}^{1} \exp[-p(\cosh t + 1)].e^{t}\,dt.$

(xiii) $\int_{-1}^{1} \exp[p(t^4 - 2t^2)].\sqrt{2+t}\,dt.$

(xiv) Repeat (13) for the range $\int_{0}^{3}$.

(xv) $\int_{-2\pi}^{2\pi} \exp(p\cos t)(4\pi + t)\,dt.$

(xvi) $\int_{1}^{2} \exp[p/(1+t)].\sqrt{3+t}\,dt.$

Selected answers

(i) $\sqrt{\frac{\pi}{p}}.$

(iii) $\sqrt{\frac{\pi}{6p}}\,e^{p}.$

(iv) $\frac{e^{5p}}{\sqrt{2}}.\frac{1}{12p}.$

(v) $\frac{1}{2}e^{-p}\sqrt{\frac{2\pi}{p}}.$

(vi) $\frac{\cos\pi/4}{\sinh\pi/4}.\frac{\exp(-p\cosh\pi/4)}{p}.$

(xiii) $\sqrt{\frac{\pi}{p}}.$

(xiv) $\frac{\sqrt{5}}{96}\frac{e^{63p}}{p}.$

(xv) $16\pi\,e^{p}\sqrt{\frac{\pi}{2p}}.$

7.5 The Riemann–Lebesgue lemma

So far in this chapter we have considered integrals whose integrands decay exponentially from their maximum value. We now turn our attention to integrands which rapidly oscillate in sign over the integration range. The contributions to the integral then tend to cancel except at end points where small contributions remain. Contributions also arise from regions where the rate of oscillation of the integrand is reduced.

Such regions are called regions of *stationary phase.* The associated asymptotic approximation technique known as the method of stationary phase will be described in the following section.

In this section the behavior of integrals of the form

$$\int_{t_1}^{t_2} \cos(pt).f(t)\mathrm{d}t \quad \text{and} \quad \int_{t_1}^{t_2} \sin(pt).f(t)\mathrm{d}t,$$

for large values of the parameter p will be studied.

Cancellation of contributions of opposite sign

Consider first the integral

$$I = \int_0^T \cos pt\,\mathrm{d}t \quad \text{as } p \to \infty. \tag{7.5.1}$$

If the upper limit of the integration happens to be π/p then the integral is zero because the contribution over the interval $0 < t < \pi/2p$ is exactly canceled by the contribution over the remaining interval $\pi/2p < t < \pi/p$. Similarly if $T = 2\pi/p$, $3\pi/p$ or $N\pi/p$ where N is any integer the integral will be zero. In the case when T is not exactly an integer number of half periods, π/p, there remains a contribution over the interval from $t = N\pi/p$ to T where N is the number of whole half periods in the integration range. This is shown in Fig. 7.3.

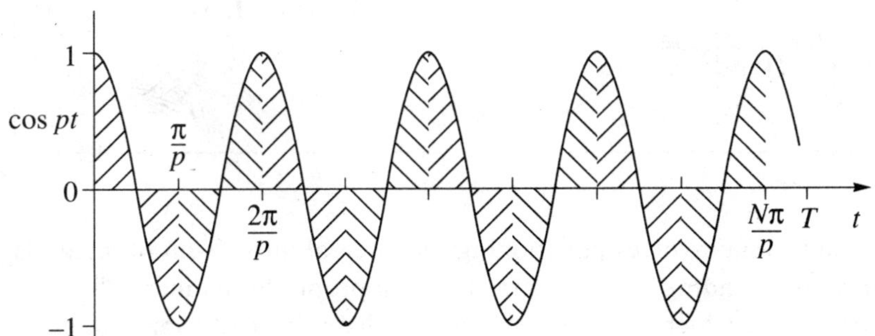

Fig. 7.3 Cancellation of contributions to the integral 7.5.1

Let $T = N\pi/p + \Delta t$, then 7.5.1 becomes

$$I = \int_0^T \cos pt\,\mathrm{d}t = \underbrace{\int_0^{N\pi/p} \cos pt\,\mathrm{d}t}_{I_1} + \underbrace{\int_{N\pi/p}^{N\pi/p+\Delta t} \cos pt\,\mathrm{d}t}_{I_2}.$$

The integral, I_1, is zero because the contributions from the integrand exactly cancel. The integral I_2 has the value $(\sin p\,\Delta t)/p$. The value of $\sin p\,\Delta t$ for large p lies in the range $-1 \leqslant \sin p\,\Delta t \leqslant 1$ depending on the value of Δt. It is bounded in value and for the purpose of describing the order of the term as p tends to infinity it has order one. Thus

$$I = I_2 = \mathrm{O}\left(\frac{1}{p}\right) \quad \text{as } p \to \infty.$$

It is a trivial extension of this result to replace the lower limit of integration in 7.5.1 by a nonzero value and argue that a contribution to the integral from this region will also be of order $1/p$.

Next we consider integrals of the form

$$I = \int_{t_1}^{t_2} \cos(pt).f(t)\mathrm{d}t \quad \text{as } p \to \infty. \qquad 7.5.2$$

Intuitively we may expect that contributions to the integral over intervals of length π/p will still tend to cancel because for large p these intervals are so short that variations in $f(t)$ will be insignificant. This is the Riemann–Lebesgue lemma.

$$\lim_{p \to \infty} \int_{t_1}^{t_2} \cos(pt).f(t)\mathrm{d}t = \lim_{p \to \infty} \int_{t_1}^{t_2} \sin(pt).f(t)\mathrm{d}t = 0 \qquad 7.5.3$$

RIEMANN–LEBESGUE LEMMA

provided that $\int_{t_1}^{t_2} |f(t)|\mathrm{d}t$ is finite. 7.5.4

The proof of the lemma involves partitioning the integral into a sum of integrals over subintervals and bounding each integral in an appropriate manner. This is a rather intricate procedure which will not be presented here. Instead some examples which verify the lemma for particular cases will be examined. The lemma is valid under very general conditions. The limits of integration may be infinite and the function f may be singular or discontinuous at a finite number of points. We require only that the condition 7.5.4 be satisfied.

As the first example consider the integral

$$I = \int_{t_1}^{t_2} t\cos(pt)\mathrm{d}t. \qquad 7.5.5$$

This satisfies the condition 7.5.4 provided t_1 and t_2 are finite. The integral can be evaluated exactly as follows:

$$I = \left[\frac{t \sin pt}{p}\right]_{t_1}^{t_2} - \frac{1}{p}\int_{t_1}^{t_2} \sin pt \, dt$$

$$= \frac{t_2 \sin pt_2 - t_1 \sin pt_1}{p} + \frac{\cos pt_2 - \cos pt_1}{p^2}.$$

Then, in the limit as $p \to \infty$, $I = 0$ thus 7.5.3 is verified. Clearly the function $f(t) = t$ in 7.5.5 can be replaced by any integer power $f(t) = t^n$ and repeated integration by parts will yield the result that $I = 0$ in the limit $p \to \infty$ again providing a verification of the lemma.

The following integral cannot be evaluated exactly,

$$I = \int_1^2 \frac{1}{t} \sin pt \, dt. \qquad 7.5.6$$

It satisfies the condition 7.5.4 of the Riemann–Lebesgue lemma so 7.5.3 asserts that I tends to zero as $p \to \infty$. We can verify this by first integrating by parts,

$$I = \left[\frac{-\cos pt}{pt}\right]_1^2 - \frac{1}{p}\int_1^2 \frac{\cos pt}{t^2} dt,$$

and bounding the integral on the right-hand side of this expression as follows:

$$\left|\int_1^2 \frac{\cos pt}{t^2} dt\right| \leqslant \int_1^2 \frac{|\cos pt|}{|t^2|} dt \leqslant \int_1^2 \frac{1}{t^2} dt$$

$$= \left[-\frac{1}{t}\right]_1^2 = \frac{1}{2}.$$

This is a coarse but sufficient bound; in fact $\int_1^2 \frac{\cos pt}{t^2} dt = \mathrm{O}\left(\frac{1}{p}\right)$ but this is not necessary for our result and the above O(1) bound is sufficient. Thus

$$I = \frac{\cos p}{p} - \frac{\cos 2p}{2p} + \mathrm{O}\left(\frac{1}{p}\right),$$

where the big O order symbol is used in its broadest sense to include the little o case. The point is that we have shown that the integral is $\mathrm{O}(1/p)$ as $p \to \infty$ and hence verified the lemma.

An important definite integral which will be used repeatedly in the next section is

$$\int_0^\infty \cos(x^2) dx = \int_0^\infty \sin(x^2) dx = \frac{1}{2}\sqrt{\frac{\pi}{2}}. \qquad 7.5.7$$

The integral can be evaluated using a complex variable technique known as Cauchy's theorem (see Churchill[20]). We will take this as a given result in subsequent applications.

The behavior of the following integral,

$$I = \int_0^1 \frac{1}{\sqrt{t}} \cos pt\, dt, \tag{7.5.8}$$

can be determined by making use of 7.5.7. Although the function $f(t) = 1/\sqrt{t}$ is singular at the lower limit, the integral exists and the condition 7.5.4 of the Riemann–Lebesgue lemma is satisfied. The lemma is verified for this example as follows. The integral may be expressed as the difference of two integrals,

$$I = \underbrace{\int_0^\infty \frac{1}{\sqrt{t}} \cos pt\, dt}_{I_1} - \underbrace{\int_1^\infty \frac{1}{\sqrt{t}} \cos pt\, dt}_{I_2}.$$

We will prove that $I_1 = O(1/\sqrt{p})$ and $I_2 = O(1/p)$ so that $I = O(1/\sqrt{p})$ as $p \to \infty$. On making the substitution $pt = x^2$ we have

$$I_1 = \int_0^\infty \frac{1}{\sqrt{t}} \cos pt\, dt = \frac{2}{\sqrt{p}} \int_0^\infty \cos(x^2) dx,$$

and so from 7.5.7 we obtain the result $I_1 = \sqrt{\pi/2p}$.

We will prove that $I_2 = O(1/p)$ as $p \to \infty$ by bounding the integral in an appropriate manner. It is not helpful to attempt a direct bound as follows:

$$|I_2| \leqslant \int_1^\infty \frac{|\cos pt|}{\sqrt{t}}\, dt \leqslant \int_1^\infty \frac{1}{\sqrt{t}}\, dt,$$

since the resulting integral is infinite in value. After first integrating by parts we are left with an integral with $t^{3/2}$ as the denominator and then such a bounding procedure is helpful.

$$I_2 = \int_1^\infty \frac{1}{\sqrt{t}} \cos pt\, dt = \left[\frac{\sin pt}{p\sqrt{t}}\right]_1^\infty + \frac{1}{2p} \int_1^\infty \frac{\sin pt}{t^{3/2}}\, dt.$$

The second integral may be bounded as follows:

$$\left|\int_1^\infty \frac{\sin pt}{t^{3/2}}\, dt\right| \leqslant \int_1^\infty \frac{|\sin pt|}{t^{3/2}}\, dt \leqslant \int_1^\infty \frac{1}{t^{3/2}}\, dt = \left[\frac{-2}{\sqrt{t}}\right]_1^\infty = 2.$$

Again this is a coarse but sufficient bound. Thus

$$I_2 = \frac{-\sin p}{p} + O\left(\frac{1}{p}\right) = O\left(\frac{1}{p}\right) \quad \text{as } p \to \infty,$$

so the order of the combined integral, 7.5.8, is determined by the dominant term I_1,

$$\int_0^1 \frac{1}{\sqrt{t}} \cos pt\, dt = O\left(\frac{1}{\sqrt{p}}\right) \quad \text{as } p \to \infty.$$

This verifies the Riemann–Lebesgue lemma for this case and provides an example where the integral is of order $1/\sqrt{p}$ rather than $1/p$ as in the previous examples.

Using the little o order symbol allows the lemma to be stated in the alternative form

$$\int_{t_1}^{t_2} \begin{matrix} \cos(pt) \\ \text{or} \\ \sin(pt) \end{matrix} . f(t)\mathrm{d}t = \mathrm{o}(1) \quad \text{as } p \to \infty,$$

provided the condition 7.5.4 is satisfied.

Construction of asymptotic expansions of integrals

The Riemann–Lebesgue lemma does not tell us how the limit zero is approached as $p \to \infty$. To determine the behavior, a combination of the lemma along with another technique (usually integration by parts) is required. For example, suppose we wish to know how the following integral,

$$I = \int_0^1 \ln(1+t)\cos pt \, \mathrm{d}t, \tag{7.5.9}$$

approaches zero as $p \to \infty$. The condition 7.5.4 is satisfied, so we know from the Riemann–Lebesgue lemma that $I = \mathrm{o}(1)$ as $p \to \infty$. To obtain a precise order relation we must first integrate by parts,

$$I = \left[\frac{\ln(1+t)\sin pt}{p}\right]_0^1 - \frac{1}{p}\int_0^1 \frac{1}{1+t}\sin pt \, \mathrm{d}t. \tag{7.5.10}$$

The integral on the right-hand side satisfies the condition 7.5.4 and hence is $\mathrm{o}(1)$. It has a multiplicative coefficient $1/p$ so the term itself is $\mathrm{o}(1/p)$ as $p \to \infty$, i.e.

$$I = \frac{\ln 2 . \sin p}{p} + \mathrm{o}\left(\frac{1}{p}\right) \quad \text{as } p \to \infty.$$

The first term on the right-hand side therefore determines the rate at which the integral approaches zero.

An asymptotic expansion can be developed by repeated integration by parts. Continuing from 7.5.10 we have

$$\begin{aligned} I &= \frac{\ln 2 . \sin p}{p} - \frac{1}{p}\left(\left[\frac{-\cos pt}{p(1+t)}\right]_0^1 - \frac{1}{p}\int_0^1 \frac{1}{(1+t)^2} . \cos pt \, \mathrm{d}t\right) \\ &= \frac{\ln 2 . \sin p}{p} + \frac{1}{p^2}\left(\frac{1}{2}\cos p - 1\right) + \mathrm{o}\left(\frac{1}{p^2}\right) \quad \text{as } p \to \infty. \end{aligned}$$

A rather more complicated example is provided by the integral

$$I = \int_0^1 \sqrt{t} \sin pt \, dt \quad \text{as } p \to \infty. \tag{7.5.11}$$

Integration by parts yields

$$I = \left[-\frac{\sqrt{t} \cos pt}{p} \right]_0^1 + \frac{1}{2p} \int_0^1 \frac{1}{\sqrt{t}} \cos pt \, dt. \tag{7.5.12}$$

The integral on the right-hand side satisfies the condition 7.5.4 and hence is o(1). Thus

$$I = -\frac{\cos p}{p} + o\left(\frac{1}{p}\right) \quad \text{as } p \to \infty.$$

So far the argument has followed the lines of the previous example. However, we cannot proceed further by integration by parts because the second member of the right-hand side of 7.5.12 is singular at $t = 0$, so straightforward integration by parts yields

$$\int_0^1 \frac{1}{\sqrt{t}} \cos pt \, dt = \left[\frac{\sin pt}{p\sqrt{t}} \right]_0^1 + \frac{1}{2p} \int_0^1 \frac{1}{t^{3/2}} \sin pt \, dt \quad \text{(WRONG)}.$$

In order to develop the asymptotic expansion further the integral in 7.5.12 may be dealt with as follows:

$$I' = \int_0^1 \frac{1}{\sqrt{t}} \cos pt \, dt = \int_0^\infty \frac{1}{\sqrt{t}} \cos pt \, dt - \int_1^\infty \frac{1}{\sqrt{t}} \cos pt \, dt. \tag{7.5.13}$$

The first member of the right-hand side has previously been shown to equal $\sqrt{\pi/2p}$. Integration by parts may now be applied to the second member because the singularity at $t = 0$ has been avoided in the new integration range, thus

$$I' = \sqrt{\frac{\pi}{2p}} - \left[\frac{\sin pt}{p\sqrt{t}} \right]_1^\infty - \frac{1}{2p} \int_1^\infty \frac{1}{t^{3/2}} \sin pt \, dt.$$

The condition of the Riemann–Lebesgue lemma is satisfied by the last integral because $\int_1^\infty \left| \frac{1}{t^{3/2}} \right| dt$ is finite.

Thus

$$I' = \sqrt{\frac{\pi}{2p}} + \frac{\sin p}{p} - o\left(\frac{1}{p}\right)$$

and substituting into 7.5.12 yields

$$\int_0^1 \sqrt{t} \sin pt \, dt = -\frac{\cos p}{p} + \frac{1}{2p}\sqrt{\frac{\pi}{2p}} + \frac{\sin p}{2p^2} + o\left(\frac{1}{p^2}\right) \quad \text{as } p \to \infty.$$

These last two examples, 7.5.9 and 7.5.11, have shown how the Riemann–Lebesgue lemma can be used to generate asymptotic expansions of integrals of the form

$$\int_{t_1}^{t_2} \begin{matrix} \cos(pt) \\ \text{or} \\ \sin(pt) \end{matrix} . f(t)\mathrm{d}t,$$

for large p. The procedure is analogous to Watson's lemma applied to integrals of the form $\int_0^T \mathrm{e}^{-pt} f(t)\mathrm{d}t$.

7.6 The method of stationary phase

In this section asymptotic approximations for integrals of the form

$$\int_{t_1}^{t_2} \cos[ph(t)].f(t)\mathrm{d}t \quad \text{and} \quad \int_{t_1}^{t_2} \sin[ph(t)].f(t)\mathrm{d}t,$$

will be obtained for large values of the parameter p.

If p is large then the trigonometric functions will oscillate rapidly as t varies through the integration range except in regions where $h(t)$ is changing slowly. These regions occur in the neighborhood of points where $\mathrm{d}h/\mathrm{d}t$ is zero. These are called points of stationary phase.

To illustrate the idea compare the behavior of the integrands of the following two integrals as p increases,

$$I_1 = \int_{-1}^{1} \cos(pt)\mathrm{d}t, \quad I_2 = \int_{-1}^{1} \cos(pt^2)\mathrm{d}t.$$

In Fig. 7.4 the integrands $\cos pt$ and $\cos pt^2$ are shown for $p = 1$ and $p = 10$. There is no point of stationary phase for $h(t) = t$ and we know from our previous considerations in Section 7.5 that the contributions to the integral will cancel except at the end points of the integration region. This is indicated in Fig. 7.4a. In contrast there is a point of stationary phase for $h(t) = t^2$; it occurs at $t = 0$. In Fig. 7.4b, regions which appear from the figure to have contributions which cancel are shown as shaded areas. There remain end point contributions and a relatively large contribution from the neighborhood of the point of stationary phase.

The order of magnitude of the contributions to the integrals can be estimated by multiplying the order of the integrand (which is unity) by the interval length over which the contribution occurs. In the case of end point contributions the interval length Δt is, for I_1, such that $p(t + \Delta t) - pt$ is of order $\pi/2$, thus $\Delta t = \mathrm{O}(1/p)$ and we expect that

$$I_1 = \mathrm{O}\left(\frac{1}{p}\right) \quad \text{as } p \to \infty.$$

Similarly for I_2 the interval over which the end point contributes is Δt such that

$$|p(t+\Delta t)^2 - pt^2| \simeq \frac{\pi}{2}, \quad \text{i.e.} \quad |p(2t\Delta t + \Delta t^2)| \simeq \frac{\pi}{2}.$$

In the region of the end points, $t = \pm 1$, the term involving Δt^2 may be neglected so that again $\Delta t = O(1/p)$ leading to an end point contribution to I_2 of $O(1/p)$ as $p \to \infty$. In the region near the point of stationary phase, $t = 0$, $\Delta t = O(1/\sqrt{p})$ leading to a contribution to I_2 of $O(1/\sqrt{p})$ as $p \to \infty$.

These ideas can be verified analytically for the two examples considered. For the first integral we have

$$\int_{-1}^{1} \cos pt \, dt = \left[\frac{\sin pt}{p}\right]_{-1}^{1} = \frac{2\sin p}{p} = O\left(\frac{1}{p}\right).$$

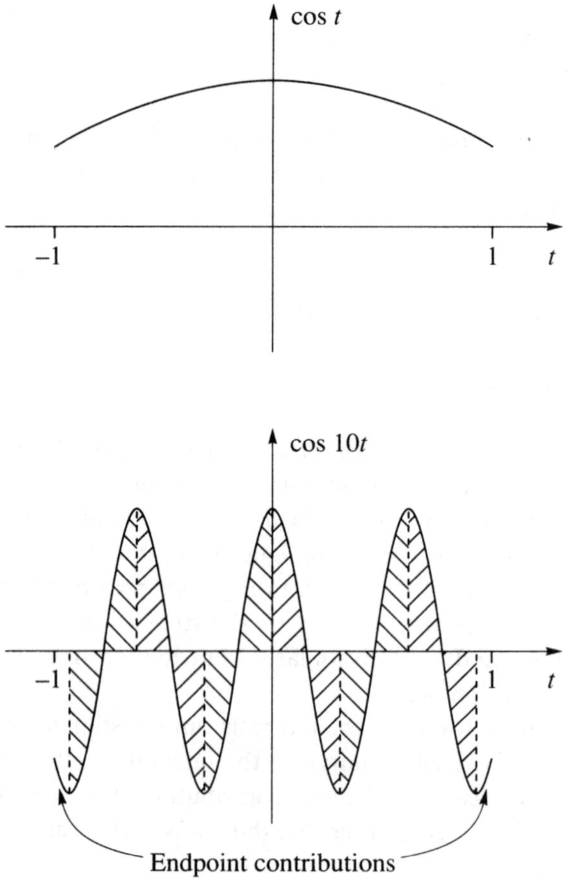

Fig. 7.4a The behavior of $\int_{-1}^{1} \cos(pt)dt$ as p increases

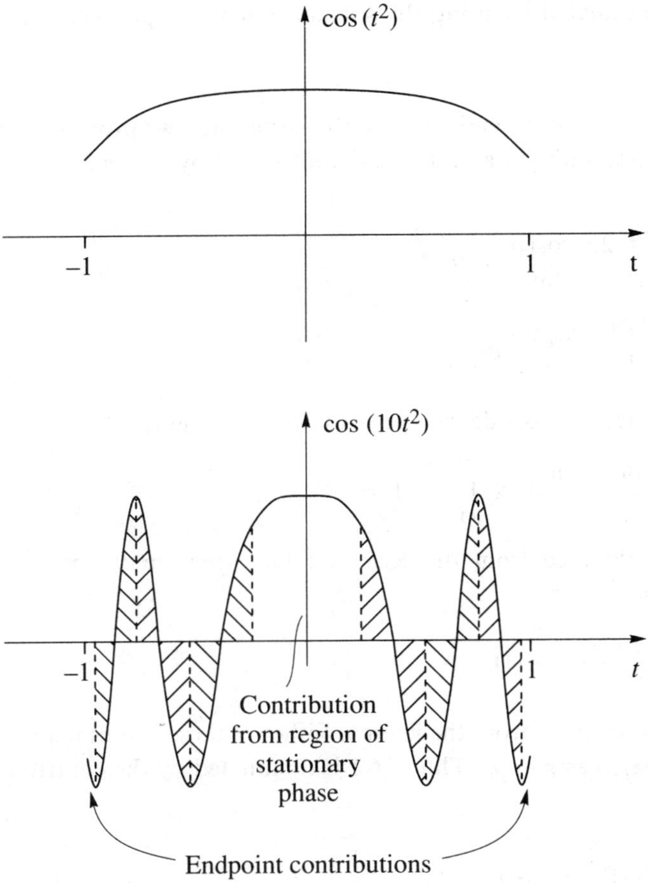

Fig. 7.4b The behavior of $\int_{-1}^{1} \cos(pt^2)dt$ as p increases

The second integral requires the result 7.5.7 namely,

$$\int_0^\infty \cos(x^2)dx = \frac{1}{2}\sqrt{\frac{\pi}{2}}. \qquad 7.5.7 \text{ (repeated)}$$

We have

$$\int_{-1}^{1} \cos pt^2\, dt = 2\int_0^1 \cos pt^2\, dt = 2\Bigg(\underbrace{\int_0^\infty \cos pt^2\, dt}_{I_1} - \underbrace{\int_1^\infty \cos pt^2\, dt}_{I_2}\Bigg). \qquad 7.6.1$$

The integral I_1 can be evaluated by using the substitution $x = \sqrt{pt}$; then from 7.5.7, $I_1 = \frac{1}{2}\sqrt{\frac{\pi}{2p}}$.

The integral I_2 is associated with the end point contributions and we prove that it is of order $1/p$ as follows. First multiply and divide the integrand by $2pt$ to facilitate integration by parts,

$$\begin{aligned} I_2 &= \int_1^\infty \cos(pt^2)\,dt = \int_1^\infty \frac{2pt\cos(pt^2)}{2pt}\,dt \\ &= \left[\frac{\sin(pt^2)}{2pt}\right]_1^\infty + \frac{1}{2p}\int_1^\infty \frac{\sin(pt^2)}{t^2}\,dt. \end{aligned}$$

The second member of the right-hand side can be bounded as follows:

$$\left|\int_1^\infty \frac{\sin(pt^2)}{t^2}\,dt\right| \leqslant \int_1^\infty \frac{|\sin(pt^2)|}{t^2}\,dt \leqslant \int_1^\infty \frac{1}{t^2}\,dt = 1.$$

A stronger bound can be obtained from the Riemann–Lebesgue lemma by first making the substitution $x = t^2$ then

$$\int_1^\infty \frac{\sin(pt^2)}{t^2}\,dt = \int_1^\infty \frac{\sin px}{2x^{3/2}}\,dx = o(1),$$

where the second equality is obtained from the lemma. Either of these routes may be taken to prove that $I_2 = O(1/p)$ as $p \to \infty$. Thus 7.6.1 is dominated by the contribution from I_1 and we have

$$\int_{-1}^1 \cos pt^2\,dt = \sqrt{\frac{\pi}{2p}} + O\left(\frac{1}{p}\right) \quad \text{as } p \to \infty.$$

Contributions from points of stationary phase

We now turn to the general form and first consider the cosine case,

$$I = \int_{t_1}^{t_2} \cos[ph(t)].f(t)\,dt, \qquad 7.6.2$$

where $h(t)$ possesses a single point of stationary phase at $t = t_0$ which is an interior point ($t_1 < t_0 < t_2$). If t_0 coincides with an end point then the corresponding approximation is halved. If there is more than one point of stationary phase then the integration range may be divided into subregions each containing only one point of stationary phase.

At a stationary point t_0, $dh/dt = 0$ and we will restrict our study to the case where d^2h/dt^2 is nonzero. The region of dominant contribution is $t_0 - \Delta t < t < t_0 + \Delta t$

where $\Delta t = O(1/\sqrt{p})$. Divide the integration range into three intervals,

$$\underbrace{\int_{t_1}^{t_2} \cdots \mathrm{d}t}_{I} = \underbrace{\int_{t_1}^{t_0-\Delta t} \cdots \mathrm{d}t}_{I_-} + \underbrace{\int_{t_0-\Delta t}^{t_0+\Delta t} \cdots \mathrm{d}t}_{I_0} + \underbrace{\int_{t_0+\Delta t}^{t_2} \cdots \mathrm{d}t}_{I_+}$$

The contributions from I_+ and I_- are of the end point type and are $O(1/p)$. The contribution from I_0 is $O(1/\sqrt{p})$. Only the leading order approximation for I will be constructed, so it is only necessary to consider I_0.

Both $f(t)$ and $h(t)$ are expanded about $t = t_0$,

$$f(t) = f(t_0) + O(t - t_0)$$

$$h(t) = h(t_0) + \frac{(t-t_0)^2}{2} h''(t_0) + O[(t-t_0)^3].$$

Introduce the variable $s = (t - t_0)\sqrt{\frac{p}{2}|h''(t_0)|}$ then

$$I_0 = \int_{-\Delta t\sqrt{p|h''(t_0)|/2}}^{\Delta t\sqrt{p|h''(t_0)|/2}} \left[f(t_0) + O\left(\frac{s}{\sqrt{p}}\right)\right] \cos\left[ph(t_0) \pm s^2 + O\left(\frac{s^3}{\sqrt{p}}\right)\right] \sqrt{\frac{2}{p|h''(t_0)|}}\,\mathrm{d}s, \qquad 7.6.3$$

where the $+$ sign is used if $h''(t_0)$ is positive and the $-$ sign used if $h''(t_0)$ is negative. The limits of integration are $O(1)$ since $\Delta t = O(1/\sqrt{p})$. Thus the two remainder terms in 7.6.3 are each $O(1/\sqrt{p})$ and may be omitted when constructing the leading order approximation. Then we have

$$I \simeq I_0 \simeq \sqrt{\frac{2}{p|h''(t_0)|}} f(t_0) \int_{-\Delta t\sqrt{p|h''(t_0)|/2}}^{\Delta t\sqrt{p|h''(t_0)|/2}} \cos[ph(t_0) \pm s^2]\,\mathrm{d}s. \qquad 7.6.4$$

At this stage the limits of integration may be replaced by $\pm\infty$ with an error of the 'end point contribution' type namely $O(1/p)$. So, to leading order, we have the approximation

$$I \simeq \sqrt{\frac{2}{p|h''(t_0)|}} f(t_0) \left(\cos[ph(t_0)] . \int_{-\infty}^{\infty} \cos(s^2)\,\mathrm{d}s \mp \sin[ph(t_0)] . \int_{-\infty}^{\infty} \sin(s^2)\,\mathrm{d}s \right).$$

The integrals occurring on the right-hand side of this expression each have the value $\sqrt{\pi/2}$ (see equation 7.5.7). Thus finally we have the stationary phase approximation for large p,

$$\int_{t_1}^{t_2} \cos[ph(t)]f(t)\mathrm{d}t \simeq \sqrt{\frac{\pi}{p|h''(t_0)|}}\, f(t_0)\{\cos[ph(t_0)] \mp \sin[ph(t_0)]\}. \qquad 7.6.5$$

If $f(t_0)$ is zero then I_0 may no longer dominate the integral. Indeed equation 7.6.3 indicates the possibility of I_0 being of order $1/p$ so that in this case all three integrals I_0, I_+ and I_- may contribute to the leading approximation. Examples of this type will not be considered in this book.

In the case of t_0 coinciding with either t_1 or t_2 one of the limits of integration in 7.6.4 will be zero. Thus only half the contribution arises and the right-hand side of equation 7.6.5 should be divided by two.

The sine version of 7.6.2 is dealt with in a similar way. The equivalent of the approximation 7.6.4 becomes

$$\int_{t_1}^{t_2} \sin[ph(t)].f(t)\mathrm{d}t \simeq \sqrt{\frac{2}{p|h''(t_0)|}}\, f(t_0) \int_{-\infty}^{\infty} \sin[ph(t_0) \pm s^2]\mathrm{d}s$$

$$\simeq \sqrt{\frac{\pi}{p|h''(t_0)|}}\, f(t_0)\{\sin[ph(t_0)] \pm \cos[ph(t_0)]\}, \qquad 7.6.6$$

where again the $+$ sign is taken if $h''(t_0)$ is positive and the minus sign if $h''(t_0)$ is negative. If the point of stationary phase coincides with an end point then the right-hand side of 7.6.6 is halved.

Contributions from end points

If there is no point of stationary phase in the interval of integration then $h'(t)$ will either be always positive or always negative. In this case the contributions will be of the 'end point' type. The leading approximation is obtained from integrating by parts as follows. Consider the cosine case and multiply and divide by $ph'(t)$,

$$I = \int_{t_1}^{t_2} ph'(t)\cos[ph(t)].\frac{f(t)}{ph'(t)}\,\mathrm{d}t$$

$$= \left[\sin[ph(t)]\frac{f(t)}{ph'(t)}\right]_{t_1}^{t_2} - \frac{1}{p}\int_{t_1}^{t_2} \sin[ph(t)]\left(\frac{f}{h'}\right)'\mathrm{d}t.$$

The Riemann–Lebesgue lemma asserts that the integral on the right-hand side of the above expression is $\mathrm{o}(1)$ so that

$$I = \frac{1}{p}\left[\frac{f(t_2)\sin[ph(t_2)]}{h'(t_2)} - \frac{f(t_1)\sin[ph(t_1)]}{h'(t_1)}\right] + \mathrm{o}\left(\frac{1}{p}\right).$$

A similar result holds for the sine case. Thus, in summary, if there is no point of stationary phase we have the following approximations for large p,

$$\int_{t_1}^{t_2} \cos[ph(t)]f(t)\mathrm{d}t \simeq \frac{1}{p}\left[\frac{f(t_2)\sin[ph(t_2)]}{h'(t_2)} - \frac{f(t_1)\sin[ph(t_1)]}{h'(t_1)}\right] \qquad 7.6.7$$

$$\int_{t_1}^{t_2} \sin[ph(t)]f(t)\mathrm{d}t \simeq \frac{1}{p}\left[\frac{-f(t_2)\cos[ph(t_2)]}{h'(t_2)} + \frac{f(t_1)\cos[ph(t_1)]}{h'(t_1)}\right]. \qquad 7.6.8$$

In practice it is not recommended that the formulae 7.6.5–7.6.8 be used directly either from memory or by reference to a text such as this. Instead the ideas which lead to these results should be applied to the integral which is being considered. This process will be illustrated by the following worked examples.

Worked examples

Obtain asymptotic approximations for large p of the following integrals,

(i) $\displaystyle\int_0^5 \frac{1}{2+t}\cos p(4t - t^4)\mathrm{d}t.$ (ii) $\displaystyle\int_0^1 \sqrt{1+t}\,\sin p(\mathrm{e}^t - t)\,\mathrm{d}t.$

(iii) $\displaystyle\int_{-1}^2 \sqrt{1+t^2}\,\sin p(t^4 - 2t^2)\,\mathrm{d}t.$ (iv) $\displaystyle\int_2^3 \ln(t).\sin pt^3\,\mathrm{d}t.$

(v) $\displaystyle\int_1^2 \frac{1}{1+t^2}\cos(p\mathrm{e}^{-t})\,\mathrm{d}t.$

Solutions

(i)

$$I = \int_0^5 \frac{1}{2+t}\cos p(4t - t^4)\mathrm{d}t.$$

We have $h = 4t - t^4$ so that $h' = 4 - 4t^3$. There is a single point of stationary phase occurring at $t = 1$. Expanding h about $t = 1$ yields

$$h = 4(t-1+1) - (t-1+1)^4 = 4 + 4(t-1) - 1 - 4(t-1) - 6(t-1)^2$$
$$+ \mathrm{O}[(t-1)^3] \simeq 3 - 6(t-1)^2.$$

Then using the standard approximations we obtain

$$I \simeq \frac{1}{2+1}\int_{-\infty}^{\infty} \cos p[3 - 6(t-1)^2]\,\mathrm{d}t.$$

Let $s = \sqrt{6p}(t-1)$ so that

$$I \simeq \frac{1}{3\sqrt{6p}}\int_{-\infty}^{\infty} \cos(3p - s^2)\,\mathrm{d}s$$

$$\simeq \frac{1}{3\sqrt{6p}}\int_{-\infty}^{\infty} (\cos 3p.\cos s^2 + \sin 3p.\sin s^2)\,\mathrm{d}s.$$

Thus

$$I \simeq \frac{1}{3}\sqrt{\frac{\pi}{12p}}(\cos 3p + \sin 3p).$$

(ii)

$$I = \int_0^1 \sqrt{1+t}\,\sin p(\mathrm{e}^t - t)\,\mathrm{d}t.$$

We have $h = \mathrm{e}^t - t$, $h' = \mathrm{e}^t - 1$ so that $h' = 0$ at $t = 0$. This is an example where the point of stationary phase occurs at an end point. Expanding h about $t = 0$ yields

$$h = \mathrm{e}^t - t = 1 + t + \frac{t^2}{2} + \mathrm{O}(t^3) - t$$

$$h \simeq 1 + \frac{t^2}{2}.$$

Then

$$I \simeq \sqrt{1+0}\int_0^\infty \sin p\left(1 + \frac{t^2}{2}\right)\mathrm{d}t.$$

Let $s = t\sqrt{\frac{p}{2}}$ so that

$$I \simeq \sqrt{\frac{2}{p}}\int_0^\infty (\sin p \cos s^2 + \cos p \sin s^2)\,\mathrm{d}s$$

$$\simeq \frac{1}{2}\sqrt{\frac{\pi}{p}}(\sin p + \cos p).$$

(iii)

$$I = \int_{-1}^2 \sqrt{1+t^2}\,\sin p(t^4 - 2t^2)\mathrm{d}t.$$

In this case $h = t^4 - 2t^2$ and $h' = 4t^3 - 4t$ so h' is zero when $t = -1, 0, 1$. The function $h(t)$ is shown in the figure.

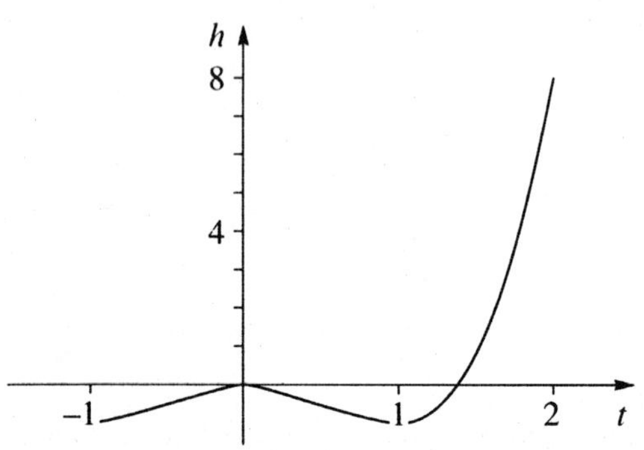

The function $h(t) = t^4 - 2t^2$ in the range $-1 < t < 2$.

There are three points of stationary phase in the integration range. They can each be dealt with separately by dividing the range into three subintervals. Any division such that there is only one stationary point in each will suffice. Thus, for example, we may choose

$$\underset{I}{\int_{-1}^{2} \cdots dt} = \underset{I_1}{\int_{-1}^{-1/2} \cdots dt} + \underset{I_2}{\int_{-1/2}^{1/2} \cdots dt} + \underset{I_3}{\int_{1/2}^{2} \cdots dt.}$$

Consider I_1 first. The point of stationary phase occurs at an end point. Expand $h(t)$ about $t = -1$,

$$\begin{aligned} h = t^4 - 2t^2 &= (t + 1 - 1)^4 - 2(t + 1 - 1)^2 \\ &= 1 - 4(t + 1) + 6(t + 1)^2 + O(t + 1)^3 - 2 + 4(t + 1) - 2(t + 1)^2 \\ &\simeq -1 + 4(t + 1)^2. \end{aligned}$$

So

$$I_1 \simeq \sqrt{1 + (-1)^2} \int_0^{\infty} \sin p[-1 + 4(t + 1)^2] dt \simeq \sqrt{2}.\frac{1}{2\sqrt{p}}\frac{1}{2}\sqrt{\frac{\pi}{2}}(-\sin p + \cos p).$$

Next consider I_2. The point of stationary phase occurs at $t = 0$ so that

$$I_2 \simeq \sqrt{1 + 0^2} \int_{-\infty}^{\infty} - \sin 2pt^2\, dt \simeq -\frac{1}{\sqrt{2p}}.\sqrt{\frac{\pi}{2}}.$$

Finally consider I_3. The point of stationary phase occurs at $t = 1$, so we expand h about $t = 1$,

$$\begin{aligned} h = t^4 - 2t^2 &= (t - 1 + 1)^4 - 2(t - 1 + 1)^2 \\ &= 1 + 4(t - 1) + 6(t - 1)^2 + O[(t - 1)^3] - 2 - 4(t - 1) - 2(t - 1)^2 \\ &\simeq -1 + 4(t - 1)^2. \end{aligned}$$

Therefore

$$I_3 \simeq \sqrt{1 + 1^2} \int_{-\infty}^{\infty} \sin p(-1 + 4(t - 1)^2)\, dt \simeq \frac{1}{2}\sqrt{\frac{\pi}{p}}(-\sin p + \cos p).$$

Thus

$$I \simeq \sqrt{\frac{\pi}{p}}\left[-\frac{1}{2} + \frac{3}{4}(\cos p - \sin p)\right].$$

(iv)

$$I = \int_2^3 \ln(t).\sin pt^3\, dt.$$

In this example $h = t^3$, so there is no point of stationary phase in the integration range.

Multiply and divide by ph' to facilitate integration by parts,

$$I = \int_2^3 \frac{\ln(t)3t^2 p \sin pt^3}{3pt^2}\,\mathrm{d}t$$

$$= \frac{1}{3p}\left[-\frac{\ln(t)}{t^2}\cos pt^3\right]_2^3 + \mathrm{o}\left(\frac{1}{p}\right)$$

$$\doteqdot \frac{1}{3p}\left[\frac{\ln 2}{4}\cos 8p - \frac{\ln 3}{9}\cos 27p\right].$$

(v)

$$I = \int_1^2 \frac{1}{1+t^2}\cos(p\mathrm{e}^{-t})\,\mathrm{d}t.$$

Here $h = \mathrm{e}^{-t}$ so again there is no point of stationary phase and

$$I = \int_1^2 \frac{-p\mathrm{e}^{-t}\cos(p\mathrm{e}^{-t})}{-p\mathrm{e}^{-t}(1+t^2)}\,\mathrm{d}t$$

$$\doteqdot \left[\frac{\sin(p\mathrm{e}^{-t})}{-p\mathrm{e}^{-t}(1+t^2)}\right]_1^2 \doteqdot \frac{1}{p}\left[\frac{\sin(p\mathrm{e}^{-1})}{2\mathrm{e}^{-1}} - \frac{\sin(p\mathrm{e}^{-2})}{5\mathrm{e}^{-2}}\right].$$

Exercises

Obtain asymptotic approximations for large p of the following integrals:

(i) $\int_0^1 \ln(2+t)\cos(pt^2)\,\mathrm{d}t$,

(ii) $\int_0^1 \ln(2+t)\cos pt\,\mathrm{d}t$,

(iii) $\int_{-10}^{10} t^2 \cos p\left(\frac{t^3}{3} - \frac{3t^2}{2} + 2t\right)\mathrm{d}t$,

(iv) $\int_0^{10} \cosh t \,.\, \sin p(2t - t^2)\,\mathrm{d}t$,

(v) $\int_0^1 \mathrm{e}^{2t}\cos p(t + t^3)\,\mathrm{d}t$,

(vi) $\int_{-10}^{10} (1+t)\cos p(t^2 - t^3)\,\mathrm{d}t$,

(vii) $\int_0^1 \sin[p\exp(t^2)]\,\mathrm{d}t$,

(viii) $\int_0^{\pi/2} \frac{\cos(p\cos t)}{\sqrt{t^2+1}}\,\mathrm{d}t$,

(ix) $\int_0^{\pi} \sin(p\cos t)\exp(-t^2)\,\mathrm{d}t$,

(x) $\int_0^1 \sin p(t^3 - t)\,\mathrm{d}t$,

(xi) $\int_0^1 \cos p(t^3 - 6t)\,\mathrm{d}t$,

(xii) $\int_0^{\infty} \frac{\cos[p(t-1)^2]}{1+t}\,\mathrm{d}t$.

Selected answers

(i) $\frac{\ln 2}{2}\sqrt{\frac{\pi}{2p}}$,

(ii) $\frac{\ln 3 \,.\, \sin p}{p}$,

(iii) $\sqrt{\frac{\pi}{p}}\left[\cos(5p/6) + \sin(5p/6) + 4[\cos(2p/3) - \sin(2p/3)]\right],$

(iv) $\sqrt{\frac{\pi}{2p}}\cosh 1\,.(\sin p - \cos p),$ (v) $\frac{e^2\,.\sin 2p}{4p},$

(vi) $\sqrt{\frac{\pi}{2p}}\left[1 + \frac{5}{3}[\cos(4p/27) + \sin(4p/27)]\right].$

7.7 Bessel and Airy functions

In this section the methods described earlier in the chapter will be used to obtain asymptotic approximations for the solution of two important equations, those of Bessel and Airy.

Bessel functions

Bessel's differential equation

$$r^2\frac{d^2y}{dr^2} + r\frac{dy}{dr} + (r^2 - \nu^2)y = 0, \tag{7.7.1}$$

often arises in the description of physical phenomena when cylindrical polar co-ordinates are used. The general solution of the equation is a linear combination of the νth order Bessel functions of the first and second kinds, $J_\nu(r)$ and $Y_\nu(r)$ respectively. $Y_\nu(r)$ is singular both at $r = 0$ and as $r \to \infty$. Consequently it is absent from the solution if the domain of the solution includes the origin or extends to infinity.

This study will be confined to Bessel functions of integer order of the first kind, $J_n(r)$. They have the following integral representation,

$$J_n(r) = \frac{1}{\pi}\int_0^\pi \cos(r\sin\theta - n\theta)d\theta. \tag{7.7.2}$$

We first verify that this expression does satisfy Bessel's equation. Differentiating 7.7.2 yields

$$\frac{dJ_n}{dr} = -\frac{1}{\pi}\int_0^\pi \sin\theta\,.\sin[\;]d\theta,$$

where [ ] represents the term $(r\sin\theta - n\theta)$. Differentiating again yields

$$\frac{d^2J_n}{dr^2} = -\frac{1}{\pi}\int_0^\pi \sin^2\theta\,.\cos[\;]d\theta.$$

Substituting these expressions into the left-hand side of equation 7.7.1 leads to the following expression,

$$\frac{1}{\pi}\int_0^{\pi}\{(r^2\cos^2\theta - n^2)\cos[\;] - r\sin\theta\sin[\;]\}d\theta. \qquad 7.7.3$$

Expressing $r^2\cos^2\theta - n^2$ as $(r\cos\theta + n)$ $(r\cos\theta - n)$ and integrating the first member of 7.7.3 by parts yields

$$\frac{1}{\pi}\Big[(r\cos\theta + n)\sin[\;]\Big]_0^{\pi} + \frac{1}{\pi}\int_0^{\pi} r\sin\theta\sin[\;]d\theta.$$

This is canceled by the second member of 7.7.3 and verifies that Bessel's equation is satisfied by $J_n(r)$.

An asymptotic approximation for the behavior of $J_n(r)$ for large r can be obtained by the method of stationary phase. We first expand the integrand in equation 7.7.2 to yield the standard form for the method

$$J_n(r) = \frac{1}{\pi}\int_0^{\pi}[\cos(r\sin\theta)\cos n\theta + \sin(r\sin\theta)\sin n\theta]\,d\theta.$$

The function $h(\theta)$ is $\sin\theta$ and the point of stationary phase occurs at $\theta = \pi/2$,

$$\sin\theta = \sin\left(\theta - \frac{\pi}{2} + \frac{\pi}{2}\right) = \cos\left(\theta - \frac{\pi}{2}\right) = 1 - \frac{1}{2}\left(\theta - \frac{\pi}{2}\right)^2 + O\left[\left(\theta - \frac{\pi}{2}\right)^4\right].$$

Thus by the usual reasoning

$$J_n(r) \simeq \frac{1}{\pi}\cos\frac{n\pi}{2}\int_{-\infty}^{\infty}\cos r\left[1 - \frac{1}{2}\left(\theta - \frac{\pi}{2}\right)^2\right]d\theta$$

$$+ \frac{1}{\pi}\sin\frac{n\pi}{2}\int_{-\infty}^{\infty}\sin r\left[1 - \frac{1}{2}\left(\theta - \frac{\pi}{2}\right)^2\right]d\theta$$

$$\simeq \frac{1}{\pi}\cos\frac{n\pi}{2}\cdot\sqrt{\frac{2}{r}}\sqrt{\frac{\pi}{2}}(\cos r + \sin r) + \frac{1}{\pi}\sin\frac{n\pi}{2}\cdot\sqrt{\frac{2}{r}}\sqrt{\frac{\pi}{2}}(\sin r - \cos r).$$

A more compact expression is obtained by noting that

$$\cos r + \sin r = \sqrt{2}\cos(r - \pi/4),$$

and

$$\sin r - \cos r = \sqrt{2}\sin(r - \pi/4).$$

Then we obtain the standard form of the large r approximation

$$J_n(r) \simeq \sqrt{\frac{2}{\pi r}}\cos(r - n\pi/2 - \pi/4).$$

Airy functions

Airy's equation

$$\frac{d^2 y}{dx^2} - xy = 0, \tag{7.7.4}$$

has a general solution consisting of a linear combination of the Airy functions $Ai(x)$ and $Bi(x)$. They have the following integral definitions,

$$Ai(x) = \frac{1}{\pi}\int_0^\infty \cos\left(\frac{1}{3}t^3 + xt\right)dt \tag{7.7.5}$$

$$Bi(x) = \frac{1}{\pi}\int_0^\infty \exp\left(-\frac{1}{3}t^3 + xt\right) + \sin\left(\frac{1}{3}t^3 + xt\right)dt. \tag{7.7.6}$$

We will show that Ai satisfies 7.7.4. Showing that Bi is also a solution follows as a straightforward extension.

If we differentiate 7.7.5 directly by differentiating the integrand twice with respect to x, a coefficient of t^2 is generated which causes the integral to be singular. The process of differentiating through the integral is invalid for this infinite integration range because the integral does not possess the property of uniform convergence.

An alternative form of the integral which is uniformly convergent can be obtained using Cauchy's theorem from the field of complex variable theory (see Churchill[20]). We first use the even property of the cosine to write 7.7.5 in the form

$$Ai(x) = \frac{1}{2\pi}\int_{-\infty}^{\infty} \cos\left(\frac{1}{3}t^3 + xt\right)dt.$$

Then let $t = -iv$ to obtain

$$\begin{aligned} Ai(x) &= \frac{1}{2\pi i}\int_{-i\infty}^{i\infty} \cos\left(\frac{iv^3}{3} - ixv\right)dv \\ &= \frac{1}{2\pi i}\int_{-i\infty}^{i\infty} \cosh\left(\frac{v^3}{3} - xv\right)dv \\ &= \frac{1}{2\pi i}\int_{-i\infty}^{i\infty} \frac{1}{2}\left[\exp(v^3/3 - xv) + \exp[-(v^3/3 - xv)]\right]dv \\ &= \frac{1}{2\pi i}\int_{-i\infty}^{i\infty} \exp\left(\frac{v^3}{3} - xv\right)dv. \end{aligned}$$

The integrand has the required analyticity property of complex variable theory to allow the integration along the imaginary v axis to be replaced by an integral along the parallel axis $v = \delta + is$ from $s = -\infty$ to $s = +\infty$, where δ is an arbitrary positive number. Then we obtain the expression

$$Ai(x)=\frac{1}{2\pi}\int_{-\infty}^{\infty}\exp\left\{\frac{\delta^3}{3}-\delta s^2-x\delta+\mathrm{i}\left(\delta^2 s-\frac{s^3}{3}-xs\right)\right\}\mathrm{d}s. \qquad 7.7.7$$

The multiplicative term $\exp(-\delta s^2)$ ensures the uniform convergence property which allows differentiation through the integral. Thus

$$\frac{\mathrm{d}Ai}{\mathrm{d}x}=\frac{1}{2\pi}\int_{-\infty}^{\infty}(-\delta-\mathrm{i}s)\exp\{\ \}\mathrm{d}s,$$

and

$$\frac{\mathrm{d}^2 Ai}{\mathrm{d}x^2}=\frac{1}{2\pi}\int_{-\infty}^{\infty}(\delta^2+2\mathrm{i}\delta s-s^2)\exp\{\ \}\mathrm{d}s.$$

The left-hand side of Airy's equation, 7.7.4, is

$$\begin{aligned}\frac{\mathrm{d}^2 Ai}{\mathrm{d}x^2}-xAi &= -\frac{\mathrm{i}}{2\pi}\int_{-\infty}^{\infty}(\mathrm{i}\delta^2-2\delta s-\mathrm{i}s^2-\mathrm{i}x)\\ &\quad\times\exp\left\{\frac{\delta^3}{3}-\delta s^2-x\delta+\mathrm{i}\left(\delta^2 s-\frac{s^3}{3}-xs\right)\right\}\mathrm{d}s\\ &= -\frac{\mathrm{i}}{2\pi}\Big[\exp\{\ \}\Big]_{-\infty}^{\infty}.\end{aligned}$$

The term $\exp(-\delta s^2)$ dominates the expression and takes the value zero at the limits $s=\pm\infty$. Thus we have shown that the function $Ai(x)$ satisfies Airy's equation.

The asymptotic approximations of the Airy functions are obtained from the integral definitions as follows. Consider Bi for x large and positive. The second member of the integrand in equation 7.7.6, $\sin(\frac{1}{3}t^3+xt)$, has no point of stationary phase in the integration region. The end point contribution will be of order $1/x$. The first member of 7.7.6 contributes an exponentially large term which dominates the second member. Thus for $Bi(x)$ we need only consider the integral

$$\frac{1}{\pi}\int_0^{\infty}\exp\left(-\frac{1}{3}t^3+xt\right)\mathrm{d}t,$$

for large positive x. The infinite integration range allows the $-\frac{1}{3}t^3$ member to compensate for the large term xt. The location of the maximum is determined by differentiating the argument, i.e.

$$\frac{\mathrm{d}}{\mathrm{d}t}\left(-\frac{1}{3}t^3+xt\right)=-t^2+x.$$

Thus the maximum contribution to the integral comes from the region $t\simeq\sqrt{x}$. The standard form for the application of Laplace's method is obtained by introducing the variable $s = t/\sqrt{x}$ so that the maximum occurs near $s = 1$.

$$Bi(x) \doteq \frac{\sqrt{x}}{\pi} \int_0^\infty \exp\left[x^{3/2}\left(-\frac{1}{3}s^3 + s \right) \right] ds.$$

Then let $s = 1 + r$ and expand the function $-\frac{1}{3}s^3 + s$ for small r,

$$-\tfrac{1}{3}s^3 + s = -\tfrac{1}{3} - r - r^2 - \tfrac{1}{3}r^3 + 1 + r \doteq \tfrac{2}{3} - r^2.$$

Then

$$Bi(x) \doteq \frac{\sqrt{x}}{\pi} \exp(\tfrac{2}{3}x^{3/2}). \int_{-\infty}^{\infty} \exp(-x^{3/2}r^2)dr,$$

so that

$$Bi(x) \doteq \frac{1}{\sqrt{\pi}} \frac{1}{x^{1/4}} \exp(\tfrac{2}{3}x^{3/2}) \quad \text{as } x \to \infty.$$

In the case of large negative x the exponential term in the definition of $Bi(x)$ is at its maximum near $x = 0$ and the integral of this term is of order $1/|x|$. The sine term has a point of stationary phase and contributes to a higher order than $1/|x|$. Thus the sine term dominates and so, for large negative x,

$$Bi(x) \doteq \frac{1}{\pi} \int_0^\infty \sin\left(\frac{1}{3}t^3 + xt \right) dt.$$

The argument $\frac{1}{3}t^3 + xt$ has a point of stationary phase when

$$\frac{d}{dt}\left(\frac{1}{3}t^3 + xt \right) = 0,$$

i.e. when $t = \sqrt{(-x)}$.

Let $s = t/\sqrt{-x}$ then

$$Bi(x) \doteq \frac{\sqrt{|x|}}{\pi} \int_0^\infty \sin\left[|x|^{3/2}\left(\frac{s^3}{3} - s \right) \right] ds,$$

where the point of stationary phase occurs at $s = 1$. Introduce the new variable r by the equation $r = s - 1$ then

$$\frac{s^3}{3} - s = +\frac{1}{3} + r + r^2 + \frac{1}{3}r^3 - 1 - r \doteq -\frac{2}{3} + r^2.$$

Therefore

$$Bi(x) \doteq \frac{\sqrt{|x|}}{\pi} \int_{-\infty}^{\infty} \sin\left[|x|^{3/2}\left(-\frac{2}{3} + r^2 \right) \right] dr$$

$$\doteq \frac{\sqrt{|x|}}{\pi} \left(\cos\frac{2}{3}|x|^{3/2} - \sin\frac{2}{3}|x|^{3/2} \right) \frac{1}{|x|^{3/4}} \sqrt{\frac{\pi}{2}},$$

i.e.

$$Bi(x) \simeq \frac{1}{\sqrt{2\pi}} \frac{1}{|x|^{1/4}} \left(\cos \frac{2}{3} |x|^{3/2} - \sin \frac{2}{3} |x|^{3/2} \right) \quad \text{as } x \to -\infty.$$

This is usually written in the following form:

$$Bi(x) \simeq \frac{1}{\sqrt{\pi}} \frac{1}{|x|^{1/4}} \cos \left(\frac{2}{3} |x|^{3/2} + \frac{\pi}{4} \right) \quad \text{as } x \to -\infty.$$

Consider next the function $Ai(x)$, defined by the integral 7.7.5, as $x \to -\infty$. This behaves in an identical way as $Bi(x)$ with the function sine replaced by cosine. The following approximation is easily obtained,

$$Ai(x) \simeq \frac{1}{\sqrt{\pi}} \frac{1}{|x|^{1/4}} \sin \left(\frac{2}{3} |x|^{3/2} + \frac{\pi}{4} \right) \quad \text{as } x \to -\infty.$$

The approximation for $Ai(x)$ as $x \to +\infty$ is more difficult to obtain than the previous results. The integrand in equation 7.7.5 has no point of stationary phase for x positive. Repeated integration by parts yields an expansion involving inverse powers of x but each coefficient is zero

$$\left(0 \times \frac{1}{x} + 0 \times \frac{1}{x^2} + 0 \times \frac{1}{x^3} + \cdots \right).$$

In fact the function is exponentially small as $x \to +\infty$. This result can be obtained using the alternative integral definition given by equation 7.7.7,

$$Ai(x) = \frac{1}{2\pi} \int_{-\infty}^{\infty} \exp \left\{ \frac{\delta^3}{3} - \delta s^2 - x\delta + \mathrm{i} \left(\delta^2 s - \frac{s^3}{3} - xs \right) \right\} \mathrm{d}s.$$

This expression is independent of the value of δ, providing it is real and positive, and a convenient choice is to let $\delta = \sqrt{x}$. Then

$$Ai(x) = \frac{1}{2\pi} \exp(-\tfrac{2}{3} x^{3/2}) \int_{-\infty}^{\infty} \mathrm{e}^{-\sqrt{x} s^2} \mathrm{e}^{-\mathrm{i}s^3/3} \, \mathrm{d}s,$$

showing that for large x the region of dominant contribution to the integral is near $s = 0$. The following asymptotic approximation is then straightforward to obtain,

$$Ai(x) \simeq \frac{1}{2\pi} \exp\left(-\frac{2}{3} x^{3/2} \right) \int_{-\infty}^{\infty} \mathrm{e}^{-\sqrt{x} s^2} \, \mathrm{d}s$$

$$= \frac{1}{2\pi x^{1/4}} \exp\left(-\frac{2}{3} x^{3/2} \right) \sqrt{\pi} = \frac{1}{2\sqrt{\pi} x^{1/4}} \exp\left(-\frac{2}{3} x^{3/2} \right) \quad \text{as } x \to \infty.$$

References

1 Abramowitz, M. and Stegun, I.A., *Handbook of Mathematical Functions*, Dover, New York 9th ed., 1965 (a) chap. 5; (b) 446; (c) chap. 6.
2 Grimshaw, R., *Nonlinear Ordinary Differential Equations*, CRC Press, Boca Raton, FL, 1990, (a) 167.
3 Poincaré, H., *New Methods of Celestial Mechanics*, Vol. I, NASA-TTF-450, Washington, D.C., 1967, chap. 3. (English translation. Original French edition 1892).
4 Lighthill, M.J., A technique for rendering approximate solutions to physical problems uniformly valid, *Phil. Mag.* 40, 1179–1201, 1949.
5 Lighthill, M.J., A technique for rendering approximate solutions to physical problems uniformly valid, *Z. Flugwiss.* 9, 267–275, 1961.
6 Nayfeh, A.H., *Perturbation Method*, Wiley, 1973, (a) chap. 3; (b) Exercise 3.34.
7 Nayfeh, A.H. and Kluwick, A., A comparison of three perturbation methods for non-linear hyperbolic waves, *J. Sound and Vibration* 48, 293–299, 1976.
8 Pritulo, M.F., On the determination of uniformly accurate solutions of differential equations by the method of perturbation of coordinates, *J. Appl. Math. Mech.* 26, 661–667, 1962.
9 Lighthill, M.J., A new approach to thin airfoil theory, *Aero. Quart.* 3, 193–210, 1951.
10 Van Dyke, M., *Perturbation Methods in Fluid Mechanics*, Parabolic Press, Palo Alto, CA, 1975.
11 Milne–Thomson, L.M., *Theoretical Hydrodynamics*, Macmillan, London, 1949.
12 Fox, P.A., On the use of co-ordinate perturbations in the solution of physical problems, Tech. Rept. No. 1, Project for Machine Method of Computation and Numerical Analysis, Massachusetts Institute of Technology, Cambridge, MA, 1953.
13 Elrod, H.G., Thin film lubrication theory for Newtonian fluids with surfaces possessing striated roughness or grooving, *Trans. ASME J. Lubric. Technol.*, 95, 484–489, 1973.
14 Krylov, N. and Bogoliubov, N.N., *Introduction to Nonlinear Mechanics*, Princeton University Press, Princeton, NJ, 1947.
15 Bluman, G.W. and Cole, J.D., *Similarity Methods for Differential Equations*, Springer-Verlag, New York, 1974.
16 Batchelor, G.K., *An Introduction to Fluid Dynamics*, Cambridge University Press, Cambridge, 1967.
17 Levich, V.G., *Physicochemical Hydrodynamics*, Prentice-Hall, Englewood Cliffs, NJ, 1962.
18 Murray, J.D., *Asymptotic Analysis*, Springer-Verlag, New York, 1984.
19 Watson, G.N., *Theory of Bessel Functions*, Cambridge University Press, Cambridge, 1952.
20 Churchill, R.V., *Complex Variables and Applications*, McGraw-Hill, New York, 1960.

Bibliography

The following books provide material for further study.

Bender, C.M. and Orszag, S.A., *Advanced Mathematical Methods for Scientists and Engineers*, McGraw-Hill, New York, 1978.

Cole, J.D., *Perturbation Methods in Applied Mathematics*, Blaisdell, Waltham, MA, 1968.

Copson, E.T., *Asymptotic Expansions*, Cambridge University Press, Cambridge, 1965.

Erdelyi, A., *Asymptotic Expansions*, Dover, New York, 1956.

Kevorkian, J. and Cole, J.D., *Perturbation Methods in Applied Mathematics*, Springer-Verlag, New York, 1981.

Murray, J.D., *Asymptotic Analysis*, Springer-Verlag, New York, 1984.

Nayfeh, A.H., *Perturbation Methods*, Wiley, New York, 1973.

Nayfeh, A.H., *Introduction to Perturbation Techniques*, Wiley, New York, 1981.

Nayfeh, A.H. and Mook, D.T., *Nonlinear Oscillations*, Wiley, New York, 1979.

Olver, F.W.J., *Asymptotics and Special Functions*, Academic Press, New York, 1974.

O'Malley, R.E., *Introduction to Singular Perturbations*, Academic Press, New York, 1974.

Schlichting, H., *Boundary Layer Theory*, McGraw-Hill, New York, 1979.

Smith, D.R., *Singular Perturbation Theory*, Cambridge University Press, Cambridge, 1985.

Van Dyke, M., *Perturbation Methods in Fluid Mechanics*, Parabolic Press, Palo Alto, CA, 1975.

Wasow, W., *Asymptotic Expansions for Ordinary Differential Equations*, Wiley, New York, 1965.

Index